IMPACT MATHEMATICS

Algebra and More

Course 2

**Developed by
Education Development Center, Inc.**

Principal Investigators: Faye Nisonoff Ruopp, E. Paul Goldenberg

Senior Project Director: Cynthia J. Orrell

Senior Curriculum Developers: Michelle Manes, Susan Janssen, Sydney Foster, Daniel Lynn Watt, Nina Arshavsky, Ricky Carter, Joan Lukas, Charles Lovitt

Curriculum Developers: Phil Lewis, Debbie Winkler

 Glencoe

New York, New York   Columbus, Ohio   Chicago, Illinois   Peoria, Illinois   Woodland Hills, California

 **Glencoe**

The McGraw·Hill Companies

The algebra content for *Impact Mathematics* was adapted from the series, *Access to Algebra*, by Neville Grace, Jayne Johnston, Barry Kissane, Ian Lowe, and Sue Willis. Permission to adapt this material was obtained from the publisher, Curriculum Corporation of Level 5, 2 Lonsdale Street, Melbourne, Australia.

Send all inquiries to:
Glencoe/McGraw-Hill
8787 Orion Place
Columbus, OH 43240-4027

ISBN 0-07-860920-8

8 9 10   079/055   13 12 11 10 09 08

Impact Mathematics Project Reviewers

Education Development Center appreciates all the feedback from the curriculum specialists and teachers who participated in review and testing.

Special thanks to:

Peter Braunfeld
Professor of Mathematics Emeritus
University of Illinois

Sherry L. Meier
Assistant Professor of Mathematics
Illinois State University

Judith Roitman
Professor of Mathematics
University of Kansas

....

Marcie Abramson
Thurston Middle School
Boston, Massachusetts

Alan Dallman
Amherst Middle School
Amherst, Massachusetts

Steven J. Fox
Bendle Middle School
Burton, Michigan

Denise Airola
Fayetteville Public Schools
Fayetteville, Arizona

Sharon DeCarlo
Sudbury Public Schools
Sudbury, Massachusetts

Kenneth L. Goodwin Jr.
Middletown Middle School
Middletown, Delaware

Chadley Anderson
Syracuse Junior High School
Syracuse, Utah

David P. DeLeon
Preston Area School
Lakewood, Pennsylvania

Fred E. Gross
Sudbury Public Schools
Sudbury, Massachusetts

Jeanne A. Arnold
Mead Junior High
Elk Grove Village, Illinois

Jacob J. Dick
Cedar Grove School
Cedar Grove, Wisconsin

Penny Hauben
Murray Avenue School
Huntingdon, Pennsylvania

Joanne J. Astin
Lincoln Middle School
Forrest City, Arkansas

Sharon Ann Dudek
Holabird Middle School
Baltimore, Maryland

Jean Hawkins
James River Day School
Lynchburg, Virginia

Jack Beard
Urbana Junior High
Urbana, Ohio

Cheryl Elisara
Centennial Middle School
Spokane, Washington

Robert Kalac
Butler Junior High
Frombell, Pennsylvania

Chad Cluver
Maroa-Forsyth Junior High
Maroa, Illinois

Patricia Elsroth
Wayne Highlands Middle School
Honesdale, Pennsylvania

Robin S. Kalder
Somers High School
Somers, New York

Robert C. Bieringer
Patchogue-Medford School Dist.
Center Moriches, New York

Dianne Fink
Bell Junior High
San Diego, California

Darrin Kamps
Lucille Umbarge Elementary
Burlington, Washington

Susan Coppleman
Nathaniel H. Wixon Middle School
South Dennis, Massachusetts

Terry Fleenore
E.B. Stanley Middle School
Abingdon, Virginia

Sandra Keller
Middletown Middle School
Middletown, Delaware

Sandi Curtiss
Gateway Middle School
Everett, Washington

Kathleen Forgac
Waring School
Massachusetts

Pat King
Holmes Junior High
Davis, California

Kim Lazarus
San Diego Jewish Academy
La Jolla, California

Ophria Levant
Webber Academy
Calgary, Alberta
Canada

Mary Lundquist
Farmington High School
Farmington, Connecticut

Ellen McDonald-Knight
San Diego Unified School District
San Diego, California

Ann Miller
Castle Rock Middle School
Castle Rock, Colorado

Julie Mootz
Ecker Hill Middle School
Park City, Utah

Jeanne Nelson
New Lisbon Junior High
New Lisbon, Wisconsin

DeAnne Oakley-Wimbush
Pulaski Middle School
Chester, Pennsylvania

Tom Patterson
Ponderosa Jr. High School
Klamath Falls, Oregon

Maria Peterson
Chenery Middle School
Belmont, Massachusetts

Lonnie Pilar
Tri-County Middle School
Howard City, Michigan

Karen Pizarek
Northern Hills Middle School
Grand Rapids, Michigan

Debbie Ryan
Overbrook Cluster
Philadelphia, Pennsylvania

Sue Saunders
Abell Jr. High School
Midland, Texas

Ivy Schram
Massachusetts Department
of Youth Services
Massachusetts

Robert Segall
Windham Public Schools
Willimantic, Connecticut

Kassandra Segars
Hubert Middle School
Savannah, Georgia

Laurie Shappee
Larson Middle School
Troy, Michigan

Sandra Silver
Windham Public Schools
Willimantic, Connecticut

Karen Smith
East Middle School
Braintree, Massachusetts

Kim Spillane
Oxford Central School
Oxford, New Jersey

Carol Struchtemeyer
Lexington R-5 Schools
Lexington, Missouri

Kathy L. Terwelp
Summit Public Schools
Summit, New Jersey

Laura Sosnoski Tracey
Somerville, Massachusetts

Marcia Uhls
Truesdale Middle School
Wichita, Kansas

Vendula Vogel
Westridge School for Girls
Pasadena, California

Judith A. Webber
Grand Blanc Middle School
Grand Blanc, Michigan

Sandy Weishaar
Woodland Junior High
Fayetteville, Arkansas

Tamara L. Weiss
Forest Hills Middle School
Forest Hills, Michigan

Kerrin Wertz
Haverford Middle School
Havertown, Pennsylvania

Anthony Williams
Jackie Robinson Middle School
Brooklyn, New York

Deborah Winkler
The Baker School
Brookline, Massachusetts

Lucy Zizka
Best Middle School
Ferndale, Michigan

CONTENTS

Chapter Three

Exploring Exponents 144

Chapter Four

Working with Signed Numbers . . 216

Chapter Five

Looking at Linear Relationships . . 298

Chapter Six

Solving Equations 382

Chapter Seven

Similarity 448

Chapter Eight

Ratio and Proportion 518

Chapter Nine

Interpreting Graphs 600

Chapter Ten

Data and Probability 664

CHAPTER 1

Understanding Expressions

Real-Life Math

Algebra in the Strangest Places You might think that algebra is a topic found only in textbooks, but you can find algebra all around you—in some of the strangest places.

Did you know there is a relationship between the speed at which ants crawl and the air temperature? If you were to find some ants outside and time them as they crawled, you could actually estimate the temperature. Here is the algebraic equation that describes this relationship.

Celsius temperature

$$t = 15s + 3$$

ant speed in centimeters per second

There are many ordinary and extraordinary places where you will encounter algebra.

Think About It What do you think is the speed of a typical ant?

Family Letter

Dear Student and Family Members,

Our class is about to begin an exciting year of *Impact Mathematics*. Some of the topics we will study include negative numbers, exponents, three-dimensional geometry, ratios, probability, and data analysis. Throughout the year, your student will also develop and refine skills in algebra.

We'll begin by looking at algebraic expressions—the combinations of numbers, letters, and mathematical symbols that form the language of algebra. We will learn about variables—letters or symbols that can change or that represent unknown quantities. For example, in the expression $b + 2$, the variable is b.

Once we're familiar with variables and expressions, we will create flow-charts to match expressions and then use the flowcharts to solve equations. We will also explore formulas used in everyday life, such as the formula used to convert degrees Celsius to degrees Fahrenheit: $F = \frac{9}{5}C + 32$.

Vocabulary Along the way, we'll be learning several new vocabulary terms:

algebraic expression	**exponent**
backtracking	**factor**
distributive property	**flowchart**
equivalent expressions	**formula**
expand	**variable**

What can you do at home?

Encourage your student to explain the kinds of problems he or she is solving in class. In addition, help him or her think about common occurrences of algebraic expressions in daily life. Your interest in your student's work helps emphasize the importance of mathematics and its usefulness in daily life.

1.1 Variables and Expressions

Every day people are confronted with problems they have to solve. Many of these problems involve such quantities as the amount of spice to add to a recipe, the cost of electricity, and interest rates. In some problem situations, it helps to have a way to record information without using a lot of words. For example, both boxes present the same idea.

To convert a Celsius temperature to a Fahrenheit temperature, find nine-fifths of the Celsius temperature and then add 32.	$F = \frac{9}{5}C + 32$

While the statement on the left may be easier to read and understand at first, the statement on the right has several advantages. It is shorter and easier to write, it shows clearly how the quantities—Celsius temperature and Fahrenheit temperature—are related, and it allows you to try different Celsius temperatures and compute their Fahrenheit equivalents.

In this lesson, you'll see that by using a few simple rules, you can write powerful algebraic expressions and equations for a variety of situations.

Think  Discuss

Shaunda, Kate, and Simon are holding bags of blocks. Isabel has just two blocks.

If you know how many blocks are in each bag, how can you figure out how many blocks there are altogether?

If you know the number of blocks in each bag, it's not hard to express the total number of blocks. For example, if there are 20 blocks in each bag, you can just add:

$$20 + 20 + 20 + 2 = 62$$

Or you can multiply and add:

$$3 \times 20 + 2 = 62$$

VOCABULARY
variable

What if you don't know the number in each bag? First, notice that, in this situation, the number of bags and the number of loose blocks don't change, but the number of blocks in each bag can change. Quantities that can change, or vary, are called **variables.**

In algebra, letters are often used to represent variables. For example, you can let the letter n stand for the number of blocks in each bag.

Now you can find the total number of blocks as you did before— by adding

$$n + n + n + 2$$

or by multiplying and adding:

$$3 \times n + 2$$

Remember
Multiplication can be shown in several ways:
$3 \times n$ $3(n)$
$3 \cdot n$ $3 * n$

In algebra, the multiplication symbol between a number and a variable is usually left out. So $3 \times n + 2$ can be written $3n + 2$.

Investigation 1 ▶ Expressions

In the bags-and-blocks situation above, you can think of $3n + 2$ as a *rule* for finding the total number of blocks when you know the number of blocks in each bag. Just substitute the number in each bag for n. For example, for 100 blocks in each bag, the total number of blocks is

$$3n + 2 = 3 \times 100 + 2 = 302$$

Rules written with numbers and symbols, such as $n + n + n + 2$ and $3n + 2$, are called **algebraic expressions.**

As you study algebra, you will work with algebraic expressions often. Using bags and blocks is a good way to start thinking about expressions. Imagining the variable as a bag that you can put any number of blocks into can help you see how the value of an expression changes as the value of the variable changes.

Problem Set A

In these problems, you will continue to explore the situation in which there are 3 bags, each containing the same number of blocks, plus 2 extra blocks.

1. Copy and complete the table.

Number of Blocks in Each Bag, n	0	1	2	3	4	5
Total Number of Blocks, $3n + 2$		5	8			

2. If $n = 7$, what is the value of $3n + 2$?

3. If $n = 25$, what is the value of $3n + 2$?

4. If there are 50 blocks in each bag, how many blocks are there altogether?

5. If there are 20 blocks altogether, how many blocks are in each bag?

6. Copy and complete the table.

n	10	5	40	25	100			22		
$3n + 2$		17		77		23	92		128	3,143

7. Compare the tables in Problems 1 and 6. Which table do you think was more difficult to complete? Why?

8. Could the total number of blocks in this situation be 18? Explain.

9. To represent the number of blocks in 3 bags plus 2 extra blocks with the expression $3n + 2$, you need to assume that all the bags contain the same number of blocks. Why?

10. The expression $3n + 2$ describes the total number of blocks in 3 bags, each with the same number of blocks, plus 2 extra blocks.

    a. Describe a bags-and-blocks situation that can be represented by the expression $5n + 6$.

    b. Explain how the expression fits your situation.

You have spent a lot of time exploring the number of blocks in three bags plus two extra blocks. Now you'll investigate some other bags-and-blocks situations.

Problem Set B

1. Here are 5 bags and 4 extra blocks.

    a. What is the total number of blocks if each bag contains 3 blocks? If each bag contains 10 blocks?

    b. Using n to represent the number of blocks in each bag, write an algebraic expression for the total number of blocks.

    c. Find the value of your expression for $n = 3$ and $n = 10$. Do you get the same answers you found in Part a?

2. Now suppose you have 4 bags, each with the same number of blocks, plus 2 extra blocks.

    a. Draw a picture of this situation.

    b. Write an expression for the total number of blocks.

3. Write an expression to represent 7 bags, each with the same number of blocks, plus 5 extra blocks.

4. Write an expression to represent 10 bags, each with the same number of blocks, plus 1 extra block.

5. Any letter can be used to stand for the number of blocks in a bag. Match each expression below with a drawing.

$$2c + 4 \qquad 4m + 2 \qquad 4y + 5 \qquad 2f + 5$$

a.

b.

c.

d.

6. Rebecca wrote the expression $3b + 1$ to describe the total number of blocks represented in this picture.

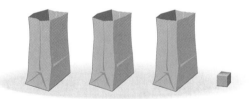

a. What does the variable b stand for in Rebecca's expression?

b. What does the 3 stand for?

c. What does the 1 stand for?

d. Complete the table for Rebecca's expression.

b	1	2	3		23		100
$3b + 1$		7		31		76	

7. Zoe thought of a new situation:

"Imagine that the total number of blocks is 2 blocks less than 3 bags' worth. This is hard to draw, but I just described it easily in words—and I can write it algebraically as $3n - 2$."

a. Describe a situation that $4n - 1$ could represent.

b. Describe a situation that $5x - 7$ could represent.

c. Describe a situation that $14 - 3p$ could represent.

8. Sascha has 1 more block than Chris, and Dean has 1 more block than Sascha. Patrick has just 2 blocks.

a. If Sascha has 6 blocks, how many blocks does each boy have? How many do they have altogether?

b. If Sascha has 15 blocks, how many blocks does each boy have? How many do they have altogether?

c. If you know how many blocks Sascha has, how can you determine the total number of blocks without figuring out how many blocks each of the other boys has?

d. If the boys have 26 blocks altogether, how many does each boy have? Explain how you arrived at your answer.

e. Let s stand for the number of blocks Sascha has. Write an expression for the number each boy has. Then write an expression for the total number of blocks.

f. Let c stand for the number of blocks Chris has. Write an expression for the number each boy has. Then write an expression for the total number of blocks.

g. Your expressions for Parts e and f both tell how many blocks the group has, and yet the expressions are different. Explain why.

Share & Summarize

1. Make a bags-and-blocks drawing.

2. Write an expression that describes your drawing.

3. Explain how you know your expression matches your drawing.

Investigation Writing Expressions

You have seen that algebraic expressions can be used to represent situations in which a quantity changes, or *varies*. So far, you've worked a lot with bags and blocks. In this investigation, you will write algebraic expressions to describe many other situations.

Problem Set C

Jay, Lyndal, Davina, Bart, Tara, and Freda live in a row of houses that are numbered 1 to 6. Use the clues below to help you solve Problems 1–3.

- Jay owns some CDs.
- Lyndal lives next to Jay and three houses from Freda. Lyndal has 2 fewer CDs than Jay.
- Davina lives on the other side of Jay and has three times as many CDs as Lyndal.
- Bart lives in House 2 and has 4 more CDs than Jay.
- If Tara had 13 more CDs, she would have four times as many as Jay.
- Jay acquired all his CDs from Freda, who lives in House 6. Freda had 17 CDs before she gave *p* of them to Jay.
- The person in House 5 owns the most CDs.

1. Figure out who lives in each house. (Hint: Focus on where the people live. Ignore the other information.)

2. Let *p* stand for the number of CDs Jay has. Write an expression containing the variable *p* for the number of CDs each person has.

3. Use your expressions to help determine how many CDs Jay has. (Hint: Experiment with different values of *p*. Only one value of *p* gives the person in House 5 the most CDs.)

In Problem Set C, the variable *p* could have only one value. The variable was used to represent an unknown quantity: the number of CDs Jay has. As you saw, it is helpful to be able to use a variable to represent an unknown quantity. Now you have two ways to think about variables: as a quantity that can change and as an unknown quantity.

Problem Set D

For these situations, the variables represent unknown quantities rather than quantities that change.

1. Esteban bought *N* apples.

    a. Granny Smith bought four times as many apples as Esteban. Write an expression to show how many apples she bought.

    b. Esteban used 11 apples in a pie. How many does he have left?

2. Suppose a bag of potatoes weighs *t* pounds. Write an expression for the number of pounds in each bag of vegetables below. Each expression should include the variable *t*.

    a. a bag of carrots weighing 3 lb more than a bag of potatoes

    b. a bag of corn weighing 2 lb less than a bag of potatoes

    c. a bag of broccoli weighing 5 lb less than a bag of corn

    d. a bag of beans weighing 3 lb more than a bag of broccoli

    e. Order the five bags from lightest to heaviest.

3. Baby Leanne is *L* inches tall. Write expressions, in terms of the variable *L*, to represent the heights in inches of these members of Leanne's family:

    a. her 4-year-old brother Tim, who is twice as tall as Leanne

    b. her 6-year-old sister Kerry, who is 5 in. taller than Tim

    c. her mother, who is 15 in. shorter than four times Leanne's height

    d. her father, who is 15 in. shorter than twice Kerry's height

    e. Baby Leanne is about 20 in. long. Check the expressions you wrote in Parts a–d by substituting this value and determining whether the heights are reasonable.

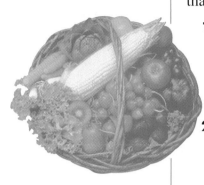

Just the facts

The average newborn is about 20 inches long. It takes people about 4 years to double their length at birth, and 8 more years to grow to three times their birth length. A person "only about 3.5 times the height of a newborn" is quite tall!

4. Maya has *k* nickels in her pocket.

  a. Write an expression for the value of Maya's nickels in cents. Find the value of the expression for *k* = 4.

  b. Maya also has 60¢ in other coins. Write an expression for the total value, in cents, of all her change.

  c. Find the total value for *k* = 5 and *k* = 12.

  d. Yesterday Maya had *k* nickels in her pocket and no other money. She spent 70¢ on a pencil. What was the total value of her coins after she bought the pencil? Could *k* be *any* whole number? Why?

5. Malik rides the city bus to school and back each weekday. The bus fare is *F* cents each way. He starts each week with $5 for bus fare. Write an expression for each of the following:

  a. the amount, in cents, Malik spends on bus fare in one day

  b. the amount he has left when he gets home on Monday

  c. the total bus fare he has spent by lunch time on Thursday

  d. the amount of bus fare he spends each week

  e. the amount he has left at the end of the week

  f. What would happen if *F* were 60?

6. You can sometimes make sense of an expression by inventing a meaning for its symbols. For example, for the expression *k* − 4, you can let *k* represent the number of kittens in a pet shop. Then *k* − 4 could stand for the number of kittens in the shop after 4 have been sold.

Interpret each expression below by making up a meaning for the symbols.

  a. *d* + 10   **b.** 3*a*   **c.** *f* − 4

Share & Summarize

Let *h* represent the number of hours Shaunda spent on homework last week.

1. Zach spent half as much time on his homework as Shaunda had on hers. Write an expression for the number of hours he spent.

2. Describe a situation that the expression $2h + 1$ might represent.

3. Write another algebraic expression containing the variable *h*, and describe a situation your expression might represent.

Investigation ▶3 Evaluating Expressions

When you study mathematics, it is important to know the shortcuts that are used for writing expressions. You already know that when you want to show a number times a variable, you can leave out the multiplication sign: instead of $6 \times t$ or $6 \cdot t$, you can write $6t$. Note that the number is normally placed before the variable.

VOCABULARY
exponent

What if you want to show that the variable *t* is multiplied by itself? You could write $t \times t$ or $t \cdot t$ or *tt*. However, an **exponent** is usually used to tell how many times a quantity is multiplied by itself. So $t \times t$ is written t^2, and $t \times t \times t$ is written t^3.

Here are some other examples.

- The expression $m \times 5 \times m$ is written $5m^2$.
- The expression $2 \cdot s \cdot s \cdot s$ is written $2s^3$.
- The expression $x \times x \times y \times 7$ is written $7x^2y$ or $7yx^2$.
- The expression $2p \cdot p \cdot p$ is written $2p^3$.

It is also important to understand the rules for *evaluating,* or finding the value of, expressions. You learned some of these rules when you studied arithmetic.

Think & Discuss

Evaluate each expression without using a calculator.

$$7 + 8 \times 5 \qquad\qquad 7 - 6 \div 2$$

Now use your calculator to evaluate each expression. Did you get the same answers with the calculator as you did without?

Most calculators will multiply and divide before they add or subtract. This *order of operations* is a convention that everyone uses to avoid confusion. If you want to indicate that the operations should be done in a different order, you need to use parentheses. Look at how the use of parentheses affects the value of the expressions below.

$$2 + 3 \cdot 5 = 2 + 15 = 17 \qquad\qquad 7 - 6 \div 2 = 7 - 3 = 4$$

$$(2 + 3) \cdot 5 = 5 \cdot 5 = 25 \qquad\qquad (7 - 6) \div 2 = 1 \div 2 = \tfrac{1}{2}$$

Which expression below means "add 4 and 6, and then multiply by n"? Which means "multiply 6 by n, and then add 4"?

$$4 + 6n \qquad\qquad (4 + 6)n$$

There are also rules for evaluating expressions involving exponents.

Think & Discuss

Who is correct, Malik or Zach?

Does "find m^2 and multiply it by 4" give the same result as "find $4m$ and square it"? Try it for $m = 2$ to see.

Think again about who is correct, Malik or Zach. Then add parentheses to the expression $4m^2$ so it will give the other boy's calculation.

Problem Set E

Rewrite each expression without using multiplication or addition signs.

1. $r + r$ **2.** $r \times r$ **3.** $t + t + t$

4. $5 \cdot g \cdot g \cdot g$ **5.** $5s + s$ **6.** $2.5m \times m$

7. Copy the expression $6 + 3 \times 4$.

     a. Insert parentheses, if necessary, so the resulting expression equals 18.

     b. Insert parentheses, if necessary, so the resulting expression equals 36.

8. Write an expression that means "take a number, multiply it by 6, and cube the result."

Evaluate each expression for $r = 3$.

9. $5r^2$ **10.** $(5r)^2$

11. $2r^4$ **12.** $(2r)^4$

13. When Kate, Jin Lee, Darnell, Zach, and Maya tried to find the value of $3t^2$ for $t = 7$, they got five different answers! Only one of their answers is correct.

     Kate: $t^2 = 49$, so $3t^2 = 349$

     Jin Lee: $3t^2 = 3 \times 7^2 = 42$

     Darnell: $3t^2 = 3 \times 7^2 = 3 \times 49 = 147$

     Zach: $3t^2 = 37^2 = 1{,}369$

     Maya: $3t^2 = 3 \times 7^2 = 21^2 = 441$

     a. Which student evaluated the expression correctly?

     b. What mistake did each of the other students make in thinking about the problem?

You have explored ways of expressing multiplication and repeated multiplication, like $3t$ and t^2. Now you will look at some ways to express division.

EXAMPLE

Three friends won a prize of P dollars in the community talent show. They want to share the prize money equally. There are several ways to express the number of dollars each student should receive.

All three expressions in the example above are correct, but the forms $\frac{1}{3}P$ and $\frac{P}{3}$ are used more commonly in algebra than $P \div 3$.

Problem Set F

1. In the prize-money example, P stands for the dollar amount of the prize, and $\frac{P}{3}$ shows each friend's share.

a. If the prize is $30, how much will each friend get?

b. Suppose $\frac{P}{3} = 25. How much prize money is there?

Find the value of each expression for $k = 5$.

2. $k^3 - 2$ **3.** $12 - k$ **4.** $\frac{k^2 - 1}{12}$ **5.** $\frac{7k}{5}$ **6.** $\frac{7}{5}k$

Problem Set G

Kate, Zach, Maya, and Darnell earn money by selling greeting cards they create on a computer.

1. In their first week in business, they sold 19 cards for *B* dollars each. They shared the money they collected equally.

   a. Write an expression that describes each friend's share.

   b. What is the value of your expression if *B* is $2.00? If *B* is $2.60? If *B* is $3.00?

   c. The friends wrote the equation $\frac{19B}{4} = 7.60$ to describe what happened their first week in business. What does the equation indicate about the money each friend made?

   d. Starting with the equation in Part c, find how much the friends charged for each card.

2. This week the friends received orders for 16 greeting cards.

   a. If they sell the cards for *B* dollars each, how much will each friend earn?

   b. The friends would like to earn $12 apiece for selling the 16 cards. How much will they have to charge for each card?

3. In Problems 1 and 2, the price of a greeting card was the only variable. Kate wants to be able to vary three amounts—the number of friends working on the cards, the number of cards, and the charge per card—and still be able to calculate how much each person would make.

Write an expression for how much each friend would earn in each situation.

   a. 3 friends sell 19 greeting cards for *B* dollars each

   b. *F* friends sell 19 cards for *B* dollars each

   c. 4 friends sell *P* cards for *B* dollars each

   d. *F* friends sell *P* cards for *B* dollars each

4. Suppose the price per card increases by $2, from *B* to (*B* + 2). Write an expression for how much *F* friends would each make after selling *P* cards at this new price.

5. Suppose the price per card increases by *K* dollars, from *B* to (*B* + *K*). Write an expression for how much *F* friends would each make after selling *P* cards at this new price.

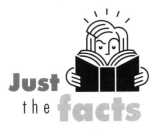

Share & Summarize

1. Write an algebraic expression that describes this set of calculations.

Step 1.   Take some number and square it.

Step 2.   Add 2 times the number you started with.

Step 3.   Divide that result by your starting number.

Step 4.   Subtract your starting number.

2. Write an expression that describes this set of calculations.

Step 1.   Take some number and triple it.

Step 2.   Subtract your result from your starting number.

Step 3.   Divide that result by the square of your starting number.

Step 4.   Subtract the result from 10.

3. Find the value of $\frac{5 + 15x^2}{13}$ when $x = 2$.

Investigation 4 Using Flowcharts

In this investigation, you will discover a tool that can help you evaluate expressions and solve equations.

In preparing for a party, Maya bought 5 bags of bagels and an extra 4 bagels.

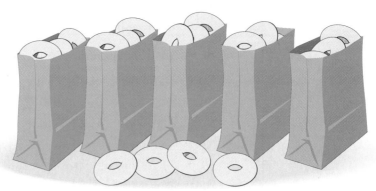

If *n* represents the number of bagels in each bag, the total number of bagels Maya purchased is $5n + 4$. To find the total number of bagels, you multiply the number in each bag by 5 and then add 4. You can make a diagram, called a **flowchart,** to show these steps.

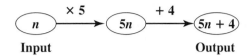

The oval at the left side of the flowchart represents the *input*—in this case, the number of bagels in each bag. Each arrow represents a *mathematical action*. The oval to the right of an arrow shows the result of a mathematical action. The oval at the far right represents the *output*—in this case, the total number of bagels.

To evaluate $5n + 4$ for a particular value of *n,* just substitute that value for the input and follow the steps until you reach the output. Here is the same flowchart, with an input—the value of *n*—of 3.

In Problem Set H you will use flowcharts to find outputs for given inputs, and you will create flowcharts to match algebraic expressions.

Problem Set H

Copy and complete each flowchart by filling the empty ovals.

1. 8 →(× 4) ◯ →(− 5) ◯

2. *n* →(× 6) ◯ →(− 1) ◯

3. 5 →(× 3) ◯ →(− 1) ◯ →(÷ 3) ◯

4. Consider the expressions $6y + 1$ and $6(y + 1)$.

   a. Make a flowchart for the expression $6y + 1$.

   b. Make a flowchart for the expression $6(y + 1)$. How is this flowchart different from the flowchart in Part a?

5. Consider the expressions $\frac{n + 2}{3}$ and $n + \frac{2}{3}$.

   a. Make a flowchart for each expression.

   b. Use your flowcharts to find outputs for three or four *n* values. Do both flowcharts give the same outputs? Explain why or why not.

At the beginning of this investigation, you saw that Maya had $5n + 4$ bagels, where n is the number of bagels in each bag. Suppose you knew she had a total of 79 bagels. How could you find the number of bagels in each bag? That is, how could you find the value of n for which $5n + 4 = 79$?

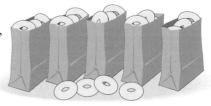

VOCABULARY
backtracking

Luis used a flowchart to find this number by **backtracking,** or working backward. This is how he did it.

Since 79 is the output, I'll put it in the last oval.

I add 4 to get 79, so the number in the second oval must be 75.

Since 5 times a number is 75, the number of bagels in each bag must be 15.

The input Luis found, 15, is the solution of the equation $5n + 4 = 79$. Now you will practice solving equations using the backtracking method.

Problem Set ▌

1. Zoe put an input into this flowchart and got the output 53. Use backtracking to find Zoe's input.

Luis solved some equations by backtracking. Problems 2–4 show the flowcharts he started with. Do Parts a and b for each flowchart.

a. Write the equation Luis was trying to solve.

b. Backtrack to find the solution.

2.

3.

4.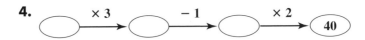

Now that you have some experience using backtracking to solve equations, you can use backtracking to solve more challenging problems.

Problem Set J

1. Lakeisha and Mateo were playing a game called *Think of a Number.*

Lakeisha must use Mateo's output to figure out his starting number.

a. Draw a flowchart to represent this game.

b. What equation does your flowchart represent?

c. Use backtracking to solve your equation from Part b. Check your solution by following Lakeisha's steps.

2. Consider this expression.

$$\frac{2(n + 1)}{3} - 1$$

a. Draw a flowchart for the expression.

b. Use backtracking to find the solution of $\frac{2(n + 1)}{3} - 1 = 5$.

Share & Summarize

Create an equation that can be solved by backtracking. Write a paragraph explaining to someone who is not in your class how to use backtracking to solve your equation.

On Your Own Exercises

Practice & Apply

For each picture, write an expression for the total number of blocks. Assume each bag contains the same number of blocks.

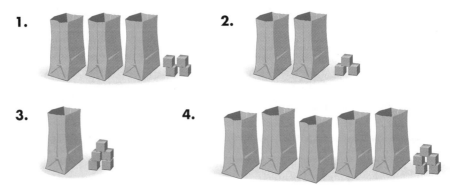

1.

2.

3.

4.

5. Consider the expression $4n + 5$.

    a. Draw a bags-and-blocks picture for this expression.

    b. Copy and complete the table.

n	0	1	2	3	26	66	
$4n + 5$		9					321

    c. If $n = 7$, what is the value of $4n + 5$?

    d. If $n = 25$, what is the value of $4n + 5$?

6. A particular bags-and-blocks situation can be represented by the expression $5n + 3$. What is the value of $5n + 3$ if there are 38 blocks in each bag?

7. Here are clues for a logic puzzle.

- Five friends—Alano, Bob, Carl, Dina, and Bonita—live in a row of houses numbered 1 to 5.
- They all keep tropical fish for pets.
- Dina got her fish from Bonita's next-door neighbor Alano. Alano had 21 fish before he gave *f* of them to Dina.
- Carl has twice as many fish as Dina, who lives in House 2. Carl does not live next to Dina.
- Bonita lives in House 4 and has 3 fewer fish than Carl.
- Bob has 4 more fish than Dina.

a. Who lives in which house?

b. Write expressions to describe the number of fish each person has.

c. If the five neighbors have a total of 57 fish, how many fish does Dina have?

8. Keshon has *s* stamps in his collection.

a. Keshon's older sister Jamila has three times as many stamps as Keshon. Write an expression for the number of stamps she has.

b. Jamila decides to give Keshon 13 of her stamps. How many stamps will she have left? How many stamps will Keshon have?

9. Franklin, a golden retriever, weighs *p* pounds. In Parts a–c, write an expression for the weight of the dog in pounds. Each expression should include the variable *p*.

a. Tatu, a pug, weighs 49 pounds less than Franklin.

b. Mia, a Chihuahua, weighs $\frac{1}{17}$ as much as Franklin.

c. Lucy, a Great Dane, weighs twice as much as Franklin, minus 15 pounds.

d. Franklin weighs 68 pounds. How much do Tatu, Mia, and Lucy weigh?

10. Mr. Karnowski has a stack of 125 sheets of graph paper. He is passing out *g* sheets to each of his students.

 a. Write an expression for the number of sheets he will pass out to the first 5 students.

 b. Write an expression for the number of sheets he will have left after he has given sheets to all 25 students in his class.

 c. What would happen if *g* were 6?

11. The Marble Emporium sells individual marbles and bags of marbles. Each bag contains the same number of marbles.

 a. Rita bought a bag of marbles, plus 1 extra marble. Is it possible that she bought an even number of marbles? Explain.

 b. Helena bought 2 bags of marbles, plus 1 extra marble. Is it possible that Helena has an even number of marbles? Explain.

 c. Challenge Let *n* stand for *any whole number*. Use what you have discovered in Parts a and b to write an algebraic expression containing *n* that means *any odd number*. Write another expression that means *any even number*.

Interpret each expression by making up a meaning for the symbols.

12. $7n$ **13.** $p - 6$ **14.** $25q - 50$

15. $2n + 1$ **16.** $3m - 6$ **17.** $5p - 3$

Rewrite each expression without using multiplication or addition signs.

18. $t + t + t + t$ **19.** $4 \times xy \times xy$

20. $6g + g + g$ **21.** $8 \cdot c \cdot c \cdot d \cdot d$

22. Find the value of $9 - 3D^2$ for $D = 1.1$.

23. Without using a calculator, find the value of $9 - 3D^2$ for $D = \frac{1}{2}$.

Evaluate each expression for $h = 4$.

24. $\frac{h^2 - 3}{13}$ **25.** $\frac{3h}{4}$ **26.** $9h \div 2$

27. Shaunda, Malik, and Luis make and sell hand-painted T-shirts. They make a profit of D dollars for each shirt they sell.

    a. If they sell 13 T-shirts, how much total profit will they make?

    b. If the three friends equally divide the profit from selling the 13 shirts, how much will each receive?

    c. Suppose each friend received $39 from selling the 13 shirts. How much profit did the students earn for each shirt?

    d. Suppose F more friends join the business. If the group sells 20 T-shirts and divides the profit equally, how much will each friend receive?

    e. If the new, larger group of friends sells T shirts, how much profit will each friend receive?

Copy and complete each flowchart.

28.

12 → ×2 → ○ → +12 → ○

29.

n → −6 → ○ → ×3 → ○

30.

7 → ×2 → ○ → +3 → ○ → ÷2 → ○

31.

○ → ×5 → ○ → −4 → 41

32. Consider the expression $3a \div 4$.

    a. Draw a flowchart to represent the expression.

    b. Use your flowchart to find the value of the expression for $a = 8$.

    c. Use your flowchart to solve $3a \div 4 = 6$. Explain each step in your solution.

33. Lehie drew this flowchart.

○ → ×16 → ○ → −4 → 28

    a. What equation was Lehie trying to solve?

    b. Copy and complete the flowchart.

34. In a game of *Think of a Number,* Jin Lee told Darnell:

    • *Think of a number.*

    • *Subtract 1 from your number.*

    • *Multiply the result by 2.*

    • *Add 6.*

    a. Draw a flowchart to represent this game.

    b. Darnell said he got 10. Write an equation you could solve to find Darnell's number.

    c. Use backtracking to solve your equation. Check your solution by following Jin Lee's steps.

35. Luis wrote the expression $9(2x + 1) + 1$.

  a. Draw a flowchart for Luis's expression.

  b. Use your flowchart to solve the equation $9(2x + 1) + 1 = 46$.

36. Consider this equation.

$$\frac{3m \times 2}{6} = 1$$

  a. Draw a flowchart to represent the equation.

  b. Use backtracking to find the solution.

37. Consider this expression.

$$3n - 2$$

  a. Why is it difficult to draw a picture of bags and blocks for this expression?

  b. Copy and complete the table.

n	15	24		38	45	60
$3n - 2$			88			

38. Owen and Noah are packing the 27 prizes left in their booth after the school fair. They have 4 boxes, and each box holds 8 prizes.

  a. How many boxes can they fill completely?

  b. After they fill all the boxes they can, will they have any prizes left over to fill another box? If so, how many prizes will be in that box?

  c. How many empty boxes will there be, if any?

39. Dario has 2 bags and a box. Each bag contains the same number of blocks. The box contains 10 more blocks than a bag contains.

  a. Draw a sketch of this situation. Label each part of your sketch with an expression showing how many blocks that part contains.

  b. Write an expression for the total number of blocks Dario has.

  c. If Dario has a total of 49 blocks, how many blocks are in each bag?

In your **own words**

Describe a situation that can be represented by the expression $7n + 4$. Explain what the 7, the 4, and the n represent in your situation.

40. Tito's Taxi charges \$3.20 for the first mile plus \$2.40 for each additional mile.

a. Which expression gives the fare, in dollars, for a trip of *d* miles? (Assume that the trip is at least 1 mile.)

$$d + 3.20 + 2.40$$

$$2.40(d - 1) + 3.20$$

$$3.20(d + 2.40)$$

$$d(2.40 + 3.20)$$

b. How much would it cost to travel 1 mile? To travel 3 miles?

c. Tito thought he might make more money if his drivers charged for every sixth of a mile, not every mile. The new rate is \$1.60 for the first $\frac{2}{6}$ of a mile, plus 40¢ for each additional $\frac{1}{6}$ mile.

If a person travels *n* sixths of a mile, which expression gives the correct fare? (Assume that the person travels at least $\frac{2}{6}$ of a mile.)

$$1.60 + 40n$$

$$1.60 + 40(n - 2)$$

$$1.60 + 0.40n$$

$$1.60 + 0.40(n - 2)$$

d. Use the formula you chose in Part c to calculate fares for trips of 1 mile and 3 miles. Compare the results to those for Part b. Does Tito's Taxi make more money with this new rate plan?

41. Challenge The members of the music club are raising money to pay for a trip to the state jazz festival. They are selling CDs and T-shirts.

a. Each CD costs the club \$2.40. The club is selling the CDs for \$12.00 apiece. Write an expression to represent how much profit the club will make for selling *c* CDs.

b. Each T-shirt costs the club \$3.50. The club is selling the T-shirts for \$14.50. Write an expression to represent how much profit the club will make for selling *t* T-shirts.

c. Now write an expression that gives the profit for selling *c* CDs and *t* T-shirts.

d. Ludwig sold 10 of each item. Johann sold 7 CDs and 12 T-shirts. Use your expression from Part c to calculate how much money each club member raised.

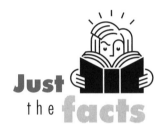

Just the facts

Profit is the amount of money a business earns. It is calculated by subtracting expenses (what was spent) from income (what was earned).

42. Preview This equation describes how the height h of a particular baseball changed over time t after it was thrown from four feet above the ground. Height is in feet, and time is in seconds.

$$h = 40t - 16t^2 + 4$$

a. How high was the ball after 0.5 second?

b. How high was the ball after 1 second?

c. Based on what you learned in Parts a and b, guess how long it took the ball to reach a height of 25 feet.

d. Substitute your guess for t in the equation to find how high the ball was at that point in time. Were you close? Is your guess too high? Too low?

43. Preview Consider the expressions $x(x - 1)$ and $x^2 - x$.

a. Choose five values for x. Evaluate both expressions for those values. Organize your results in a table.

b. What appears to be the relationship between the two expressions?

44. Challenge In this problem, you will compare x^2 and x for different values of x.

a. Find three values for x such that $x^2 > x$.

b. Find three values for x such that $x^2 < x$.

c. Find a value for x such that $x^2 = x$. Can you find more than one?

45. Geometry Franklin's garden is rectangular, with a length of L ft and an area of 80 square feet.

a. Write an equation you could solve to find the garden's width, W.

b. How wide is Franklin's garden if its length is 15 feet?

c. If the length of Franklin's garden is 12 feet, how far does Franklin travel when he walks the perimeter of his garden?

46. In a game of *Think of a Number,* the input was 3 and the output was 2. Tell whether each of the following could have been the game's rule.

  a. Think of a number. Double it. Add 3 to the result. Then divide by 5.

  b. Think of a number. Subtract 2 from it. Multiply the result by 4. Then divide by 2.

  c. Think of a number. Divide it by 3. Add 5 to the result. Then divide by 3 again.

47. Solve the equation $10 = 0.25x - 0.25$. You may want to draw a flowchart and use backtracking.

48. Challenge Mr. Frazier has a part-time job as a telemarketer. He earns $4.00 per hour, plus 50¢ for every customer he contacts.

  a. Mr. Frazier works a 3-hour shift and contacts c customers. Write an equation for computing how many dollars he earns, e, during his shift.

  b. Use your equation to find Mr. Frazier's earnings if he contacts 39 customers during a 3-hour shift.

  c. What total amount did Mr. Frazier make per hour, on average, for this shift?

Mixed Review

Find each product without using a calculator.

  49. 98×54 **50.** $256 * 67$ **51.** $2,692 \cdot 53$

Find each quotient.

  52. $88 \div 11$ **53.** $261 \div 9$ **54.** $2,064 \div 86$

55. Order these numbers from least to greatest: $^-7, \ ^-2.5, \ ^-3.8, \ ^-3, 0$.

56. Order these numbers from least to greatest: $^-7, 6, \ ^-2, \ ^-\frac{1}{2}, 0$.

Find each sum.

  57. $\frac{1}{4} + \frac{3}{4}$ **58.** $\frac{1}{3} + \frac{5}{3}$ **59.** $\frac{1}{6} + \frac{1}{6}$

Fill in each blank to make a true statement.

  60. $\frac{1}{7} + \underline{\hspace{1cm}} = 1$

  61. $\frac{9}{3} - \underline{\hspace{1cm}} = 1$

  62. $\frac{1}{9} \times \underline{\hspace{1cm}} = \frac{1}{81}$

Find the value of *a* in each equation.

63. $3a = 12$ **64.** $2a + 4 = 10$ **65.** $5 - a = 0$

66. Geometry Approximate the area of the spilled ink on the grid. Each square represents 1 square meter.

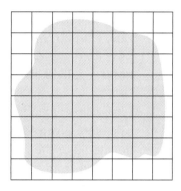

67. Statistics Kyle kept a list of the selling prices of the homes sold in his town during September.

$128,000	$85,500	$220,000	$105,000	$135,500
$259,000	$97,000	$98,500	$100,600	$263,000
$96,600	$175,000	$259,000	$187,000	$190,000

a. Find the mean, median, and mode of Kyle's data.

b. Which of these three measures do you think best represents the average selling price of a home in Kyle's town in September?

c. In November, Kyle read that one more home had sold in his town during September: a sprawling mansion, with a sale price of $1,800,000. Will this new piece of data change any of the measures you calculated in Part a? If so, how will it change them?

d. Does the new information in Part c change your answer to Part b?

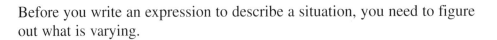

1.2 Expressions and Formulas

Before you write an expression to describe a situation, you need to figure out what is varying.

Zoe, Darnell, Maya, and Zach are members of a community group that raises money for various charities. Every year, the community group holds a calendar sale.

Last year's calendars pictured different animals, and all the profits from the sale were given to a wildlife preservation fund. The students agreed to each donate $2 of their own money to the fund in addition to the money they collected from the calendar sale. They sold the calendars for $12 each.

Think & Discuss

How much money did each student collect for the fund? Consider both the money they earned for selling the calendars and the money they donated themselves.

What is varying in this situation?

Write an expression for the total amount of money given to the fund by each student. Tell what your variable stands for.

For their summer project, the community group hosted a walkathon to raise money for cancer research. A successful radio campaign prompted a lot of people to request more information. The community group responded by sending informational packets.

Zoe, Darnell, Maya, and Zach each spent time one weekend preparing packets, including addressing them. It takes about an hour for one student to prepare 80 packets. On Monday, after school, they each completed another 50 packets.

Zoe, Darnell, and Maya discussed how much time they spent on the project.

Think & Discuss

About how many packets did each of the three students prepare? Consider both the packets prepared over the weekend and those completed on Monday.

When you calculated how many packets each student prepared, what was varying?

Write an expression for the total number of packets prepared by a volunteer working for *t* hours.

Zach prepared 690 packets. How long did he spend preparing packets over the weekend?

Investigation 1 ▶ What's the Variable?

In this investigation, you will write algebraic expressions to match situations, and you'll figure out which of several situations match a given expression. You will also make up your own situations based on an expression and a given meaning for the variable.

Problem Set A

1. A movie ticket costs $6, and a box of popcorn costs $2.

   a. Luis was meeting friends at the theater to see a movie. He arrived first and bought one ticket. He decided to buy a box of popcorn for each of his friends. Write an expression for the total amount Luis spent. Choose a letter for the variable in your expression, and tell what the variable represents.

   b. Another group of friends was meeting to see a movie. They planned to share one box of popcorn among them. Write an expression for the total amount the group spent, and tell what the variable represents.

2. A pancake recipe calls for flour. To make thicker pancakes, Mr. Lopez adds an extra cup of flour. He then divides the flour into three equal portions so that he can make blueberry pancakes, banana pancakes, and plain pancakes. Write an expression for how much flour is in each portion, and tell what your variable represents.

3. Luis made two round trips to the beach and then traveled another 600 yards to the concession stand.

   a. Write an expression to represent the total distance Luis traveled. Be sure to say what your variable stands for.

   b. If Luis walked a total of 1,584 yards, how far is it to the beach?

4. Ms. Franklin wants to break ground to add some new sections to her garden: three square sections with the same side length, and another section with an area of 7 square meters.

   a. Write an expression to represent the total area of the sections Ms. Franklin needs to till.

   b. Ms. Franklin's son Jahmall offers to help her. If they share the work equally, how many square meters will each have to dig up?

5. Simon received $60 for his birthday. Each week, for x weeks, he spent $4 of the money to go ice skating.

    a. How much money, in dollars, did Simon have left after x weeks?

    b. Does *any* number make sense for the value of x? If not, describe the values that don't make sense.

A single expression can represent lots of situations, depending on what the variable stands for.

EXAMPLE

Consider the expression $c + 10$.

- If c is the number of cents in Jin Lee's piggy bank, $c + 10$ could represent the number of cents in the bank after she drops in another dime.

- If c is the number of gallons of gas left in Ms. Lopez's gas tank, $c + 10$ could represent the number of gallons after she adds 10 gallons.

- If there are c members in the science club, $c + 10$ could represent the number of members after 10 new students join.

Can you think of some other situations $c + 10$ could represent?

Problem Set B

In Problems 1–6, decide whether the expression $2d + 5$ can represent the answer to the question. If it can, explain what d stands for in that situation.

1. Lidia bought two tickets to a symphony concert and five tickets to a movie. How much did she pay?

2. Sam bought several pens for $2 each and a notebook for $5. How much did he spend altogether?

3. A herd of zebras walked a certain distance to a watering hole. On the return trip, a detour added an extra 5 km. How many kilometers did the herd walk altogether?

4. There are two dogs in the house and five more in the yard. How many dogs are there altogether?

5. Because their parents were sick, Maya and Santo spent twice as many hours as usual doing housework during the week plus an extra 5 hours on the weekend. How many hours of housework did they do?

6. Chau gets paid $5 for every hour he baby-sits, plus a $2 bonus on weekends. How much money does he earn if he baby-sits on Saturday?

Describe a situation that can be represented by each expression.

7. $4m - 3$, if m stands for the number of pages in a book

8. $4m - 3$, if m stands for the distance in kilometers from home to school

9. $4m - 3$, if m stands for the number of eggs in a waffle recipe

10. $4m - 3$, if m stands for the number of grams of water in a beaker of water

11. Describe two situations that can be represented by the expression $10 - x^2$. Discuss with your partner how your situations match the expression.

Share & Summarize

Describe two situations that can be represented by this expression:

$$3t + 7$$

Explain how your situations match the expression. Check by trying some values for the variable t and seeing whether the solutions make sense.

Investigation ▶2▶ Using Formulas

Weather reports in most countries give temperatures in degrees Celsius. If you are like most Americans, however, a temperature of 20°C is something you've heard of, but you wouldn't know offhand whether it is very hot, pleasantly warm, or fairly chilly.

Think & Discuss

Test your sense of how warm these Celsius temperatures are.

- Which outdoor attire do you think is most appropriate when the temperature is 33°C: a winter coat, jeans and a sweatshirt, or shorts and a T-shirt?

- Which outdoor activity do you think is most appropriate in 10°C weather: swimming, soccer, or skiing?

VOCABULARY
formula

There is a **formula**—an algebraic "recipe"—for converting Celsius temperatures to the Fahrenheit system.

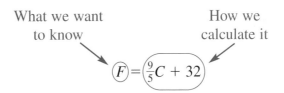

What we want to know   How we calculate it

$$F = \frac{9}{5}C + 32$$

In the formula, the variable C represents the temperature in degrees Celsius, and the variable F represents the temperature in degrees Fahrenheit.

This formula can also be written like this:

$$F = \frac{9C}{5} + 32$$

Use the formula to find the Fahrenheit equivalent of 20°C.

$$F = \frac{9}{5}C + 32 \qquad \text{Start with the formula.}$$

$$F = \frac{9}{5} \times 20 + 32 \qquad \text{Substitute 20 for } C.$$

$$F = 68 \qquad \text{Simplify.}$$

So 20°C is the same as 68°F.

Following the example above, find the Fahrenheit equivalents of 33°C and 15°C. Were your answers to the Think & Discuss questions on page 37 reasonable?

Problem Set C

1. Most pastry is made from flour, shortening, and water. There are different types of pastry. In *short pastry,* the relationship between the amount of flour F and the amount of shortening S is given by the formula

$$S = \frac{1}{2}F \qquad \text{or} \qquad S = \frac{F}{2}$$

   a. Alice wants to make lemon tarts with short pastry using exactly 800 grams of flour. How much shortening should she use?

   b. Daryl wants to make short pastry for cheese sticks, but he has only 250 g of shortening. What is the maximum amount of flour he can use?

2. The relationship between flour and shortening in *flaky pastry* is given by the formula

$$S = \frac{3}{4}F \qquad \text{or} \qquad S = \frac{3F}{4}$$

   How much shortening would you need to add to 200 g of flour to make flaky pastry?

3. Which type of pastry—short or flaky—needs more shortening for a given amount of flour? Check your answer by testing the amount of shortening for two amounts of flour.

4. You saw that the relationship between temperature in degrees Celsius and degrees Fahrenheit is given by the formula

$$F = \tfrac{9}{5}C + 32$$

a. Convert 38°C to degrees Fahrenheit.

b. Water freezes at 0°C. How many degrees Fahrenheit is this?

c. Water boils at 100°C. How many degrees Fahrenheit is this?

d. Convert 50°F to degrees Celsius.

5. The formula for the area of a circle is $A = \pi r^2$, where r is the radius. The Greek letter π is not a variable. It is a number equal to the circumference of any circle divided by its diameter.

a. Your calculator probably has a button that automatically gives a good approximation for π. Do you recall some approximations you have used before?

b. What is the value of πr^2 for $r = 5.6$?

c. Tetromino's Pizza makes a pizza with a diameter of 12 in. Use the formula to find the area of a 12-in. pizza.

d. Tetromino's 12-in. pizza costs $7.50. They also make a pizza with a diameter of 24 in., which sells for $25. Which size is the better buy? Explain your answer.

e. Tetromino's smallest round pizza has an area of 78.5 in.^{2}. Write the equation you would need to solve to find the radius of the pizza. Solve your equation.

Remember

The *diameter* is the longest measurement across a circle and equals twice the radius.

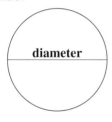

diameter

Now you will use formulas to help you analyze a race.

Problem Set D

Three tortoises entered a 10-meter race.

 Tortoise 1 was especially slow. He moved at only 0.9 meter per minute, so he was given a 3.1-meter head start.

 Tortoise 2 roared along at 1.3 meters per minute and received no head start.

 Tortoise 3 also got no head start. Her distance was equal to the square of the time, in minutes, since the race began, multiplied by 0.165.

1. In investigating this problem, it would be useful to have a formula that tells each tortoise's distance from the starting line at any time after the race began. For each tortoise, choose the formula below that describes its distance, D, from the starting point m minutes after the race began.

$$D = 3.1m + 0.9 \qquad D = 1.3m \qquad D = 0.165 + m^2$$

$$D = 3.1 + 0.9m \qquad D = 0.165m^2 \qquad D = 10 - 1.3m$$

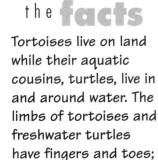

Just the facts

Tortoises live on land while their aquatic cousins, turtles, live in and around water. The limbs of tortoises and freshwater turtles have fingers and toes; marine turtles have flat flippers.

You can use the formulas you found in Problem 1 to figure out the tortoises' positions at any time during the race. A table or a graph showing the positions of the tortoises at various times can help you understand how the race progressed.

2. Make a table like the one below that shows the tortoises' positions at various times during the race. To find a tortoise's distance at a given time, substitute the time value into the tortoise's formula. In your table, include whole-number minutes and any other time values that help you understand what happens during the race.

The Tortoise Race

Time (min)	Distance from Start (m)			Comments
	Tortoise 1	Tortoise 2	Tortoise 3	
0	3.1	0	0	They're off!
1	4.0	1.3	0.17	Tortoise 1 is well in front.
2	4.9	2.6	0.66	Tortoise 3 looks hopeless!
3				

3. On a single set of axes, make graphs that show the position of each tortoise at the times you have included in your table. Use a different color or symbol for each tortoise.

4. Use your table and graph to determine which tortoise won the race. You may need to add values to your table or extend your graph to find the winner.

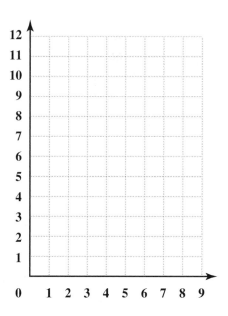

Share & Summarize

A *cube* is a block with six identical square faces.

1. The *surface area* of a cube is the total area of all six faces. Write a formula for the surface area *S* of a cube with edge length *e*. Explain why your formula is correct.

2. Use your formula to find the surface area of a cube with edges of length 8 cm.

Lab Investigation ▶ Formulas and Spreadsheets

MATERIALS
computer with spreadsheet software
(1 per group)

Jo has started her own business selling hand-painted note cards. She designs and paints the front of the cards, and then she bundles the finished cards, with envelopes, in packs of 10.

Using her own Web site, Jo has started to get orders from many specialty shops. She charges $15 for each pack of cards. By charging $7 for shipping regardless of the size of the order, she hopes to encourage large orders.

To help her calculate how much to charge for the various orders she gets, Jo set up a spreadsheet on her computer. She started by entering headings in the first three columns.

	A	B	C
1	Number of Packs	Price	Price with Shipping
2			
3			
4			

Each box is called a *cell*. This is Cell A2.

When she gets a new order, Jo enters the number of packs ordered into Column A, Row 2. This is Cell A2.

To have the spreadsheet do the calculations automatically, Jo entered formulas into Cells B2 and C2 to tell the spreadsheet what to do. When she entered the formulas, she used the name of Cell A2 as the variable. This tells the spreadsheet to use the value in Cell A2 for its calculations.

	A	B	C
1	Number of Packs	Price	Price with Shipping
2		=A2*15	=A2*15+7
3			
4			

The = sign tells the spreadsheet that the entry is a formula to evaluate.

Look at the formula in Cell B2. It tells the spreadsheet to calculate the price of the number of packs entered in Cell A2: "Multiply the value in A2 by 15." The formula in Cell C2 calculates the price with shipping: "Multiply the value in A2 by 15, and then add 7."

Try It Out

Set up a spreadsheet just like Jo's.

1. Without highlighting the cells, look at Cells B2 and C2. What do they display? Why do you think this may have happened?

2. Jo received an order from Carl's Cards and Gifts for 5 packs of cards. Use your spreadsheet to find how much Jo charges for 5 packs, both with and without shipping. To do this, select Cell A2. Then type "5" and press the return key or an arrow key.

3. Suppose Jo gets three more orders: 8 packs, 10 packs, and 30 packs. Find the charges with shipping for these orders.

Just the facts

When you start a spreadsheet program, Cell A1 is usually highlighted in some way. You can choose cells by using the arrows keys to move the highlight or by using a mouse to click on the cell you want.

Try It Again

After a while, Jo realized that most of her orders were in multiples of 5: 5 packs, 10 packs, 15 packs, 20 packs, and so on. Rather than changing her spreadsheet for every order, she wanted a more efficient way to compute the charges. A friend suggested she make a chart listing the charges for orders of various sizes. She could print the chart and then wouldn't have to turn to her computer so often.

To make her chart, Jo used the Fill Down command. To do this, she selected all the cells in Columns B and C, from Row 2 to Row 21.

	A	**B**	**C**
1	Number of Packs	Price	Price with Shipping
2		0	7
3			
4			
5			
6			
7			
8			
9			
10			
11			
12			
13			
14			
15			
16			
17			
18			
19			
20			
21			
22			

The Fill Down command took the formulas in the top cells, B2 and C2, and copied them into Cells B3–B21 and C3–C21. The formulas in Column B all looked like the formula in Cell B2, and the formulas in Column C all looked like the one in Cell C2.

4. Copy the formulas in your spreadsheet just like Jo did. Then look at the formulas in Cells B14 and C14. They don't look *exactly* like the formulas in Cells B2 and C2. For example, the formula in Cell B14 should be A14*15. What changed? Why do you think this happened?

To finish her chart, Jo needed to enter the order amounts into Column A—the numbers from 5 to 100 by fives—but she didn't want to take the time to type each number into its cell. She thought of an easier way to do it.

Jo entered the first number, 5, into Cell A2. Then she typed the formula A2+5 into Cell A3.

5. What number would the spreadsheet calculate for Cell A3?

Jo used the Fill Down command to fill the cells in Column A, from Row 4 to Row 21, with her formula for Cell A3.

6. Do you think the formula in Cell A4 is *exactly* the same as the formula in Cell A3? If not, what do you think the formula is?

7. What value would be calculated for Cell A4?

8. Try it with your spreadsheet. What happened? What does Column A look like now?

9. What numbers are given in Row 14 of your spreadsheet? What do these numbers mean?

10. Look at your spreadsheet to find what Jo should charge, with shipping, for an order of 45 packs of cards.

11. Design a spreadsheet of your own. Use a formula or equation from this chapter, or from a science book. Show a reasonable range of values, using an increment in the inputs to give about 20 to 30 rows of calculations. Label the columns clearly so someone could understand what your chart shows.

What Did You Learn?

12. A spreadsheet cell contains the formula D3*4–1.

a. What is the variable in this formula?

b. The formula in this cell is copied to the cell below it. What is the formula in the new cell?

13. Name at least one way in which spreadsheets can be helpful.

On Your Own Exercises

1. A ticket to the symphony costs $17 for adults and $8.50 for children.

    a. Write an expression for the total ticket cost in dollars if A adults and three children attend the symphony.

    b. Write an expression for the total ticket cost if two adults and C children go to the symphony.

2. Anica and her friends are organizing a trip to the state skateboarding championships. The bus they will rent to take them there and back will cost $250, and Anica estimates that each person will spend $40 for meals and souvenirs.

    a. If p people go on the trip, what will be the total cost for travel, meals, and souvenirs?

    b. What will be the cost per person if p people go on the trip?

    c. What will be the total cost per person if 25 people go on the trip?

3. A full storage tank contains 2,160 gallons of water. Every 24 hours, 4.5 gallons leak out.

    a. How many gallons of water will be in the tank H hours after it is filled?

    b. How many gallons of water will be in the tank after D days?

    c. How many gallons of water will be in the tank after W weeks?

In Exercises 4–6, determine whether the expression $9m - 4$ can be used to represent the answer to the question. If it can, explain what m stands for.

4. The ski club bought nine lift tickets and received a $4 discount on the total price. What was the total cost for the tickets?

5. Of the nine people on a baseball field, four are girls. How many are boys?

6. Last Saturday, Kendra listened to her new CD nine times, but the ninth time she played it, she skipped a 4-minute song she didn't like. How much time did Kendra spend listening to her CD last Saturday?

In Exercises 7–9, describe a situation that the expression can represent.

7. $3p + 2$, if p is the price of a pepperoni pizza in dollars

8. $2x - 8$, if x is the number of ounces in a pitcher of lemonade

9. $\frac{L^2}{2}$, if L is the length of a square piece of paper

10. Three friends, Aisha, Mika, and Caitlin, have electronic robot toys that move at different speeds. They decide to have a 10-meter race.

- Aisha's robot travels 1 meter per second.

- Mika's robot moves 0.9 meter per second.

- Caitlin's robot's speed is 1.3 meters per second.

a. If they all start together, whose robot will win? How long will each robot take to reach the finish line?

b. To make the next race more interesting, the friends agree to give Mika's and Aisha's robots a head start.

- Mika's robot has a 3-meter head start, so it has to travel only 7 meters.

- Aisha's robot has a 2-meter head start, so it has to travel only 8 meters.

Whose robot wins this race? Whose comes in second? Whose comes in last?

c. The girls decide they would like the robots to finish as close to *exactly* together as possible. How much of a head start should Mika's and Aisha's robots have for this to happen?

Hint: Use your calculator. Caitlin's robot takes $10 \div 1.3 = 7.6923077$ seconds to travel 10 meters. How can you position the other two robots so that they take this same amount of time?

11. **Measurement** By measuring certain bones, forensic scientists can estimate a person's height. These formulas show the approximate relationship between the length of the tibia (shin bone) t and height h for males and females. The measurements are in centimeters.

$$\text{males: } h = 81.688 + 2.392t$$

$$\text{females: } h = 72.572 + 2.533t$$

a. How tall is a male if his tibia is 38 cm long? Give your answer in centimeters, and then use the facts that 2.54 cm is about 1 in. and 12 in. equals 1 ft to find the height in feet.

b. How tall is a female if her tibia is 38 cm long? Give your answer in centimeters and in feet.

c. If a woman is 160 cm tall (about 5.25 ft), how long is her tibia?

12. **Physics** The distance in meters an object falls in T seconds after it is dropped—not taking into account air resistance—is given by the formula $D = 4.9T^2$.

a. A stone is dropped from a high cliff. How far will it have fallen after 3 s?

b. After dropping from a plane, a parachutist counts slowly to 10 before pulling the ripcord to unfold her parachute. How far does she fall while she is counting?

c. Shelley dropped a pebble from a cliff and timed it as it fell. The pebble hit the water 3.7 s after she dropped it. How high is the cliff above the water?

d. The watch Shelley used was accurate only to the nearest 0.1 s. This means that the actual time it took the pebble to fall may have been 0.1 s less or 0.1 s more than what Shelley timed. Find a range of heights for the cliff to allow for this error in timing the pebble's fall.

13. **Geology** Geologists estimate that Earth's temperature rises about 10°C for every kilometer below Earth's surface.

a. If it is 50°C on the surface of Earth, what is the formula for the temperature T in degrees Celsius at a depth of k km?

b. If it is 50°C on the surface, what is the temperature at a depth of 15 km?

14. A rental car costs $35 per day plus $.10 per mile.

a. What is the cost, in dollars, to rent the car for 5 days if you drive a total of *M* miles?

b. What is the cost to rent the car for 5 days if you drive *M* miles each day?

c. What is the cost to rent the car for *D* days if you drive a total of 85 miles?

d. What is the cost to rent the car for *D* days if you drive 85 miles each day?

e. What would be the cost per person if three people share the cost of renting the car in Part d?

15. The diagram shows the square floor of a store. A square display case with sides of length 3 ft stands in a corner of the store. The manager wants the floor area painted. Assume the display case can't be moved.

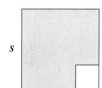

a. How many square feet need to be painted?

b. The manager hires two people to paint the floor. How many square feet would each person have to paint if they share the job equally?

c. Are there any restrictions on the value of *s*? If so, explain why and tell what the restrictions are. If not, explain why not.

16. The formula below helps scuba divers figure out how long they can stay under water.

$$T = \frac{120V}{d}$$

T: approximate maximum time, in minutes, a diver can stay under water

V: volume of air in the diver's tank, in cubic meters, before compression

d: depth of the water, in meters

a. Machiko has 1 cubic meter of air compressed in her tank. She is 4 meters under water. How long can she stay down?

b. How long can Machiko stay 8 meters under water with 1 cubic meter of air in her tank?

c. If Machiko wanted to stay 4 meters under water for 1 hour, how much air would she need?

17. Geometry The formula for the area of a trapezoid is $A = \frac{h(B + b)}{2}$, where B and b are the lengths of the two parallel sides and h is the height.

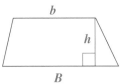

a. Find the area of a trapezoid with parallel sides of length 6 cm and 7 cm and height 5 cm.

b. The area of a trapezoid is 6 cm². What might the values of h, B, and b be?

18. Sports A baseball player's slugging percentage, P, can be computed with this formula:

$$P = \frac{S + 2D + 3T + 4H}{A}$$

where S is the number of singles, D is the number of doubles, T is the number of triples, H is the number of home runs, and A is the number of official at bats.

a. In 1998, Sammy Sosa of the Chicago Cubs hit 112 singles, 20 doubles, 0 triples, and 66 home runs in 643 at bats. What was his slugging percentage?

b. In 1998, Mark McGwire of the St. Louis Cardinals hit 61 singles, 21 doubles, 0 triples, and 70 home runs in 509 at bats. What was his slugging percentage?

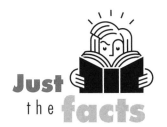

Just the facts

A *single* hit enables the batter to reach first base; a *double*, to reach second base, a *triple*, to reach third base; and a *home run*, to make a complete circuit of the bases and score a run.

A batting average and a slugging percentage are different. A *batting average* is a measure of the number of hits a player makes for every time at bat. A *slugging percentage* figures in how many bases the batter ran for every time at bat.

Sammy Sosa

Mixed Review

Evaluate each expression.

19. $\frac{7}{8} + \frac{1}{4}$ **20.** $\frac{8}{11} - \frac{1}{2}$ **21.** $\frac{1}{10} \cdot \frac{1}{3}$ **22.** $\frac{2}{5} \div \frac{1}{2}$

Find each percentage.

23. 5% of 100   **24.** 100% of 5   **25.** 5% of 10

Find the value of b in each equation.

26. $\frac{3}{4}b = 6$ **27.** $\frac{5}{2} - b = \frac{3}{2}$ **28.** $2.2b - 4.2 = 2.4$

29. Draw a factor tree to find all the prime factors of 100.

30. Copy and complete the chart.

Fraction	Decimal	Percent
$\frac{1}{2}$	0.5	50%
		45%
$\frac{1}{8}$		
	0.05	

31. How many triangles are there in this figure? Be sure to look for different sizes!

32. Probability Kyle made up a game for the school carnival. He put 9 blocks in a container—3 red, 3 blue, and 3 yellow. To play his game, you choose one of the three colors and then—without looking into the container—select a block. If the block matches your color, you win 10 points.

a. What are the chances of selecting a blue block? A red block? A yellow block?

b. Meela chose the color yellow and then reached into the container and pulled out a yellow block. Kyle says that if she picks one of the two remaining yellow blocks on her next draw, she will get 20 more points. What are the chances that Meela's second block will be yellow?

c. Meela does it again! Now Kyle says that if she chooses the last yellow block on her third draw, she will earn 30 more points. What are the chances that Meela's third block will be yellow?

The Distributive Property

If you have ever solved a multiplication problem like 4 × 24 by thinking, "It's 4 × 20 plus 4 × 4" or "It's 4 less than 4 × 25," you were using a mathematical property called the *distributive property.* Using this property, you can change the way you think about how numbers are grouped. For example, rather than think about 4 groups of 24, you can think about 4 groups of 20 added to 4 groups of 4. As you'll see, this property is helpful for more than just mental arithmetic.

You have used bags and blocks to help you think about algebraic expressions. Bags and blocks can help you look more closely at the distributive property, too.

Think & Discuss

Shaunda, Kate, and Malik are each holding one bag and two extra blocks.

Find the total number of blocks if each bag contains

• 5 blocks

• 20 blocks

• 100 blocks

• *b* blocks

How did you find your answers?

Here's how Jin Lee and Luis found the total number of blocks when there were 20 in each bag.

Jin Lee's method:

"Each person has 22 blocks, so I multiplied 22 by 3 and got 66."

Luis's method:

"There are 3 bags—that's 60 blocks—and 3 sets of 2 leftover blocks—that's 6 more. So, there are 66 blocks."

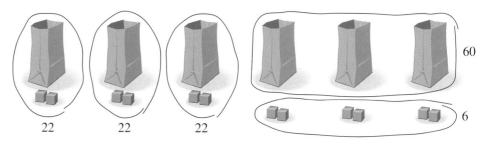

$$3(20 + 2) = 3 \cdot 22 = 66$$

$$3 \cdot 20 + 3 \cdot 2 = 60 + 6 = 66$$

Jin Lee's and Luis's methods both work no matter how many blocks are in each bag. You can express their methods in symbols for b blocks in each bag.

Jin Lee's method	**Luis's method**
$3(b + 2)$	$3b + 3 \cdot 2 = 3b + 6$

Which way do *you* like to think about this situation? Or do you have another way?

Whichever method you prefer, the important thing to understand is that these two ways of looking at the problem give the same answer. That is,

$$3(b + 2) = 3b + 6$$

Investigation Grouping Bags and Blocks

One way to see how different expressions can describe the same situation is to consider different groupings of a given number of bags and blocks. Each grouping will have the same total number of blocks, so the expressions you create must represent the same quantity.

Problem Set A

1. Brigitte placed 3 blocks in front of each of 4 bags. For each situation below, show two ways of finding the total number of blocks she has. If you need help, look back at Jin Lee's and Luis's methods.

a. 6 blocks in each bag   **b.** 15 blocks in each bag

c. 100 blocks in each bag   **d.** *b* blocks in each bag

2. Keenan set 4 blocks in front of three sets of 2 bags. For each situation, show two ways of finding the total number of blocks he has.

a. 7 blocks in each bag   **b.** 11 blocks in each bag

c. 100 blocks in each bag   **d.** *b* blocks in each bag

3. Flowcharts can also help you see different ways to express a quantity.

   a. Think about how you might create a flowchart to calculate the total number of blocks in the situation from Problem 2. Draw and label a flowchart that has four ovals.

   b. Use an input of *b* for your flowchart. Find the output expression.

   c. Draw another flowchart to find the number of blocks in this situation, but this time use only three ovals. Use an input of *b* and find the output expression.

4. Solana placed 1 block in front of each of 4 bags. Find two expressions for the total number of blocks she has.

5. Simon and Zoe's teacher held up 3 bags and told the class that each contained the same number of blocks. She removed 2 blocks from each bag. How can you express the number of blocks still in the bags?

To answer this question, Simon and Zoe decided to experiment by starting with 7 blocks in each bag.

I would have the 3 bags of 7 that I started with (which is 3 x 7), and I would remove 2 from each of the bags (which is 3 x 2). So I would have 3 x 7 − 3 x 2 blocks. That's the same as 21 − 6 = 15 blocks.

I would have 3 bags, and each bag would have 7 − 2 blocks in it. So I would have 3 groups of 7 − 2, which is 3(7 − 2) blocks. That's the same as 3 x 5 = 15 blocks.

a. If b blocks are in each bag, Zoe's reasoning can be expressed in symbols as $3b - 6$. Write an expression that fits Simon's reasoning.

b. Copy and complete the table to show the results of using Zoe's and Simon's methods for various numbers of blocks.

Number of Blocks	b	2	4	7	10	18	
Zoe's Method	$3b - 6$						60
Simon's Method				15			

Share & Summarize

Dante put 5 blocks in front of each of 2 bags.

1. Describe, in words, two ways of finding the total number of blocks if you know the number of blocks in each bag.

2. Write two rules, in symbols, for finding the total number of blocks if there are s blocks in each bag.

Investigation 2 ▶ The Same and Different

In the last investigation, you discovered more than one way to find the total number of blocks in a bags-and-blocks situation. You can also use tables and flowcharts to help figure out why two expressions can *look* different but produce the same outputs.

Kaya, Maria, and Luis played the game *What's My Rule?* Luis made up a rule for finding the output for any input. Then, Kaya and Maria gave Luis several inputs, and he told them the outputs his rule would produce. The girls organized the input/output values in a table.

Input	1	3	6	4	5
Output	8	12	18	14	16

Kaya and Maria used their data to guess Luis's rule. When they gave their rules, Luis had a problem.

Kaya and Maria described two rules that fit the table. If the input is represented by K, Kaya's rule is written in symbols as $2K + 6$. Maria's rule is written as $2(K + 3)$.

It is reasonable to believe that the calculation $2K + 6$ will always give the same result as the calculation $2(K + 3)$. To help you see *why* this works—and why it *must* work no matter what K is—put back the missing multiplication signs.

Start with $2(K + 3)$, and put back the missing multiplication sign:

$$2 \times (K + 3)$$

This is two groups of $(K + 3)$, which can be written

$$K + 3 + K + 3$$

This is the same as $K + K + 3 + 3$, or

$$2K + 6$$

VOCABULARY
equivalent expressions

So, $2K + 6$ and $2(K + 3)$ are **equivalent expressions.** That means they must give the same result for every value of K.

Problem Set B

1. Let's look more closely at the expressions $2K + 6$ and $2(K + 3)$.

   a. Draw a flowchart for each expression.

   b. For one of the expressions, you multiply first and then add. For the other, you add first and then multiply. Which expression is which?

2. You have seen that $2K + 6$ and $2(K + 3)$ are two ways of writing the same thing. And remember, Luis was thinking of a rule that was different from both of these but gave the same outputs.

   The rules below are also equivalent to $2K + 6$ and $2(K + 3)$. Using K for the input, write an expression for each rule, and show that your expression is equivalent to $2K + 6$.

   a. Add the input to 6 more than the input.

   b. Add 2 more than the input to 4 more than the input.

   c. Add 2 to the input, and double the sum. Add 2 to the result.

   d. Add 5 to the input, and double the sum. Subtract 4 from the result.

3. Write another rule, in words, that is equivalent to $2K + 6$.

You have seen that in the game *What's My Rule?* the rule can often be written in more than one way. You will now look at input/output tables from the game and express each rule in two ways.

Problem Set C

1. Here's an input/output table for a game of *What's My Rule?*

Input	0	1	2	3	4
Output	6	8	10	12	14

a. Use symbols to write two rules for this game. One rule should involve multiplying first and then adding, and the other should involve adding first and then multiplying. Substitute some other inputs to check that the two rules give the same output.

b. Why do the two rules give the same output for a given input?

Use symbols to write two rules for the data in each table. One rule should involve multiplying first and then adding or subtracting, and the other should involve adding or subtracting first and then multiplying.

2.
Input	2	3	4	5	6
Output	0	5	10	15	20

3.
Input	0	1	2	3	4
Output	4	4.5	5	5.5	6

Copy and complete each table. Then write another expression that gives the same outputs as the given expression. Check your expressions by substituting input values from your tables.

4.
j	0	1		8		20
$4(j + 2)$	8		20		76	

5.
x	2	3	4	5		18		
$5x + 5$	15				35		215	415

Share & Summarize

1. Write three expressions that are equivalent to $3P + 12$.

2. Use symbols to write two rules for the data in the table. One rule should involve multiplying first and then adding or subtracting; the other, adding or subtracting first and then multiplying.

Input, k	3	4	13	24	42	50
Output	0	4	40	84	156	188

Investigation ▶ 3 Grouping with the Distributive Property

For the bags-and-blocks situation on page 53, Jin Lee and Luis found two ways to calculate the total number of blocks. In the *What's My Rule?* game on page 56, Kaya and Maria found two rules for determining the output for a given input.

One way to find different rules is to look at different groupings of quantities, as you did with bags and blocks in Investigation 1. For example, you can think of the diagram below as a single rectangular array of dots or as two rectangular arrays put together.

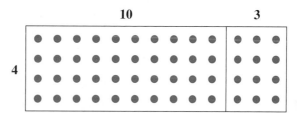

These two ways of thinking about the diagram lead to two ways of calculating the total number of dots in the diagram.

- The total number of dots can be found by noting that there are 4 rows with 10 + 3 dots each:

$$4(10 + 3)$$

- The total number of dots can be found by adding the number of dots in the left rectangle to the number of dots in the right rectangle:

$$4 \times 10 + 4 \times 3$$

Since both $4(10 + 3)$ and $4 \times 10 + 4 \times 3$ describe the total number of dots in the diagram,

$$4(10 + 3) = 4 \times 10 + 4 \times 3$$

Problem Set D

1. Describe two ways to find the number of dots in this diagram. Write an expression for each method.

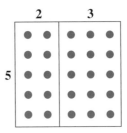

2. Create a dot diagram to show that $3(4 + 5) = 3 \times 4 + 3 \times 5$.

In the diagrams below, the dots are not shown, but the total number of dots is given and labels indicate the number of rows and columns. Use the clues to determine the value of each variable.

3.

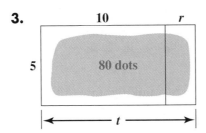

4.

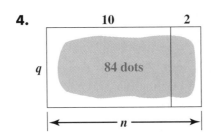

5. Jane wrote $5(10 + 6) = 5 \times 10 + 6$ in her notebook.

   a. Find the value of the expression on each side to show that Jane's statement is incorrect.

   b. Make a dot diagram you could use to explain to Jane why her statement doesn't make sense.

6. Describe two ways to find the number of dots in this diagram. Write an expression for each method.

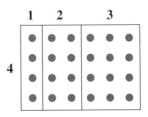

Dot diagrams help you see how different groupings can give equivalent expressions. In this lesson, you have seen many pairs of equivalent expressions, like $3(10 + 9)$ and $3 \times 10 + 3 \times 9$, or $2(n - 3)$ and $2n - 2 \times 3$. When you rewrite an expression like $3(10 + 9)$ or $3(20 - 1)$ as a sum or difference of products, you are using the *distributive property.*

Think & Discuss

What does it mean to *distribute* something? (A dictionary might help.)

What is being distributed in this equation:

$$3(10 + 9) = 3 \times 10 + 3 \times 9$$

At the beginning of this lesson, you read about a shortcut that can be used to compute 4×24 mentally. Such shortcuts are examples of the distributive property.

Shortcut in Words	Shortcut in Symbols
"It's 4×20 plus 4×4."	$4 \times 24 = 4(20 + 4) = 4 \times 20 + 4 \times 4$
"It's 4 less than 4×25."	$4 \times 24 = 4(25 - 1) = 4 \times 25 - 4 \times 1$

Sometimes a calculation can be simplified by using the distributive property in reverse.

EXAMPLE

Find a shortcut for calculating $12 \times 77 + 12 \times 23$.

$$12 \times 77 + 12 \times 23 = 12(77 + 23)$$
$$= 12(100)$$
$$= 1{,}200$$

Problem Set E

Use the distributive property to help you do each calculation mentally. Write the grouping that shows the method you used.

1. $5 \cdot 17$ **2.** $6 \cdot 41$

3. $4 \cdot 19$ **4.** $7 \cdot 27$

5. $6 \cdot 45$ **6.** $9 \cdot 38$

Copy each equation, inserting parentheses if needed to make the equation true.

7. $4 \times 8 + 3 = 44$ **8.** $4 \times 8 + 3 = 35$

9. $3 \times 7 + 4 = 25$ **10.** $3 \times 7 + 4 = 33$

Find a shortcut for doing each calculation. Use parentheses to show your shortcut.

11. $9 \times 2 + 9 \times 8$ **12.** $19 \times 2 + 19 \times 8$

13. $12 \times 4 + 12 \times 6$ **14.** $7 \times \frac{3}{5} + 3 \times \frac{3}{5}$

You have been rewriting expressions as sums or differences of products using two versions of the **distributive property.** Each version has its own name.

When addition is involved, you use the *distributive property of multiplication over addition.* The general form of this property states that for any numbers *n*, *a*, and *b*,

$$n(a + b) = na + nb$$

The distributive property you have used to write an expression as a difference of products is the *distributive property of multiplication over subtraction.* The general form of this property states that for any numbers *n*, *a*, and *b*,

$$n(a - b) = na - nb$$

Each of these more specific names mentions two operations: multiplication, and either addition or subtraction. You distribute the number that multiplies the sum or difference to each part of the sum or difference.

In the next problem set, you will explore whether distribution works for several combinations of operations.

Problem Set F

1. The expressions in the following statement involve division rather than multiplication.

$$\frac{a + b}{c} = \frac{a}{c} + \frac{b}{c}$$

Choose some values for a, b, and c, and test the statement to see whether it is true. For example, you might try $a = 2$, $b = 5$, and $c = 7$. Try several values for each variable. Do you think the statement is true for all values of a, b, and c?

2. The expressions in the statement below are like those in Problem 1, but they involve multiplication rather than addition.

$$\frac{ab}{c} = \frac{a}{c} \times \frac{b}{c}$$

Choose some values for a, b, and c, and test the statement to see whether it is true. Try several values for each variable. Do you think the statement is true for all values of a, b, and c?

3. Choose some values for a and b, and test this statement:

$$(a + b)^2 = a^2 + b^2$$

Do you think the statement is true for all values of a and b?

Share & Summarize

1. Make a dot diagram to show that $6 \times 3 + 6 \times 2 = 6(3 + 2)$.

2. Give examples of calculations that look difficult but are easy to do mentally by using the distributive property. For each example, explain how the distributive property can be used to simplify the calculation.

Investigation ▶ 4 ▶ Removing and Inserting Parentheses

VOCABULARY
expand
factor

The distributive property explains how you can write expressions in different ways. Using the distributive property to remove parentheses is called **expanding.** Using the distributive property to insert parentheses is called **factoring.**

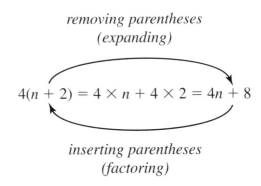

removing parentheses
(expanding)

$$4(n + 2) = 4 \times n + 4 \times 2 = 4n + 8$$

inserting parentheses
(factoring)

Expanding and factoring allow you to change the *form* of an expression (what it looks like) without changing the output values it gives.

You have seen that rewriting an expression in a different form can be useful for simplifying calculations and comparing expressions. Later, you will see that rewriting expressions can help you solve equations.

In this investigation, you will practice using the distributive property to rewrite expressions. First, you'll focus on expanding expressions, or removing parentheses.

Problem Set G

1. Consider the equation $4(n + 2) = 4n + 8$.

   a. Is the equation true for $n = 2$?

   b. It is impossible to check that $4(n + 2) = 4n + 8$ for every value of n. How do you know that this equation is true for any value of n?

Decide whether each equation is true for all values of n. Explain how you decided.

2. $5(n + 6) = 5n + 30$ **3.** $7(n + 3) = 7n + 3$

4. $(n + 12) \cdot 9 = 9n + 108$ **5.** $1.5(n + 2.5) = 1.5n + 3.75$

6. $2(n + 4) = 2n + 4$ **7.** $(n + 9) \cdot 8 = n \cdot 8 + 72$

Expand each expression. (That is, rewrite it to remove the parentheses.)
Check the resulting expression by making sure it gives the same values as
the original expression for at least three values of x.

8. $4(2x + 3)$

9. $4(3x + 7)$

10. $0.5(2x + 8)$

Expand each expression.

11. $3(n - 4)$ **12.** $4(2n + 7)$ **13.** $8(3n - 3)$

14. $1(4n - 7)$ **15.** $2(2n - 3.5)$ **16.** $10(n + 10)$

17. $1(n + 1)$ **18.** $0(n + 9{,}999)$ **19.** $2(n + 0)$

The expressions below look a bit different from those in Problems 11–19,
but the distributive property still works. Expand each expression. If you
are uncertain about a result, test it by substituting values for the variables.

20. $r(n + 1)$ **21.** $n(2 + 7)$ **22.** $x(4 - 3n)$

23. $2(n + 9n)$ **24.** $n(2 + 18)$ **25.** $2(1.5n + 2.5n)$

Now you will use the distributive property to insert parentheses.

Just the facts

In your study of mathematics, you will sometimes use more than just standard parentheses. Here are a few styles of parentheses, brackets, and braces you might encounter.

$$\{\,[\,(\{\,[\,(\ )\,]\,\}\,)\,]\,\}$$

EXAMPLE

To rewrite $3d + 21$ by inserting parentheses, think about how the
distributive property could have been used to produce the expression.

Both parts of the expression, $3d$ and 21, can be rewritten as 3 times
something.

$$3d + 21 = 3 \times d + 3 \times 7$$
$$= 3(d + 7)$$

You can check this result by using the distributive property to expand
$3(d + 7)$.

$$3(d + 7) = 3 \times d + 3 \times 7$$
$$= 3d + 21$$

Rewriting an expression by inserting parentheses is called *factoring* because the resulting expression is the product of factors.

You could factor $3d + 21$ in other ways. For example, you could write $\frac{1}{2}(6d + 42)$ or $0.1(30d + 210)$. However, working with these expressions is a lot more complicated. Unless the expression you are factoring contains fractions or decimals, use only whole numbers in your factors.

The examples below show how some other expressions can be factored.

$$7 + 7x = 7 \times 1 + 7 \times x = 7(1 + x)$$

$$7t - 3t = 7 \times t - 3 \times t = (7 - 3)t = (4)t = 4t$$

$$6m + 3m^2 = 3m \times 2 + 3m \times m = 3m(2 + m)$$

$$s + ms + nms = s + m \times s + n \times m \times s = s(1 + m + nm)$$

After you factor, look carefully at the expression inside the parentheses. You may find that more factoring can be done.

EXAMPLE

Factor the expression $27p^2 + 18p$.

Since 9 divides both parts of the expression, it can be rewritten as follows:

$$27p^2 + 18p = 9(3p^2 + 2p)$$

Now look at the expression inside the parentheses. Since p divides both $3p^2$ and $2p$, the expression can be factored further.

$$9(3p^2 + 2p) = 9p(3p + 2)$$

So, $27p^2 + 18p = 9p(3p + 2)$.

Not all expressions can be easily factored. For example, no whole number or variable evenly divides both parts of $5x + 7$.

Problem Set H

Determine whether there is a whole number or variable that divides both parts of each expression. If there is, use the distributive property to rewrite the expression using parentheses. Check the resulting expression by expanding it.

1. $4a + 8$ **2.** $4b + 12$ **3.** $4c + 17$

4. $3g - 15$ **5.** $5f + 13$ **6.** $8h - 24$

Factor each expression. You may need to look closely to see how to do it. Check the resulting expression by expanding it.

7. $22s + 33$ **8.** $34t - 4$

9. $45m + 25k$ **10.** $7j^2 + 3j$

11. $4t + 9t$ **12.** $8g^2 + 12g$

13. $10m + 15t + 25$ **14.** $8 - 16h^2 + 20h$

15. Every morning Tonisha and her dog Rex run to the local park, around the park, and back home. It is 400 meters to the park and x meters around the park. Write two expressions for the total distance, in meters, they run in a week. One of your expressions should involve parentheses.

16. Explain why $4t + 9t$ and $6k + 9m$ can be factored but $4k + 9m$ cannot.

17. Prove It! Show that $2k + 3k = 5k$ by first factoring $2k + 3k$. Then check that $2k + 3k = 5k$ for $k = 7$ and $k = 12$.

Share & Summarize

1. Create an expression in factored form. Give it to your partner to expand.

2. Create an expression that can be factored. Give it to your partner to factor.

3. Explain how expanding and factoring are related.

On Your Own Exercises

Practice &
Apply

1. Suppose there are 5 bags with *n* blocks in each bag, and 3 extra blocks beside each bag.

    a. Write two expressions for the total number of blocks, as you did in Problem Set A.

    b. Check that your two expressions give the same total for $n = 8$, $n = 12$, and $n = 25$.

    c. Explain—using a picture, if you like—why your two expressions must give the same total number of blocks for any value of *n*.

2. Suppose there are four sets of 3 bags, with 3 blocks in front of each set. For each situation, show two ways to find the total number of blocks.

    a. 5 blocks in each bag

    b. 100 blocks in each bag

    c. *b* blocks in each bag

3. Shaunda has 5 bags, with the same number of blocks in each. She removes 2 blocks from each bag. For each situation below, show two ways of finding the total number of blocks in the bags.

    a. The bags start with 7 blocks in each.

    b. The bags start with 50 blocks in each.

    c. The bags start with *p* blocks in each.

4. Consider the expressions $3(T - 1)$ and $3T - 3$.

    a. Copy the table, and complete the first four columns. In the last four columns, choose your own input values and calculate the output for both rules. Use fractions for at least two of your input values.

T	4		100				
$3T - 3$		30		87			
$3(T - 1)$	9	30					

    b. Show that $3(T - 1)$ and $3T - 3$ are equivalent expressions.

    c. Find two more expressions that are equivalent to $3(T - 1)$ and $3T - 3$.

 impactmath.com/self_check_quiz

Copy and complete each table. Then, for each table, write another expression that gives the same outputs. Check your new expression to make sure it generates the same values.

5.

h	0	1	2		42	
$2(h + 3)$	6			20		100

6.

r		5	10	11		23
$5(r - 2)$	5				100	

7.

n	4	5	6	8		
$\frac{n-4}{4}$	0				10	24

For each *What's My Rule?* table, find two ways to write the rule in symbols. Choose your own variable.

8.

Input	0	1	2	3	4
Output	8	10	12	14	16

9.

Input	0	1	2	3	4
Output	6	6.5	7	7.5	8

Use the clues on each dot diagram to find the unknown values.

10.

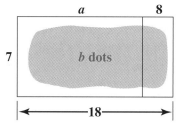

11.

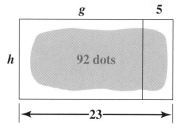

Copy each equation, inserting parentheses when needed to make the equation true.

12. $5 \times 2 + 3 = 25$

13. $12 + 3 \times 7 = 105$

14. $11 + 8 \times 4 = 43$

15. $0.2 + 0.2 \times 0.2 = 0.08$

Use the distributive property to help you do each calculation mentally. Write the grouping that shows the method you used.

16. $17 \cdot 2 + 17 \cdot 8$

17. $16 \cdot 4 - 4 \cdot 4$

18. $11 \cdot 5 + 5 \cdot 9$

19. $\frac{20}{87} + \frac{80}{87}$

Decide whether each equation is true for all values of the variable. Justify your answers.

20. $6(W + 2) = 6W + 2$

21. $(Y + 176) \div 8 = \frac{Y}{8} + \frac{176}{8}$

22. $2.5(B + 12) = 2.5B + 30$

23. $(a + 3) \times 7 = a \times 7 + 3$

Expand each expression. Check the resulting expression by making sure it gives the same values as the original expression for several values of the variable.

24. $5(3j + 4)$

25. $0.2(4k + 9)$

26. $2(n - 7)$

27. $n(n - 6)$

28. $3(3n + 3)$

29. $0(n + 875)$

Factor each expression, if possible.

30. $4a + 8$

31. $18 + 5b$

32. $3g - 15$

33. $18 - 2A$

34. $11v + 3v$

35. $5z^2 + 2z$

Remember

Expanding an expression means rewriting it to remove the parentheses.

Connect & Extend

36. Joshua has 4 bags of blocks. He removes 3 blocks from each bag.

    a. Write two expressions for the total number of blocks in Joshua's bags now.

    b. Explain—using a picture, if you like—why your two expressions must give the same total number of blocks for any value of the variable.

37. The rock band "The Accidents" brought five boxes of their new CD, *Waiting to Happen,* on their tour. Each box holds the same number of CDs. They hoped to sell all the CDs, but they sold only 20 CDs from each box.

    a. Write an expression that describes the total number of unsold CDs.

    b. Write an expression that describes how much the band will earn, in dollars, by selling the remaining CDs for $15 each.

38. In this pattern of toothpicks, the width of the figure increases from one figure to the next. Using symbols, write two rules for finding the number of toothpicks in a figure. Use parentheses in one of your rules but not in the other. Use *w* to represent the width of a figure.

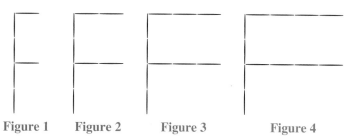

Figure 1   Figure 2   Figure 3   Figure 4

39. This dot diagram is missing so much information that many sets of numbers will work. Find at least three sets of values for *a, b,* and *c*.

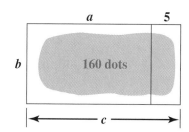

40. You have learned that multiplication distributes over addition. Do you think addition distributes over multiplication? That is, does $a + (b \times c) = (a + b) \times (a + c)$? Support your idea with numerical examples.

41. Marcus says he knows a shortcut for multiplying by 99 in his head. He claims he can mentally multiply any number by 99 within 5 seconds.

   a. Find Marcus's shortcut for multiplying by 99.

   b. Using symbols, explain why his shortcut works. (Hint: His shortcut uses the distributive property.)

Preview For each *What's My Rule?* table, fill in the missing numbers and find the rule. Hint: Consider exponents when you are looking for the rules.

42.

Input	0	1	2	3	5		10	
Output	0	1	4	9		64		625

43.

Input	0	1	2	3	4		7	
Output	3	4	7		28			103

In y o u r **own words**

Assume you are talking to a student two years younger than you are. Explain why $3(b + 4) = 3b + 12$, not $3b + 4$. You might want to draw a picture to help explain the idea.

Expand each expression. Check your answer by substituting 1, 2, and 10 for the variable.

44. $7Q(Q + 8)$

45. $9L(L + 3L^2)$

46. $8R(\frac{1}{2} - R)$

47. $3D(D^2 - \frac{D}{2})$

Factor each expression.

48. $3m^3 + 6m$

49. $2t^3 + 4t^2$

50. $4L + 3L^2$

51. $3P + mP$

52. Challenge When you take any whole number, add $\frac{1}{2}$, and then multiply the sum by 2, the result is an odd number. Explain why.

53. Challenge This fraction contains variables.

$$\frac{8x + 4z + 2}{2}$$

To simplify this fraction, you might first think of it as

$$\frac{1}{2}(8x + 4z + 2)$$

Then you can apply the distributive property.

$$\frac{1}{2}(8x + 4z + 2) = \frac{1}{2}(8x) + \frac{1}{2}(4z) + \frac{1}{2}(2)$$

$$= 4x + 2z + 1$$

Use this method to simplify each fraction.

a. $\dfrac{12x + 6z + 3}{3}$

b. $\dfrac{10x^2 + 5x}{5}$

54. World Cultures In some countries, long multiplication is taught to students using a grid. Here's how to multiply 15 and 31 using a grid.

×	10	5
30	300	150
1	10	5

The product of 15 and 31 is 465, the total of the four numbers inside the grid. Use the grid method to find each product.

a. 14×56

b. 23×23

c. 45×21

Remember

A *prime number* has only two factors: itself and 1.

Mixed Review

55. Challenge You can use factoring to prove some interesting facts about numbers.

    a. If you start with a whole number greater than 1 and add it to its square, the result will never be a prime number. Explain why not. (Hint: Use x to stand for the whole number. Then x^2 is its square, and $x^2 + x$ is the whole number added to its square.)

    b. If you start with a whole number greater than 2 and subtract it from its square, the result can never be a prime number. Explain.

Find each product.

56. 0.08×0.2 **57.** 0.15×0.4 **58.** 0.65×0.2

59. 0.08×0.08 **60.** 0.03×3 **61.** 0.06×0.02

Find each percentage.

62. 70% of 1   **63.** 1% of 70   **64.** 16% of 50

Write each expression without using multiplication or addition signs.

65. $r + r + r$ **66.** $4.2 \cdot r \cdot r \cdot r$ **67.** $5st + 2.7st$

Find the value of each expression for $k = 1.1$.

68. $k^3 - 1^3$ **69.** $\frac{7k}{5}$ **70.** $\frac{7}{5}k$

Geometry Find the area and perimeter of each figure.

71.
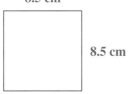
8.5 cm
8.5 cm

72.
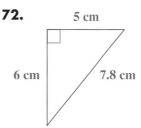
5 cm
6 cm
7.8 cm

73.
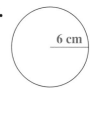
6 cm

74. The grid contains four types of squares: white, shaded, striped, and squares with stars.

    a. What percent of the squares in the grid are striped?

    b. What percent of the squares are shaded?

    c. What percent of the squares have stripes or stars?

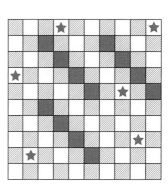

Chapter Summary

In this chapter, you learned that many situations can be described with algebraic expressions. In an *algebraic expression,* symbols—usually letters—are used as variables. *Variables* can be quantities that change or unknown quantities. By investigating different values for a variable, you can explore what happens in a situation as the variable changes.

You learned the standard ways of indicating multiplication, division, and repeated multiplication in algebraic expressions. You created flowcharts to match expressions, and you used flowcharts to backtrack to solve equations.

You found that the same situation can often be described with several *equivalent expressions,* and that sometimes one expression is more useful than another. You saw that you could change expressions into equivalent expressions by using the *distributive property* to *expand* and *factor* them.

Strategies and Applications

The questions in this section will help you review and apply the important ideas and strategies developed in this chapter.

Matching expressions and situations

1. Draw a bags-and-blocks picture to match the expression $2b + 6$. If there are 5 blocks in each bag, what is the total number of blocks?

2. Hector has h baseball caps.

   a. Rachel has 4 more than three times the number of caps Hector has. Write an expression for the number of caps Rachel has.

   b. Hector paid $10 for each of his caps. Write an expression for the amount he paid, in pennies, for all his caps.

Using formulas and evaluating expressions

3. The formula for the circumference of a circle is $C = 2\pi r$, where r is the radius. If the radius of a circle is 4.3 cm, what is its circumference?

Evaluate each expression for $r = 2$.

4. $\frac{r}{2}$ 5. $\frac{3}{2}r$

6. $\frac{3r}{2}$ 7. $3r^2$

Solving equations by backtracking

8. Consider the expression $4(2n - 1) + 5$.

   a. Draw a flowchart to represent the expression.

   b. Use backtracking to solve the equation $4(2n - 1) + 5 = 21$.

9. Find the length of a rectangle with width 4.2 cm and area 98 cm^2.

Using the distributive property

Use the distributive property to rewrite each expression.

10. $3(0.5r + 7)$ **11.** $\frac{10x - 6}{2}$ **12.** $\frac{1}{2}(24r - 1)$

13. $14p + 21$ **14.** $3g + 9g^2$ **15.** $49s - 14s$

16. Find two rules, using symbols, that could produce this table.

Input	0	1	2	3	4
Output	4	8	12	16	20

Demonstrating Skills

Rewrite each expression without using multiplication or addition signs.

17. $s + s + s + s$ **18.** $7 \cdot b \cdot b \cdot b$ **19.** $7g + 3g$

20. Find the value of $6y^2$ for $y = 4$.

Insert parentheses to make each equation true.

21. $3 \times 5 - 2 = 9$ **22.** $4 + 7 \times 3 = 33$

Use parentheses to show a shortcut for doing each calculation.

23. $17 \times 6 + 17 \times 4$ **24.** $8 \times 26 + 8 \times 4$

Use the distributive property to expand each expression.

25. $5(x + 4)$ **26.** $3(a + 2b + 3c)$ **27.** $3y(y - \frac{1}{3})$

Factor each expression. Check the resulting expression by expanding it.

28. $25r - 50$ **29.** $16h - 4h^2$ **30.** $9x - 81y + 27z$

CHAPTER 2

Geometry in Three Dimensions

Real-Life Math

Patterns and Plans Some of the most powerful examples of using two-dimensional drawings to represent three-dimensional objects can be found in different types of designing. Architects must be very skilled at drawing three-dimensional objects so there is no confusion about what they represent.

Before a house is built, an architect makes drawings of what the house will look like. The drawings include *elevations* that show how the house will look from two sides.

The plans for a house also include blueprints, which show how the interior will be divided into rooms. Blueprints also show details such as where doors are, which way they open, and where to place items like sinks, the stove, and the bathtub.

Think About It Can you think of other careers or companies that use two-dimensional drawings to represent three-dimensional objects?

Family Letter

Dear Student and Family Members,

In Chapter 2, our class will begin studying three-dimensional geometry, including the measurement of surface area and volume. We will build three-dimensional patterns and use words or algebraic expressions to describe them. We will also use blocks to build complex structures that match two-dimensional drawings, as well as make drawings to describe structures we build.

Here is one example of a geometric block pattern that we will build. We will be identifying the pattern and deciding how many cubes there will be in Stage 4, Stage 5, and so on.

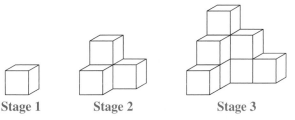

Another way to draw this pattern is called a *top-count* view. Imagine the view as you are looking down on the block patterns from the top. The first three *top-count* views look like this. The numbers in the drawings show how many blocks are in each stack.

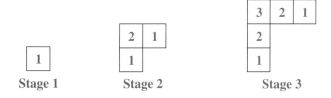

Eventually we will try to determine how many blocks are in Stage 93 if there are 8,464 blocks in Stage 92.

Vocabulary Along the way, we'll be learning about these new vocabulary terms:

base   prism
cylinder   surface area
net   volume

What can you do at home?

As we near completion of Chapter 2, you and your student might enjoy collecting cans with each of you predicting the volumes of the cans. Then find the actual volume of each can and check to see how close each prediction was to the actual volume. Who was the better predictor?

2.1 Block Patterns

Using algebra lets you communicate information in a concise way. In this lesson, you will see how you can write algebraic rules to describe geometric patterns made with blocks.

MATERIALS

cubes

Explore

The block pattern below grows from one stage to the next. The square in each stage is larger than the square in the previous stage.

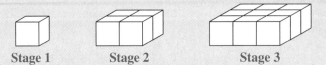

Stage 1   Stage 2   Stage 3

With your own blocks, build Stage 1. Add blocks to build Stage 2. Then add blocks again to build Stage 3.

Continue this pattern. Use your blocks to build Stages 4, 5, and 6.

- When you go from Stage 1 to Stage 2, how many blocks do you add? How do you arrange them?

- When you go from Stage 2 to Stage 3, how many blocks do you add? How do you arrange them?

- What would you do to go from Stage 3 to Stage 4? From Stage 4 to Stage 5? From Stage 5 to Stage 6?

- In general, how many blocks do you add to go from one stage to the next? In other words, how many blocks do you add to Stage s to make Stage $s + 1$? How do you arrange them?

- How many blocks in total are used in Stage 1? In Stage 2? In Stage 3? In Stage 4? In Stage 5? In Stage 6?

- Write an expression for the number of blocks in Stage s.

Investigation 1 ▶ A Staircase Pattern

Here is another block pattern to investigate.

Stage 1   Stage 2   Stage 3

Problem Set A

Build the next three stages of this "staircase" pattern.

1. Describe how you add blocks to go from one stage to the next. That is, tell how many blocks you add, and how you arrange them. Draw pictures if they help to explain your thinking. You may want to use a table like the one below to organize your ideas.

To Create Stage	2	3	4	5	6
Add This Many Blocks	2				

2. Marty thinks there are 37,401 blocks in Stage 273. If he's right, how many blocks are in Stage 275?

3. How many blocks do you add to Stage *s* to make Stage *s* + 1? Write an expression to describe the number of blocks you add, and tell how you would arrange them.

Look at the number of blocks in each stage. If you made a table for Problem 1, you may want to add a row to your table to help you find a pattern to solve the next few problems.

To Create Stage	2	3	4	5	6
Add This Many Blocks	2				
Total Number of Blocks Needed	3				

4. How many blocks does it take to build Stage 10?

5. How many blocks does it take to build Stage 100?

6. Write an expression to describe the number of blocks you need to build Stage *s*.

7. Prove It! Find a way to explain why the expression you wrote for Problem 6 is correct, based on how you built each stage of the staircase pattern. To help you get started, you might try these ideas:

• Think about building two copies of one staircase and putting them together.

• Try breaking up a staircase into smaller pieces.

• Rewrite your expression in a different way, and try to fit both expressions to how the staircase grows.

Share & Summarize

Describe how you found your answers to Problems 3 and 6.

Investigation 2 ▶ Other Block Patterns

Here is another block pattern to build and investigate.

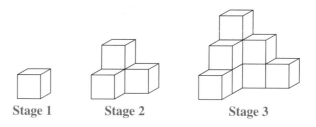

Another way to represent this "double staircase" pattern is to use a *top-count view*. Imagine you are looking down on the buildings from above. The numbers in the drawings below show how many blocks are in each stack.

1		

Stage 1

2	**1**
1	

Stage 2

3	**2**	**1**
2		
1		

Stage 3

Problem Set B

1. Build the next three stages of this pattern. Draw a top-count view of each stage.

2. Think about how you go from one stage to the next.

   a. How many blocks do you add to Stage 1 to make Stage 2? How do you arrange them?

   b. How do you add blocks to Stage 2 to make Stage 3?

   c. How do you add blocks to Stage 3 to make Stage 4?

3. If Stage 92 has 8,464 blocks, how many blocks are in Stage 93?

4. How many blocks do you add to Stage s to make Stage $s + 1$? Write an expression to describe the number of blocks you add, and tell how you would arrange them.

5. How many blocks does it take to build Stage 3? Stage 4? Stage 5?

6. How many blocks would it take to build Stage 25? Stage 1,000?

7. There are 100 blocks in a particular stage of this pattern. Which stage is it?

8. If you have 79 blocks, which is the largest stage of the pattern you could build?

9. Write an expression to describe how many blocks you need to build Stage s. Explain why your expression works.

Remember
A *cube* is a three-dimensional figure with six square sides, or faces.

Think & Discuss

Suppose you were making a pattern of larger and larger cubes. Stage 1 would look like this:

Stage 1

Which of these figures is Stage 2 in a growing pattern of *cubes*?

A   B   C

---

1. Build Stages 3 and 4 of the growing pattern of cubes. Draw them using top-count views.

2. How many blocks did you use to build Stage 2? Stage 3? Stage 4?

3. Without building it, predict how many blocks you would need to build Stage 5.

4. Write an expression to describe how many blocks you need to build Stage *s*. Explain how you found your answer.

Share & Summarize

Compare all the block patterns you have seen. Thinking only about the numbers of blocks needed to build each pattern, which of the patterns grow more quickly? Which grow more slowly? Do any grow in the same way, using the same numbers of blocks at each stage?

Investigation 3 Building Block Patterns

Zoe noticed that the square pattern and the double-staircase pattern grow the same way numerically.

1 block   4 blocks   9 blocks

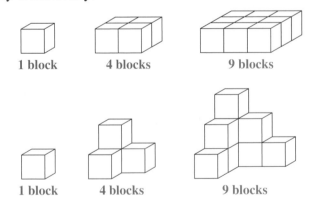

1 block   4 blocks   9 blocks

Think & Discuss

The cartoon on the next page shows how Zoe thought about creating the square pattern from the staircase pattern. Can you work the other way? That is, can you find a way to rearrange the blocks from the squares to form the double staircase? Describe your method.

Remember

To be a pattern, blocks must be added in a *predictable* way—both how many blocks you add and how you arrange them.

Problem Set D

1. Work with your group to build a third block pattern that grows according to the same numerical pattern as the squares and the double staircase. Stage 1 will have 1 block, Stage 2 will have 4 blocks, and Stage s will have s^2 blocks. Draw the first three stages of your pattern using top-count views.

2. Now build another pattern that grows like the staircase pattern from Investigation 1: Stage 1 will have 1 block, Stage 2 will have $1 + 2$ blocks, Stage 3 will have $1 + 2 + 3$ blocks, and so on. Draw the first three stages of your pattern using any method you like.

3. Build a pattern with $s^3 - 1$ blocks in Stage s. Stage 1 will have no blocks, Stage 2 will have 7 blocks, Stage 3 will have 26 blocks, and so on.

Share & Summarize

Create your own block pattern.

1. Draw at least the first three stages of your pattern.

2. Describe how your pattern grows from one stage to the next.

3. Write an expression for the number of blocks in each stage of your pattern.

On Your Own Exercises

1. Squares can be used to make growing patterns. Here is a growing pattern of L-shapes.

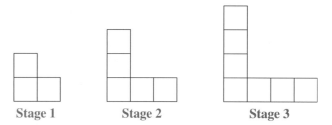

Stage 1   Stage 2   Stage 3

a. How many squares are added to Stage 1 to make Stage 2? To Stage 2 to make Stage 3?

b. How many squares do you add to go from Stage s to Stage $s + 1$?

c. How many squares are used in Stage 1? In Stage 2? In Stage 3?

d. Write an expression for the number of squares in Stage s.

2. Here is a growing pattern of H-shapes.

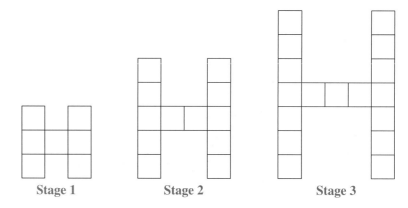

Stage 1   Stage 2   Stage 3

a. How many squares are added to Stage 1 to make Stage 2? To Stage 2 to make Stage 3?

b. How many squares do you add to go from Stage s to Stage $s + 1$?

c. How many squares are used in Stage 1? In Stage 2? In Stage 3?

d. Write an expression for the number of squares in Stage s.

3. Choose a letter of the alphabet that you can draw with squares. Some good choices are A, C, E, F, I, O, T, U, and X.

a. Draw Stage 1, Stage 2, and Stage 3 versions of your letter. Make sure the letter grows predictably from one stage to the next.

b. How many squares are added from Stage 1 to Stage 2? From Stage 2 to Stage 3?

c. How many squares do you add to go from Stage s to Stage $s + 1$?

impactmath.com/self_check_quiz

d. How many squares are in Stage 1? In Stage 2? In Stage 3?

e. Write an expression for the number of squares in Stage *s* of your letter.

4. Here is a new block pattern.

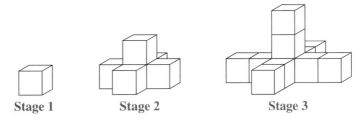

Stage 1 Stage 2 Stage 3

The top-count views of Stages 1 and 2 are shown at left.

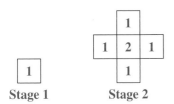

Stage 1 Stage 2

a. Draw the top-count views of Stages 3 and 4.

b. How many blocks are added to Stage 1 to make Stage 2? To Stage 2 to make Stage 3?

c. How many blocks do you add to go from Stage *s* to Stage *s* + 1?

d. How many blocks are used in Stage 1? In Stage 2? In Stage 3?

e. Write an expression for the number of blocks in Stage *s*.

5. Look at the block pattern in Exercise 4.

a. Describe how you could modify the pattern so that only 4 blocks are added from one stage to the next.

b. Use any method you like to draw the first three stages of your pattern.

6. Here is another block pattern.

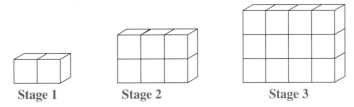

Stage 1 Stage 2 Stage 3

a. How many blocks are added to Stage 1 to make Stage 2? To Stage 2 to make Stage 3?

b. How many blocks are added to get from Stage *s* to Stage *s* + 1?

c. How many blocks are used to make Stage 1? Stage 2? Stage 3?

d. Write an expression for the number of blocks in Stage *s*.

Remember

To be a pattern, blocks must be added in a predictable way.

7. Draw the first three stages of another pattern that uses the same number of blocks in each stage as the pattern in Exercise 6. You may want to use top-count views.

8. Challenge To build Stage 2 of this block pattern, you could make its base (the bottom layer) and set Stage 1 on top of it. To build Stage 3, you could make its base and set Stage 2 on top of it.

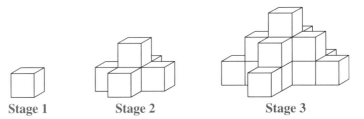

Stage 1   Stage 2   Stage 3

a. There are 5 blocks in the base of Stage 2. How many blocks are in the base of Stage 3?

b. How many blocks are in the base of Stage s?

c. How many blocks are used in Stage 1? In Stage 2? In Stage 3?

d. Describe a rule for the number of blocks in Stage s. Write an expression for the rule, or describe it in words.

Connect & Extend

9. Jamie found that the number of blocks in Stage s of his block pattern was $2s$.

a. Copy and complete this table for Jamie's pattern.

Stage	1	2	3
Blocks			

b. What might Jamie's block pattern look like? Draw the first three stages.

c. How many blocks are used in Stage 10 of Jamie's pattern?

d. How many blocks does Jamie add to Stage 1 to make Stage 2? To Stage 2 to make Stage 3?

e. How many blocks does Jamie add to one stage to build the next stage in his pattern?

10. Kate found that the number of blocks in Stage s of her block pattern was $s^2 - 1$.

a. Copy and complete this table for Kate's pattern.

Stage	1	2	3
Blocks			

b. What might Kate's block pattern look like? Draw the first three stages.

c. How many blocks are used in Stage 10 of Kate's pattern?

d. How many blocks does Kate add to Stage 1 to make Stage 2? To Stage 2 to make Stage 3?

e. Describe how many blocks Kate adds to one stage to build the next.

11. Keisha drew a growing line pattern on dot paper.

Remember

To find the area of a figure on dot paper, you can count the squares inside the figure. Sometimes you have to count half squares.

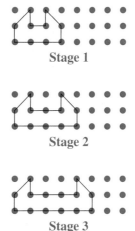

Stage 1

Stage 2

Stage 3

a. Make a table showing the stage, the number of dots on the perimeter of the figure, and the area of the figure.

Stage	1	2	3	4	5
Dots on Perimeter	10				
Area (square units)	4				

b. Write an expression for the number of dots on the perimeter in Stage s.

c. Write an expression for the area of Stage s.

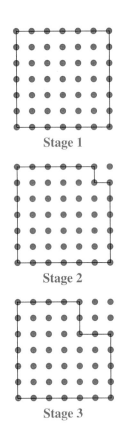

Stage 1

Stage 2

Stage 3

12. Keisha drew the shrinking line pattern shown at left on dot paper.

 a. Make a table showing the stage, the number of dots on the perimeter of the figure, and the area of the figure.

Stage	1	2	3	4	5
Dots on Perimeter	24				
Area (square units)	36				

 b. Write an expression for the number of dots on the perimeter in Stage *s*.

 c. Write an expression for the area of Stage *s,* or describe in words the area that is left.

 d. At what stage will the figure have area 0? Does it make sense to continue the pattern beyond that stage?

13. Every morning, Minowa helps raise the flag up the school's flagpole. She hooks the flag onto the rope at a height of 3 ft. She found that with each pull on the rope, the flag rises 2 ft.

 a. How high is the flag after three pulls?

 b. Write an expression to describe the flag's height after *p* pulls.

 c. If the flagpole is 40 ft high, how many pulls does it take Minowa to get the flag to the top?

14. At a popular clothing store, clothes go on sale when they have hung on the rack too long. When an item is first put on sale, the store marks the price down 10%. Every week after that, the store takes an additional 10% off the original price. So one week after the first markdown, you get 20% off; after 2 weeks, you get 30% off; and so on.

 a. If some shoes are regularly priced at $50, how much will they cost after the first discount?

 b. How much will the shoes cost 2 weeks after the first discount?

 c. When will the shoes be half price?

 d. When will the shoes cost $30?

 e. Write an expression for the price of the shoes *w* weeks after the first discount is taken.

 f. When an item is discounted to 100% off, the store donates it to charity. If the shoes don't sell, how long will they stay in the store before they are donated? Does your answer depend on the original price of the item?

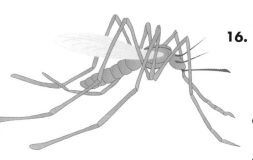

15. Joaquin owes his sister $30 for taking him to a concert. He earns $5 each week for doing chores. He promised to give his sister half his earnings every week until the $30 is paid.

    a. How much does Joaquin owe his sister after 1 week?

    b. Write an expression for how many dollars Joaquin owes his sister after w weeks.

    c. How many weeks will it take Joaquin to pay off his debt?

16. Preview Biologists take water samples from a particular swamp in the beginning of spring. They estimate there are 100 mosquito larvae per square meter in the swamp. The area has had a very rainy spring, so the mosquito population is growing quickly, tripling every month.

    a. How many mosquito larvae are there per square meter 1 month after the scientists first measured? 2 months after?

    b. If there are x larvae in a given month, how many are there in the next month? In the month after that?

    c. How can you find the number of mosquito larvae in a square meter m months after the scientists first measured?

    d. After how many months will there be more than 5,000 mosquito larvae per square meter?

    e. Would this growth continue forever? Explain your answer.

17. Here is a shrinking pattern that starts with a square sheet of paper. Each stage is the same height as, and half as wide as, the previous stage.

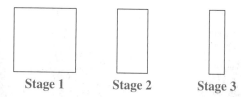

Stage 1   Stage 2   Stage 3

    a. What happens to the area of the paper from one stage to the next?

    b. What happens to the perimeter of the paper from one stage to the next?

    c. Will the paper ever disappear (have an area of 0 units)? Explain your answer.

18. Challenge Malik found that the number of blocks in Stage s of his block pattern is $s^3 + 3$.

  a. How many blocks are in Stage 10 of Malik's pattern?

  b. How many blocks does Malik add to go from Stage 1 to Stage 2? From Stage 2 to Stage 3? From Stage 3 to Stage 4?

  c. Without calculating the number of blocks in each stage, predict how many blocks Malik will add to go from Stage 4 to 5 and from Stage 5 to 6. Explain how you found your answer.

Mixed Review

Rewrite each expression using exponents.

19. $t \times t \times t$ **20.** $\pi \times r \times r$ **21.** $y \times y \times y \times y \times y$

Order each group of numbers from least to greatest.

22. $\frac{1}{3}, \frac{2}{3}, 1, \frac{5}{3}, 0$

23. $\frac{2}{6}, \frac{1}{6}, -\frac{1}{2}, -1, -\frac{3}{6}, 1$

24. $4, 0, 0.4, -0.04, -\frac{1}{4}, \frac{1}{4}$

25. $1\frac{3}{8}, 0.125, \frac{3}{8}, \frac{9}{8}, 0$

26. $3, 30, 0.3, 0.03, \frac{3}{10}$

27. $-1, 10, -0.1, \frac{1}{5}, 0, 0.005$

Fill in the blanks to make true statements.

28. $\frac{1}{8} +$ _____ $= 1$

29. $\frac{8}{5} -$ _____ $= 1$

30. $0.5 +$ _____ $= 1$

31. $0.5 + 0.25 +$ _____ $= 1$

32. $\frac{9}{12} + \frac{3}{12} +$ _____ $= 1$

33. $\frac{3}{10} + \frac{7}{10} -$ _____ $= 1$

Find the product or quotient.

34. $156 \cdot 3$

35. $12 \cdot 25$

36. $52 \div 13$

37. $352 \div 11$

38. Kiran wants to display his collection of 30 first-class stamps in groups of equal size. What group sizes are possible?

Visualizing and Measuring Block Structures

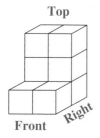

Top

Front   Right

One way to describe a structure is to draw how it looks from different viewpoints. For example, here are six views that might describe the block structure at left.

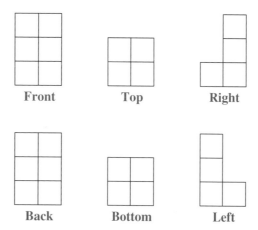

Front   Top   Right

Back   Bottom   Left

These flat views are like shadows that would be cast against a wall or a floor by the structure, but they also show the lines between the blocks. The top view is a "bird's-eye view," looking down from above. For the bottom view, imagine building the structure on a glass table and looking at it from underneath. The left and right views show how the structure looks when viewed from those sides.

MATERIALS
- cubes
- graph paper or dot paper

Explore

Build this block structure. Then draw the six views of it using graph paper or dot paper. If you build your structure on a sheet of paper, you can write the labels *Front, Back, Left,* and *Right* on the paper. This will make it easier for you to remember which side is which.

Front   Right

Which of your views look exactly the same?

Investigation Seeing All the Angles

In this lesson, you will consider only block structures in which at least one face of each block matches up with a face of another block. (Creating views like those in the Explore on page 91 can be difficult for other kinds of buildings.) Keep this in mind as you build your structures.

Good Structure

Not Good

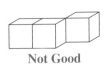
Not Good

Since you will build these structures on your desk, the laws of gravity apply. Blocks need to rest on the table or on other blocks, so you can't make a structure like the one at left.

Not Good

MATERIALS
- cubes
- graph paper or dot paper

Problem Set A

Work with your partner to build three different block structures. Use 6 to 10 blocks for each of them, and follow the rules of good structures. Keep the structures in front of you for this set of problems.

1. First, focus on top and bottom views.

    a. For each of your structures, draw top and bottom views. Be sure to label which is which.

    b. Describe in words the relationship between the top and bottom views of a structure.

    c. Does that relationship hold for *any* block structure, or just for those you built?

2. Now focus on left and right views.

    a. For each structure, draw and label left and right views.

    b. Describe the relationship between the left and right views of a structure.

    c. Does that relationship hold for *any* block structure?

3. Finally, focus on front and back views.

    a. For each structure, draw and label front and back views.

    b. Describe the relationship between the front and back views of a structure.

    c. Does that relationship hold for *any* block structure?

Problem Set B

Use the relationships you discovered in Problem Set A to complete these problems.

1. This is the right view of a structure. Draw the left view.

2. This is the top view of a structure. Draw the bottom view.

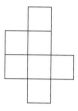

3. This is the front view of a structure. Draw the back view.

Share & Summarize

1. Suppose you know the top view of a structure. Would you learn anything more about the structure from the bottom view? Explain your answer.

2. If you know the front, top, and right views of a structure, would any other view give you more information about the structure? Explain.

Investigation ▶2▶ Different Views

Remember
The *top-count view*
shows a top view of
the structure and the
number of blocks in
each stack.

MATERIALS

• cubes

• graph paper or
  dot paper

Think & Discuss

In many problems in this chapter, you have used
top-count views to describe block structures.
For this top-count view, is there more than one
possible structure?

2	2	2
2	1	2
1	1	1

Is there *any* top-count view that could fit more than one structure?
Explain your answer.

Problem Set C

You will now explore this question: Can more than one structure fit the
same front, top, and right views?

1. Working on your own, use 10 to 15 blocks to build a block structure.

   a. Draw a top-count view for your structure.

   b. On another sheet of paper, draw front, top, and right views. Then
   take your structure apart.

   c. Exchange front, top, and right views with a partner, and build a
   structure that matches your partner's drawings. Did you build the
   structure your partner had in mind? (Use the top-count view to
   check.)

2. Here are three views of a block structure.

   Front   Top   Right

   a. Build a structure that fits these views. Draw a top-count view for
   your structure.

   b. Build a different structure that fits these views, and draw a top-
   count view.

   c. Do you think there are other possible structures? Why or why
   not?

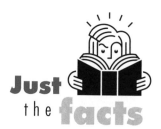

Although top-count views show exactly how the blocks are arranged in a block structure, they aren't really useful to engineers and architects because things in real life aren't usually made of blocks.

Creating a structure from three given views is an interesting puzzle, but if three views represent lots of different structures, they are not very useful. If you were designing a house, for instance, you would want to be very specific in your drawings so that what you imagine is what gets built.

Engineers and architects often need to make drawings of structures, but the types of drawings they make must give more precise information about a structure. The drawings you will work with next can give more information about the block structures they represent.

EXAMPLE

Remember this structure?

Below is another type of three-view drawing that describes it. This type of drawing shows the different levels of the block structure but not the lines between blocks.

Front   Top   Right

The line in the front view shows that there are two different levels of blocks in the structure. In this structure, the base extends beyond the tall part.

The line in the top view shows that there are two different heights of blocks in the structure. The right view shows that the part in front is lower.

These three views together often contain more information than the front, top, and right views you have used before. We will refer to the earlier kind of views as *regular views* and the new kind as *engineering views*.

It is sometimes hard to picture a structure from just looking at the engineering views. Practice will help!

MATERIALS

- cubes
- graph paper or dot paper

Problem Set D

Build the structure described by each set of three views. Draw a top-count view to record each structure you build.

1.

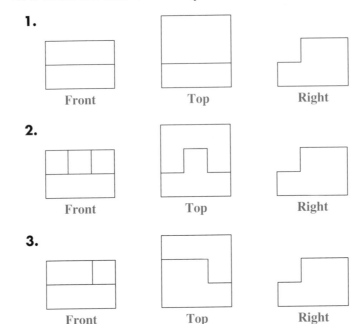

Front   Top   Right

Just **the facts**

These types of views are technically known as *orthographic drawings.*

2.

Front   Top   Right

3.

Front   Top   Right

Share & Summarize

1. Can there be more than one structure with the same *regular* front, top, and right views? Explain your answer by giving an example.

2. The three structures in Problem Set D all fit the same regular front, top, and right views. How do the engineering views show the differences among the structures?

Investigation 3 Creating Engineering Views

In Problem Set D, you learned to interpret engineering views and to show the structures they represent with top-count views. Now you will draw engineering views yourself.

Problem Set E

Work with a partner to draw the three engineering views for each structure shown. It will probably help if you build the structures.

1.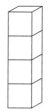

2.

3.

4.

5.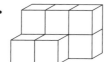

6.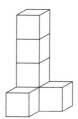

7. Working on your own, use 10 to 15 blocks to build a block structure.

  a. Draw a top-count view for your structure.

  b. On another sheet of paper, draw front, top, and right engineering views. Then take your structure apart.

  c. Exchange engineering views with a partner, and build a structure that matches your partner's drawings. Did you build the structure your partner had in mind? (Use the top-count view to check.)

Share & Summarize

1. Use 6 to 10 blocks to build a new structure. Then draw each of the following:

  - the top-count view

  - the regular front, top, and right views

  - the engineering front, top, and right views

2. Which do you find easier to work with: regular front, top, and right views; engineering views; or top-count views? Why?

Area, which is the space inside a two-dimensional figure, is measured in *square units*. To find the area of an irregular figure like the one shown here, you can count squares (shown with dashed lines).

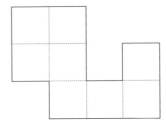

You can say that the area of this figure is 8 square units. Or, if you know the size of the squares, you can use it to state the area exactly. For example, these squares have a side length of 1 centimeter, so the area of the figure is 8 square centimeters. This can be written 8 cm^2.

VOCABULARY
surface area

The **surface area** of a three-dimensional object is the area of the region covering the object's surface. If you could open up the object and flatten it so you could see all sides at once, the area of the flat figure would be the surface area. (Don't forget to count the bottom surface!) Surface area is also measured in square units.

VOCABULARY
volume

Volume, the space inside a three-dimensional object, is measured in *cubic units*. If the blocks you build with are each 1 cubic unit, then the volume of a block structure is equal to the number of blocks in the structure. For example, a structure made from eight blocks has a volume of 8 cubic units. If the blocks have an edge length of 1 cm, the structure's volume is 8 cm^3.

In the problems that follow, the blocks each have an edge length of 1 unit, faces of area 1 square unit, and a volume of 1 cubic unit.

Remember
The volume of a one-block structure is 1 cubic unit.

Think & Discuss

What is the surface area of a single block in square units?

If the edge lengths of a block are 2 cm, what is the block's surface area?

What is the volume of the structure at the right in cubic units?

What is the surface area of the structure above in square units? (Remember: Count only the squares on the *outside* of the structure.)

- cubes
- graph paper or dot paper

Problem Set F

1. Find the volume and the surface area of each three-block structure.

a.

b.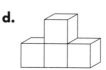

2. Find the volume and the surface area of each four-block structure.

a.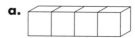

b.

c.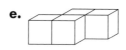

d.

e.

f.

3. Do the structures in Problem 2 all have the same volume? Explain your answer.

4. Which of the structures in Problem 2 have the greatest surface area? Which has the least surface area?

5. Build two block structures with at least six blocks each that have the same volume but different surface areas.

a. For each structure, draw a top-count view.

b. Record the volume and the surface area of each structure.

Share & Summarize

Make a block structure with the same volume as this structure but with less surface area. How did you decide how to rearrange the blocks?

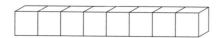

On Your Own Exercises

Practice & Apply

1. Here is a block structure and four views. Some of the views match the structure, but at least one does not. The top-count view is given to help you know how the blocks are arranged. For each view, write *top, front, right,* or *not possible.*

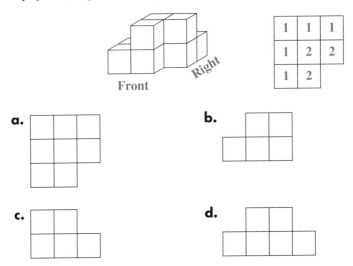

a.

b.

c.

d.

2. Here is a block structure and five views. Some of the views match the structure, but at least one does not. The top-count view is given to help you know how the blocks are arranged. For each view, write *top, front, right,* or *not possible.*

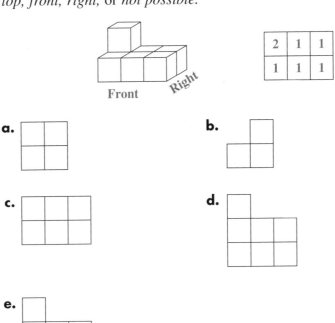

a.

b.

c.

d.

e.

impactmath.com/self_check_quiz

3. Here is Felipe's block structure.

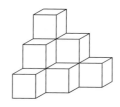

    a. Draw a top view of Felipe's structure.

    b. Draw a front view of Felipe's structure.

    c. Draw a right view of Felipe's structure.

4. Here are three views for a structure.

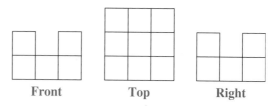

            Front   Top   Right

Choose which of the following structures fit these views. Does more than one fit? The top-count views are given to help you know how the blocks are arranged.

a.

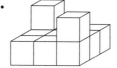

2	1	1
1	1	1
1	1	2

b.

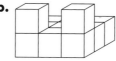

1	1	1
1	1	1
2	1	2

c.

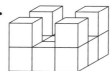

2	1	2
1	1	1
2	1	2

5. Here is a block structure and four engineering views. Some of the views match the structure, but at least one does not. For each engineering view, write *top, front, right,* or *not possible*.

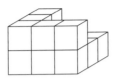
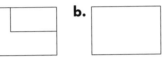

a.   b. c. d.

6. Here is Maya's block structure.

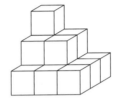

a. Draw an engineering top view of Maya's structure.

b. Draw an engineering front view of Maya's structure.

c. Draw an engineering right view of Maya's structure.

7. Here are three engineering views for a structure. Choose which of the three structures shown fit these views. Does more than one fit?

Front   Top   Right

a.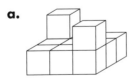

b.

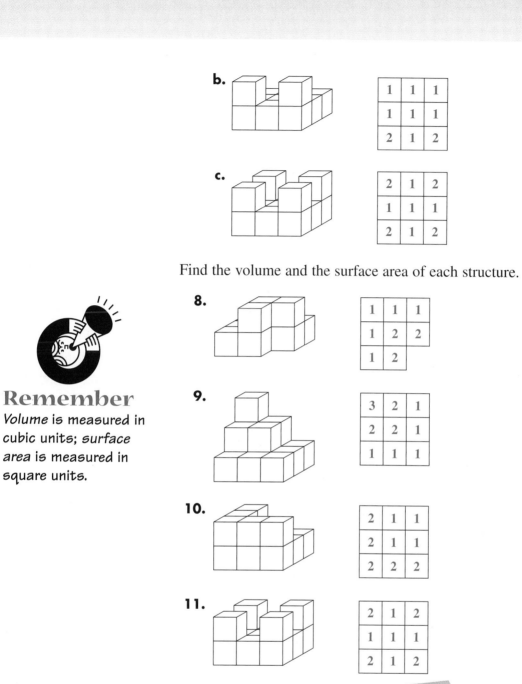

1	1	1
1	1	1
2	1	2

c.

2	1	2
1	1	1
2	1	2

Find the volume and the surface area of each structure.

8.

1	1	1
1	2	2
1	2	

9.

3	2	1
2	2	1
1	1	1

10.

2	1	1
2	1	1
2	2	2

11.

2	1	2
1	1	1
2	1	2

Remember

Volume is measured in cubic units; *surface area* is measured in square units.

Below are top-count views for four block structures. Find the volume and the surface area of each structure.

12.

3	2	1
3	2	1

13.

2	2	2
2	2	2

14.

1	1	1	1	1	1
1	1	1	1	1	1

15.

2	2	2
2	0	2
2	2	2

16. Draw top-count views for three structures that each have a volume of 9 cubic units. Find the surface area of each structure.

Connect &
Extend

17. If possible, draw a top view for a block structure so that the bottom view would be identical. If you have blocks available, you may want to build the structure.

18. If possible, draw a right view for a block structure so that the left view would *not* be identical. If you have blocks available, you may want to build the structure.

19. If possible, draw a front view of a block structure so that the back view would show a different number of squares.

Polyominoes are figures that are formed by joining squares with edges lined up exactly.

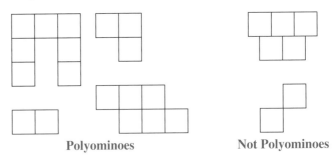

Polyominoes   Not Polyominoes

Two polyominoes are the same if you can cut them out of the paper and fit them exactly on top of each other.

20. Which of these are polyominoes?

a.

b.

c.

d.

e.

21. Which of the polyominoes below are the same?

a.

b.

c.

d.

e.

f.

g.

h.

i.

22. You have probably played the game of dominoes. In the game, dominoes are tiles that have two squares containing dots. In mathematics, *dominoes* are just polyominoes made of two squares. Considering the *mathematical* definition, how many different dominoes are there? Draw them.

23. *Trominoes* are built with three squares. How many different trominoes are there? Draw them.

24. A popular computer game is played with tetrominoes (polyominoes with four squares). These pieces are used:

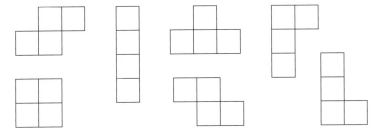

As the pieces fall from the top of the screen, you can move them left or right and rotate them clockwise and counterclockwise, trying to create solid rows (with no gaps) when the pieces land near the bottom of the screen.

a. Are all the pieces different tetrominoes, or are some of them the same?

b. Why must this game include more than one piece to represent the same tetromino?

25. Preview Hexominoes are built with six squares.

a. Draw at least six different hexominoes.

b. You can cut out some hexominoes along the outside and fold them along the lines on the inside to make a cube.

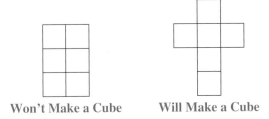

Won't Make a Cube   Will Make a Cube

Group the hexominoes you found in Part a by whether or not they will make a cube. (You can draw them on graph paper and cut them out to test.)

26. You can draw three engineering views for objects other than block structures. For example, here are front, top, and right views for a roll of tape.

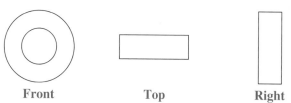

Front   Top   Right

Draw front, top, and right engineering views for two other common objects. Name each object.

27. Draw front, top, and right views of a sphere. It might help to look at a ball.

28. Some engineers draw mechanical devices, like gears. Draw front, top, and right engineering views for this simple gear.

Preview Suppose you had a different kind of block to build with, a half cube. You could build structures like those below. Find the volume, in cubic units, of each structure.

29. ◸   **30.** △   **31.** ▱

32. Estimate the volume of a room in your home in 1-inch blocks. Describe your method.

Mixed Review

Find the indicated percentage.

33. 50% of 12   **34.** 12% of 50   **35.** 25% of 12

36. 12% of 25   **37.** 75% of 12   **38.** 12% of 75

39. Which of the following are factors of 36?

1   2   3   4   5   6   7   8
9   10   11   12   13   14   24   36

The number pair "2, 10" is a *factor pair* of 20 because $2 \times 10 = 20$. For each number below, list all the factor pairs. (The factor pairs "2, 10" and "10, 2" are the same.)

40. 5 **41.** 10 **42.** 25 **43.** 18

Evaluate each expression.

44. $\frac{3}{7} - \frac{2}{7}$

45. $\frac{8}{7} - 1$

46. $\frac{3}{8} + \frac{4}{8} - \frac{1}{8}$

47. $\frac{1}{2} \times \frac{9}{3}$

48. $\frac{1}{8} \times \frac{3}{5}$

49. $\frac{4}{7} \times \frac{1}{4}$

50. $\frac{3}{5} \div \frac{1}{2}$

51. $\frac{1}{4} \div \frac{3}{4}$

52. $\frac{1}{8} \div \frac{3}{8}$

Sketch a graph to match each story.

53. Paul charted his height from when he was 3 until he turned 12.

54. The space shuttle was launched on Monday, orbited the planet for 10 days, and then returned to Earth.

55. Batai threw the ball into the air and watched as it rose, fell, and then hit the ground.

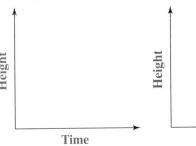

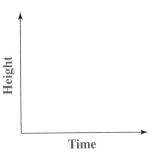

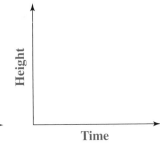

56. Write each of the numbers 1 through 35 in a diagram like the one below. If a number does not fit in any of the circles, write it outside the diagram.

Multiple of 3

Multiple of 2

Multiple of 5

Surface Area and Volume

These figures are *prisms*.

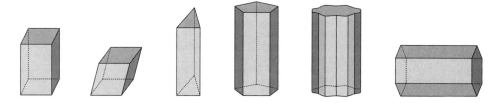

These figures are not prisms.

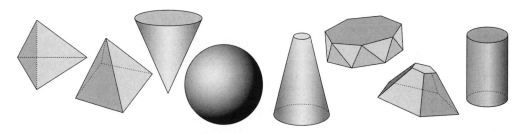

Think & Discuss

What do all the prisms have in common?

How are the nonprisms different from the prisms?

VOCABULARY
prism

All **prisms** have two identical, parallel faces. These two faces are always polygons. A prism's other faces are always parallelograms.

A prism is sometimes referred to by the shape of the two identical faces on its ends. For example, a *triangular prism* has triangular faces on its ends, and a *rectangular prism* has rectangular faces on its ends.

Triangular Prism   Rectangular Prism

Investigation Finding Volumes of Block Structures

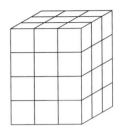

If a block structure has a constant height—that is, has the same number of blocks in every column—that structure is a prism. If the top view of such a structure is a rectangle, the structure is a rectangular prism.

MATERIALS

cubes

Problem Set A

By using its dimensions, you can describe a rectangular prism exactly. For example, a prism with edge lengths 3 units, 2 units, and 4 units is a $3 \times 2 \times 4$ prism.

1. Make all the rectangular prisms you can that contain 8 blocks.

   a. Record the dimensions, volume, and surface area of each prism you make.

   b. Do the 8-block rectangular prisms all have the same volume?

   c. Which of the 8-block rectangular prisms has the greatest surface area? Give its dimensions.

   d. Which of the prisms has the least surface area?

2. Make all the rectangular prisms you can that have a volume of 12 cubic units.

   a. Record the dimensions and surface area of each prism you make.

   b. Which prism has the greatest surface area?

   c. Which prism has the least surface area?

3. Now find all the rectangular prisms that have a volume of 20 cubic units. Try to do it without using your blocks.

   a. Record the dimensions and surface area of each prism.

   b. Which prism has the greatest surface area?

   c. Which prism has the least surface area?

Problem Set B

cubes

Remember
Your block has an edge length of 1 unit and a volume of 1 cubic unit. Each face has an area of 1 square unit.

1. Here is a top view of a prism.

   a. Build a prism 1 unit high with this top view. What is its volume?

   b. Build a prism 2 units high with this top view. What is its volume?

   c. What would be the volume of a prism 10 units high with this top view?

   d. Write an expression for the volume of a prism with this top view and height h.

   e. What is the area of this top view?

2. Here is another top view.

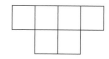

   a. Build a prism 1 unit high with this top view. What is its volume?

   b. Build a prism 3 units high with this top view. What is its volume?

   c. Suppose you built a prism 25 units high with this top view. What would its volume be?

   d. Write an expression for the prism's volume, using h for height.

   e. What is the area of this top view?

3. Here is a third top view.

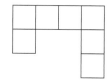

   a. Build a prism 1 unit high with this top view. What is its volume?

   b. Build a prism 5 units high with this top view. What is its volume?

   c. Suppose you cut your blocks in half to build a structure half a unit high with this top view. What would its volume be?

Two Half Blocks

   d. Write an expression for the prism's volume, using h for height.

   e. What is the area of this top view?

Share & Summarize

1. If someone gave you some blocks, how could you use all of them to build a rectangular prism with the greatest surface area? How could you use all of them to build a rectangular prism with the least surface area?

2. Suppose the top-count view of a prism contains 8 squares. What is the volume of a prism that is

   a. 1 unit high?　　　　　　**b.** 10 units high?

   c. $\frac{1}{2}$ unit high?　　　　　**d.** h units high?

3. Write a general rule for finding the volume of a prism made from blocks.

Investigation ▶2 Finding Other Volumes

Will your method for finding the volume of a prism work for prisms that are not block structures? The problems that follow will help you find out.

V O C A B U L A R Y
base

Remember that a prism has two identical, parallel faces that can be any type of polygon. These faces are called the **bases** of the prism.

Problem Set C

1. This triangle is half of the face of one of your blocks.

   a. What is the area of this triangle?

   b. If you cut one of your blocks in half as shown here, you could build a structure 1 unit high with the triangle as its base. What would the volume of this structure be?

   c. If you built a structure 10 units high that had the triangle as its base, what would its volume be?

Remember
Your block has an edge length of 1 unit and a volume of 1 cubic unit. Each face has an area of 1 square unit.

2. The dashed lines on the parallelogram below show the relationship between the parallelogram and a face of one of your blocks.

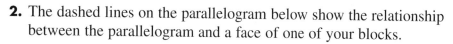

a. What is the area of the parallelogram?

b. If you had blocks like the one shown here, you could build a structure 2 units high that had the parallelogram as its base. What would its volume be?

c. If you could build a structure h units high with this parallelogram as its base, what would its volume be?

A **cylinder** is like a prism, but its two bases are circles. These are all cylinders.

3. The dashed lines on the circle below show the relationship between the circle and a face of one of your blocks.

a. What is the area of the circle?

b. If you had blocks like this one, you could build a structure 3 units high with the circle as its base. What would its volume be?

c. If you could build a structure 1,000 units high with this base, what would its volume be?

VOCABULARY
cylinder

Remember
The area of a circle is πr^2, where r is the radius and π is about 3.14.

4. Luis drew a floor plan for a playroom.

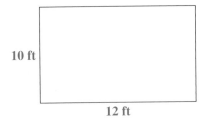

10 ft

12 ft

a. What is the area of the floor?

b. If the playroom has 8-ft ceilings, what is its volume?

c. If the playroom has 10-ft ceilings, what is its volume?

5. Here is a design for a large oil tank.

9 ft

a. What is the area of the base of the tank?

b. If the tank is 50 ft high, how much water will it hold (in cubic feet)?

c. If the tank is *h* ft high, how much water will it hold?

Think & Discuss

Look back over your answers for Problem Set C.

• Describe a single strategy that you think will work for finding the volume of *any* prism. Test your strategy with the prisms in Problem Set C.

• Will the same strategy work for the volume of any *cylinder*? If not, can you modify the strategy so that it *will* work?

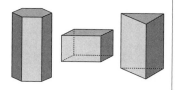

Right Prisms

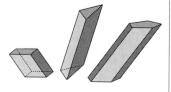

Oblique Prisms

height

Problem Set D

Some prisms, called *oblique prisms,* are slanted. However, you have only been finding the volumes of prisms with sides that are straight up and down, which are called *right prisms.*

1. This right prism has been sliced into very thin pieces, like a deck of cards. These "cards" have then been pushed into an oblique prism, but the cards themselves haven't changed.

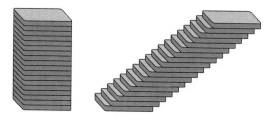

a. If each card is 5 cm wide and 8 cm long, and the stack is 15 cm high, what is the volume of the first "deck"?

b. Is the volume of the second deck *the same as, greater than,* or *less than* the volume of the first deck? Explain your answer.

c. Is the base of the second deck *the same as, larger than,* or *smaller than* the base of the first deck? Explain.

d. The diagram at left shows how to measure the height of an oblique prism. Is the height of the second deck *the same as, greater than,* or *less than* the height of the first deck? Explain.

2. Now think about a deck of circular cards.

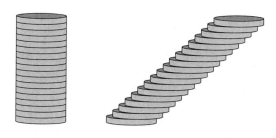

a. If the cards each have radius 4 cm, and the stack is 9 cm high, what is the volume of the first deck?

b. Is the volume of the second deck *the same as, greater than,* or *less than* the volume of the first deck? Explain.

c. Is the base of the second deck *the same as, larger than,* or *smaller than* the base of the first deck? Explain.

d. Is the height of the second deck *the same as, greater than,* or *less than* the height of the first deck? Explain.

3. Will the formula *area of base* × *height* give you the volume of an oblique prism? Explain.

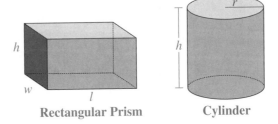
Investigation 3 ▶ Modeling with Block Structures

Many adults find hot tubs, saunas, steam rooms, and hot baths relaxing. In public places, though, these are off-limits to babies and young children, because children are affected much more quickly by heat than adults are. Why do you think this is true?

Your body uses energy, in the form of calories taken from food you have eaten, to keep its temperature relatively constant. When your body grows too hot, you start to sweat. As the sweat evaporates, it carries heat away from your skin, and you feel cooler.

Sweat is composed mostly of water and minerals. The more you sweat, the more your body cools—and the more water it loses. When your body is hot—because of the weather, because you are active, or because you are relaxing in a hot bath—you need to drink fluids to replace the water you are losing through sweat. Otherwise, you can become dehydrated and very sick. Because you sweat only through your skin, how fast you cool off—and how fast you lose water—depends on how much skin you have. That is, it depends on your surface area.

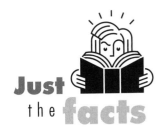

Just the facts

Pigs don't sweat. After they roll around in mud, the mud dries and cools them. So "sweating like a pig" means not sweating at all!

Think & Discuss

Build models of two different 8-block "animals." Design them so that one would cool more quickly than the other.

Explain the reasoning you used to construct your animals.

Building animals from blocks may sound silly. After all, blocks are nothing like animals. However, these models actually work quite well.

When mathematicians build a model, they ignore less-important details in order to simplify the situation they are modeling. Using blocks, you built two shapes that represented the *relative* sizes of the animals. You used the same number of blocks for each animal so that your animals had the same volume. Then you were able to compare just their surface areas.

Problem Set E

Mathematical models of a human baby and a human adult can help you make more comparisons between surface area and volume. For these problems, use a single block to model a baby and a $2 \times 2 \times 2$ cube to model an adult.

1. Given that the baby is represented by 1 block, consider whether the model for the adult is a good model.

   a. If the baby weighs 15 lb, how much does the adult weigh? Does that seem about right?

   b. Are adults really twice as large as babies in every dimension (length, width, and height)?

2. What is the volume of the model of a baby? What is its surface area?

3. What is the volume of the model of an adult? What is its surface area?

4. For every 1 cubic unit of the adult's volume, how many square units of surface area are there? Is this more or less than the surface area of the baby's 1 cubic unit?

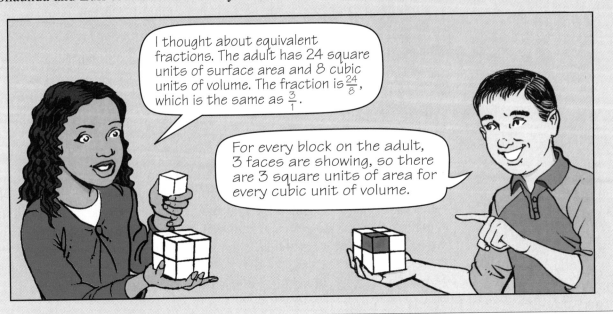

Problem Set F

MATERIALS

cubes

As you do these problems, consider both Shaunda's and Luis's reasoning.

1. For each structure, answer this question: For every 1 cubic unit of volume, how many square units of surface area are there?

    a.    b.   c.

2. Continue to use the single block as a model for a baby, but build a new block model for an adult that you think is better than a $2 \times 2 \times 2$ cube. (It does not have to be a cube, a prism, or any other special shape, but it should be about the right size and shape compared to the model of a baby.) Draw a top-count view of your model.

3. Explain why you think your model of an adult is a better model than a $2 \times 2 \times 2$ cube.

4. What is the volume of your adult?

5. What is the surface area of your adult?

6. For every block in your adult model, how many square units of surface area are there?

7. For its size, does your adult have relatively more or less surface area than the baby has?

Share & Summarize

1. It is not safe to leave a baby in a closed car on a hot day because it can quickly become dehydrated (lose water)—much more quickly than an adult would. Explain why this is true.

2. In chilly weather, babies get colder much more quickly than adults do, so they must be wrapped warmly. Explain why.

Lab Investigation ▶ The Soft Drink Promotion

MATERIALS

materials for making models of soft drink containers

Bursting Bubbles soft drink company wants to attract attention to its products at an upcoming convention. The company normally packages its soft drinks in 350-milliliter cylindrical cans that are 15 centimeters high.

Bursting Bubbles wants to show creativity at the convention by packaging its soft drinks in new containers of different shapes. The company has commissioned you, a mathematician, to investigate some other sizes and shapes and to make a recommendation. Company representatives tell you that the new containers can be any height at all—but they must have a volume of 350 milliliters. It's now up to you to create some attention-getting containers.

15 cm

The company gave the required volume in milliliters (mL). You probably already know that there are 1,000 mL in 1 liter (L). To do this problem, you also need to know that 1 milliliter has the same volume as 1 cubic centimeter.

Make a Prediction

You decide to design some cylindrical containers first. As the height of the can changes, the base area must also change to keep the volume fixed at 350 milliliters.

1. You could design a really tall container—even as tall as 1 m (100 cm)! To keep the volume 350 mL, would the radius of the base circle have to increase or decrease as the height of the can increased?

2. Suppose you designed a very short can—even as short as 2 cm. To keep the volume 350 mL, would the radius of the base circle have to increase or decrease as the height decreased?

Try It Out

3. Bursting Bubble's standard cans are 15 cm high. What is the radius of the circular base of these cans? To find the radius, you can use a strategy of systematic trial and error. The table shows the volume for a base radius of 3 cm and of 2.5 cm. Use a calculator to find the radius that would give a volume of 350 mL. Keep searching until you get within 1 mL of 350 mL.

Remember

The number π, pronounced "pie" and spelled pi, is about 3.14. Your calculator probably has a π key to make these calculations easier.

Base Radius (cm)	Base Area (cm²) ($A = \pi \times r \times r$)	Volume (mL, want 350) (base area × height)
3	28.27	424.05 mL (too large)
2.5	19.635	294.525 mL (too small)

4. Choose five heights from the table below and make your own table. Complete your table with radius values that give a volume within 1 mL of 350 mL.

Height (cm)	Radius of Base Circle (cm)	Height (cm)	Radius of Base Circle (cm)
1		30	
2		50	
3		75	
5		90	
10		100	
20		150	

5. Were your predictions from Questions 1 and 2 correct?

Try It Again

The committee wants you to consider other shapes, not just cylinders, that might get attention at the convention. You can think about cones, spheres, other prisms—anything you want!

Here are two shapes, with formulas for their volumes. Use this information to answer the next set of questions.

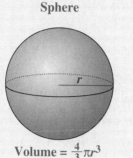

Sphere

Volume = $\frac{4}{3}\pi r^3$

Cone

Volume = $\frac{1}{3}$ base area × height, or $\frac{1}{3}\pi r^2 h$

6. Try the cone shape first. Choose at least three heights for the cone. For each height, find the radius needed to give a cone a volume of 350 mL. Get within 1 mL of 350 mL.

7. Using the formula for the volume of a sphere, find as many spheres as you can with a volume of 350 mL.

8. Find at least two other shapes for containers with a volume of 350 mL.

What Did You Learn?

Choose one design to present to the company.

9. Describe your design to the company, including its dimensions.

10. Explain why your design might be successful at the convention.

11. Make a model of your design to show the company.

On Your Own Exercises

Practice & Apply

1. Consider all the rectangular prisms that can be made with 27 blocks.

   a. Give the dimensions of each prism.

   b. Which of your 27-block prisms has the greatest surface area? Which has the least surface area?

2. Think of a top view made of squares that is different from the top views you have seen so far.

   a. Draw your top view.

   b. Suppose you wanted to build a prism 2.5 units high with that top view. What would its volume be?

   c. Write an expression for the volume of the prism, using h for height.

Remember

The formula for the area of a parallelogram is *bh*, where *b* is the base and *h* is the height.

3. Here is a parallelogram.

   a. What is the area of the parallelogram?

   b. If you build a prism 1 cm high using this parallelogram as a base, what will its volume be?

   c. Draw two other bases for containers that would have the same volume for a 1-cm height.

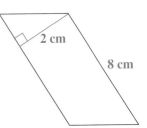

4. Six equilateral triangles are joined to form a hexagon.

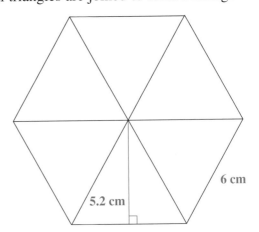

Remember

The formula for the area of a triangle is $\frac{1}{2}bh$, where *b* is the base and *h* is the height.

   a. What is the area of the hexagon? Explain how you found your answer.

   b. If you built a prism on this base with a height of h cm, what would its volume be?

impactmath.com/self_check_quiz

5. The base of a cylinder has radius *r* meters.

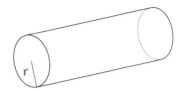

    a. What is the area of the cylinder's base?

    b. If the height of the cylinder is 1 m, what is its volume (in m^3)?

    c. If the height is 10 m, what is the volume?

    d. What is the cylinder's volume if its height is *h* meters?

6. You can find, or at least make a good estimate of, the volume of a room in your home.

    a. Draw the base (floor) of the room, and show the measurements in feet or meters. If you measured exactly, say so. If you estimated, describe how you made your estimates.

    b. What is the height of the room? Do you know exactly, or did you approximate it?

    c. What is the volume of the room?

7. Give the dimensions of four different containers that each have a volume of 360 cubic centimeters. They should not all be rectangular prisms.

8. For each structure, answer this question: For each 1 cubic unit of volume, how many square units of surface area are there?

    a.

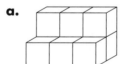

    b.

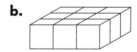

    c.

9. World Cultures Countries keep statistics on the average weights of newborns and adults.

   a. Babies born in the United States have an average weight of 7.25 lb. Adults in the United States have an average weight of 160 lb. How heavy is an adult compared to a newborn (in other words, how many times as heavy)?

   b. If a block is your model of a newborn, how many blocks would be in your model of an adult? How would you arrange them?

   c. Babies born in the United States have an average length of 20 in. Adults in the United States have an average height of 5 ft 7 in. How tall is an adult compared to a baby (how many times as tall)?

   d. Use the information from Part c to check your answer for Part b. Now how would you arrange the blocks?

10. Suppose you want to design five different covered boxes that each hold the same amount of sand. You want each box to hold exactly 500 cm^3.

   a. Describe five different containers with this volume. Draw a top-count view of each structure, or draw the base (including the dimensions) and tell how high each structure will be.

   b. Which of your boxes has the least surface area? Is that the least surface area possible for a 500-cm^3 box?

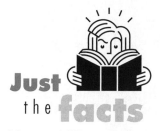

11. A standard playing card is about 5.7 cm wide and 8.9 cm long. A stack of 52 cards—a whole deck—is about 1.5 cm high.

a. What is the volume of a deck of cards?

b. Use your answer to Part a to find the volume of a single card.

12. Many cereal boxes have dimensions of about 6 cm by 20 cm for the base and are about 27 cm high.

a. What volume of cereal could this shape box hold?

b. Give the dimensions (in cm) of four other rectangular boxes that have this same volume.

c. Give the dimensions of two other containers that have this same volume. They do not have to be rectangular boxes.

13. Here are three cylinders that you probably have in your house. Pick one and find its volume.

- a penny (It is hard to measure the height of a single penny, but it's not hard to measure the height of 10 pennies and then divide.)

- a soup can

- a strand of spaghetti

14. Which of these containers holds the most water?

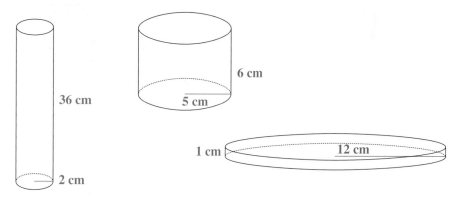

15. Draw two glasses that look different but hold the same amount of water. Show their dimensions.

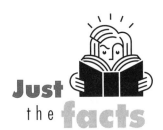
16. Tino has two Alaskan malamutes, Trooper and Scooter. Scooter is Trooper's puppy.

    a. Does Trooper have more or less total surface area than Scooter?

    b. Does Trooper have more or less surface area for a given unit of volume than Scooter?

    c. Which dog will cool off more quickly, Trooper or Scooter?

17. Life Science One of the plants pictured here lives in the desert and must conserve water. The other plant lives in the rain forest and does not need to conserve water. Describe how surface area and volume relationships could help you determine which plant is which.

Mixed Review

Find each fractional amount.

18. $\frac{3}{4}$ of 12   **19.** $\frac{1}{2}$ of 50   **20.** $\frac{3}{5}$ of 15

21. $\frac{2}{3}$ of 12   **22.** $\frac{4}{5}$ of 100   **23.** $\frac{5}{4}$ of 100

Complete each table.

24.

Fraction	Decimal	Percent
$\frac{1}{2}$	0.5	50%
$\frac{3}{4}$		
		30%
	0.25	

25.

Fraction	Decimal	Percent
$\frac{1}{2}$	0.5	50%
$\frac{1}{8}$		
	0.8	
		5%

26. If there are *n* blocks in each bag, write an expression for the total number of blocks.

27. If there are *n* blocks in each bag, write an expression for the total number of blocks.

28. Complete the table.

n	0	1	2	3	10	50	
n + 9							110

29. The formula for the perimeter of a rectangle is $2L + 2W$, where *L* is the length of the rectangle and *W* is the width. Find the perimeter of a rectangle with length 6 cm and width 2 cm.

30. This table is from a game of *What's My Rule?* Find two ways of writing the rule in symbols. Choose your own variable.

In	0	1	2	3	4
Out	3	6	9	12	15

31. **Probability** On a scale from 0 (impossible) to 100 (very likely), rate the chances of each event happening today.

   a. You do homework.

   b. Someone in your state has a baby.

   c. Someone in your city has a baby.

   d. You learn to drive.

   e. It snows somewhere in your country.

32. Zoe wants to build a doghouse for her dogs Trooper and Scooter. She made this sketch to work from.

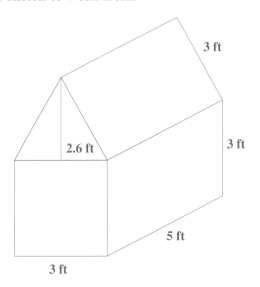

a. Zoe plans to make the doghouse out of plywood, and she wants to include a floor. How many square feet of plywood will she need?

b. The hardware store sells scrap plywood for 25¢ a square foot, in increments of 10 square feet. How much will Zoe spend on plywood?

c. Zoe plans to cut a circular door in the front of the doghouse, and then paint the outside of the house (but not the floor). If the door has a diameter of 2 ft, how much area will Zoe need to paint to give the house two coats?

d. If a small can of paint covers 30 square yards, how many cans will Zoe need to buy?

e. The hardware store sells discount carpeting for $3 per square yard. How much will it cost Zoe to buy carpet for the floor of her doghouse?

2.4 Nets and Solids

VOCABULARY
net

A **net** is a flat figure that can be folded to form a closed, three-dimensional object. Such an object is called a *solid*.

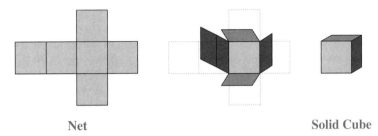

Net   Solid Cube

MATERIALS
- scissors
- graph paper

Explore

- Which of these figures are also nets for a cube? That is, which will fold into a cube? Cut out a copy of each figure and try to fold it into a cube.

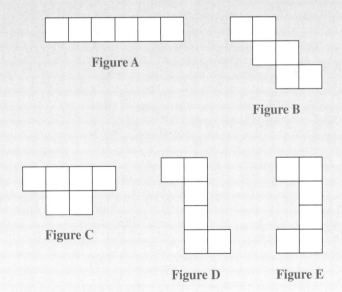

Figure A

Figure B

Figure C

Figure D   Figure E

- Find and draw three other nets for a cube.

With your class, compile all the different nets for a cube that were found. How many nets are there? Do you think your class found all of them?

Investigation 1 ▸ Will It Fold?

For a net to form a closed solid, certain lengths have to match. For example, in the nets for cubes, the side lengths of all the squares must be the same. You will now investigate whether other nets will form closed solids. Pay close attention to the measurements that need to match.

MATERIALS

scissors

Problem Set A

Decide whether each figure is a net—that is, whether it will fold into a solid. You might be able to decide just by looking at the figure. You can also cut out a copy of the figure and try to fold it. If the figure is a net, describe the shape it creates. If it isn't a net, tell what goes wrong.

1.

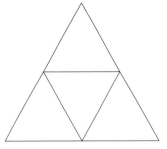

2.

3.

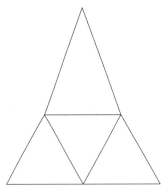

4.

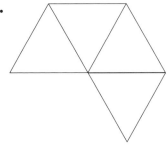

5.

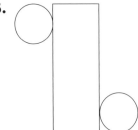

6.

7.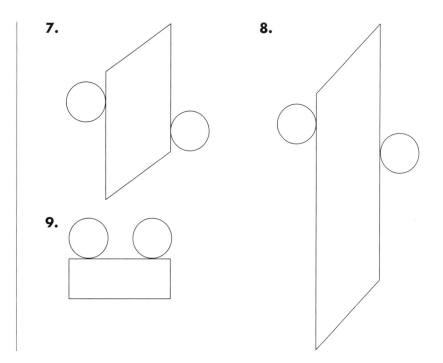

8.

9.

Share & Summarize

1. For a net to fold into a cylinder, the circumference of the circles must be the same as what other length?

2. Draw a net for a figure different from those in Problem Set A. Explain how you know your net will fold to form a solid.

Investigation ▶2 Using Nets to Investigate Solids

You have calculated surface area by counting the squares on the outside of a block structure. If the faces of a solid are not squares—like the figures below—you can find the solid's surface area by adding the areas of all its faces.

  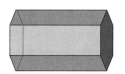

EXAMPLE

This net folds to form a cylinder.

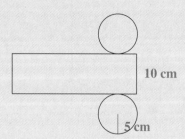

10 cm

5 cm

Because the net folds into a cylinder, the circles must be the same size. The area of each circle is πr^2 cm^2, or 25π cm^2. This is about 78.5 cm^2.

The length of the rectangle must be equal to the circumference of each circle, which is 10π cm, or about 31.4 cm. So the area of the rectangle is about 314 cm^2.

To find the cylinder's surface area, just add the three areas:

$$78.5 \text{ cm}^2 + 78.5 \text{ cm}^2 + 314 \text{ cm}^2 = 471 \text{ cm}^2$$

M A T E R I A L S

scissors

Remember

The formula for the circumference of a circle is $2\pi r$, where r is the radius; or πd, where d is the diameter.

Problem Set B

Each net shown will fold to form a closed solid. Use the net's measurements to find the *surface area* and *volume* of the solid. If you can't find the exact volume, approximate it. It may be helpful to cut out and fold a copy of the net.

1.

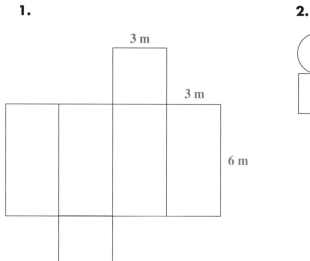

3 m

3 m

6 m

2.

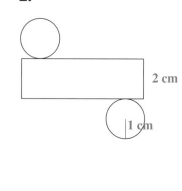

2 cm

1 cm

Remember

The formula for the area of a triangle is $\frac{1}{2}bh$, where b is the base and h is the height.

3.

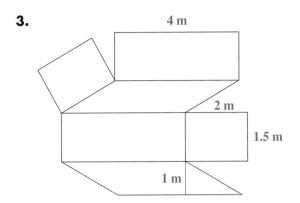

4.

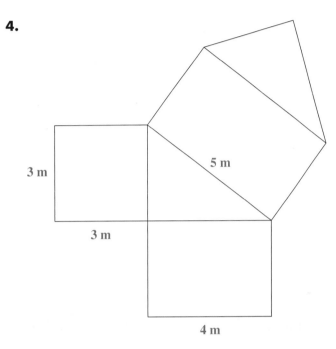

5.

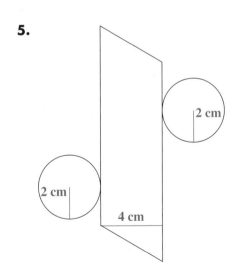

Share & Summarize

1. Explain how to find the surface area of a solid from the net for that solid.

2. Here is a net for a rectangular solid. Take whatever measurements you think are necessary to find the solid's *volume* and *surface area*. Explain what measurements you took and what you did with them.

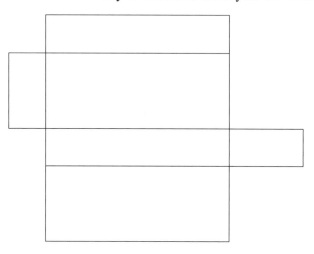

Investigation 3 Is Today's Soft Drink Can the Best Shape?

Bursting Bubbles soft drink company is trying to reduce their manufacturing costs in order to increase their profits. The president of the company wonders if there is a way to use less material to make a soft drink can.

How could Bursting Bubbles reduce the amount of aluminum it takes to make a can? One way is to design a can that has the minimum surface area for the volume of beverage it contains.

Problem Set C

To get started, inspect a 12-oz soft drink can. Get an idea of its dimensions, volume, and surface area.

1. Find the area in square centimeters of the base of the can. (If the can you are using is indented at the bottom, assume that it isn't.)

2. Measure the can's height in centimeters.

The volume printed on the can is given in milliliters (abbreviated mL). A milliliter has the same volume as a cubic centimeter, so 1 mL = 1 cm^3.

3. Does multiplying the area of the base by the height give the volume that is printed on the can? If not, why might the measures differ?

4. What is the surface area of the can in square centimeters? How did you find it? (Hint: Imagine what a net for the can would look like.)

Problem Set D

Now consider how the can design might be changed. For these problems, use the volume you *calculated* in Problem Set C, rather than the volume printed on the can.

1. Design and describe five other cans that have approximately the same volume as your can. Sketch a net for each can, and record the radius of the base and the height. Include cans that are both shorter and taller than a regular soft drink can. (Hint: First choose the height of the can, and then find the radius.)

2. Calculate the surface area of each can you designed. Record your group's data on the board for the class to see.

3. Compare the surface areas of the cans you and your classmates found. Which can has the greatest surface area? Which has the least surface area?

Bursting Bubbles wants to make a can using the least amount of aluminum possible. They can't do it by evaluating every possible can, like the five you found, and choosing the best. There are too many—an infinite number!—and they could never be sure they had found the one using the *least* material.

One problem-solving strategy that mathematicians use is to gather data, as you did in Problem 1, and look for patterns in the data. For a problem like this one, they would also try to show that a particular solution is the best in a way that doesn't involve testing every case.

4. Use what you know about volume and any patterns you found in Problems 1 and 2 to recommend a can with the least possible surface area to the president of Bursting Bubbles. Describe your can completely. Explain why you believe it has the least surface area.

Just the facts

The first "soft drinks" were mineral waters from natural springs. As water bubbles up through the earth, it collects minerals and carbon dioxide. In the 1700s, chemists perfected a method of adding carbonation and minerals to plain water and bottling the result.

Share & Summarize

Does the shape of standard soft drink cans use the minimum surface area for the volume they contain? If not, what might be some reasons companies use the shape they do?

Practice & Apply

Decide whether each figure is a net—that is, whether it will fold into a closed solid. One way to decide is to cut out a copy of the figure and try to fold it. If the figure is a net, describe the shape it creates. If it isn't a net, tell what goes wrong.

1. 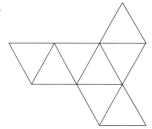   **2.**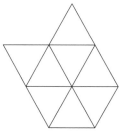

3.

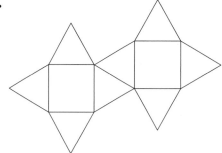

4. A *square pyramid* has a square base and triangular faces that meet at a vertex. Find and draw at least three nets for a square pyramid.

Square Pyramid

In Exercises 5–8, find the surface area of the solid that can be created from the net. Show how you found the surface area.

5. Each triangle is equilateral, with sides 5 cm and height 4.3 cm.

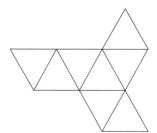

impactmath.com/self_check_quiz

6. Each triangle is equilateral, with sides 8 cm and height 7 cm. The other shapes are squares.

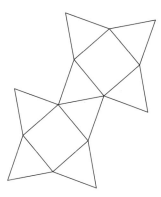

7. The radius of each circle is 2 cm. The height of the parallelogram is 2 cm.

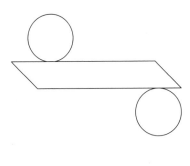

8. The pentagons have side lengths of 3 cm. If you divide each pentagon into five isosceles triangles, the height of each triangle is 2 cm. The other shapes are squares.

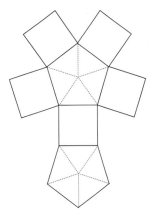

9. An unopened box of tissues has length 24 cm, width 12 cm, and height 10.5 cm.

a. What is the volume of the box?

b. What is the surface area of the box?

10. A juice pitcher is shaped like a cylinder. It is 30 cm tall and has a base with radius 6 cm. It has a flat lid.

a. How much juice will the pitcher hold?

b. What is the surface area of the pitcher?

11. **Challenge** This net will fold into a *triangular pyramid*. Find the surface area of the pyramid, and estimate its volume. You may want to fold a copy of the net into the pyramid.

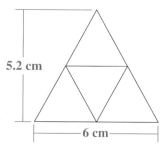

5.2 cm

6 cm

12. Choose a container in your house. You might choose a can (but not a soft drink can), a rectangular box, or some other object.

   a. Draw a net for the object.

   b. Find the surface area and volume of the object by measuring its dimensions.

   c. Draw nets for three other objects with the same volume as your container, and find the surface area of each. The new objects do not have to be real containers from your home.

   d. Is there a prism or cylinder with less surface area but the same volume as your container? If so, draw a net for it.

Connect & **Extend**

13. Imagine removing a label from a soup can.

   a. What is the shape of the label?

   b. If the radius of the soup can is *r* and the height is *h*, what are the dimensions of the label?

Create a net for each box.

14.

15.

16. **Challenge**

17. **Challenge** A cone has a circular base and a vertex some height away from that base. Make a net for a cone.

Tetrahedron

18. Challenge A *tetrahedron* is a solid with four triangular faces. You have made tetrahedrons from nets in this lesson. Nets for a tetrahedron contain only triangles. Can you make a net for another figure using only triangles? (Hint: You might try taping paper triangles together to form a new solid.)

19. A circular pond is 100 ft in diameter. In the middle of winter, the ice on the pond is 6 inches thick.

a. Think of the ice on the pond as a cylinder. What is the volume of ice?

b. What is the surface area of the floating cylinder of ice?

c. What would be the edge length of a cube of ice that had the same volume?

d. What would be the surface area of a cube of ice that had the same volume?

e. Which has greater surface area: the cylinder of ice or the cube of ice? How much more?

20. Kinu's Ice Cream Parlor packs ice cream into cylindrical tubs that are 20 cm in diameter and 30 cm tall.

a. How much ice cream does a tub hold?

b. Kinu's Ice Cream Parlor sells ice cream in cylindrical scoops. If each scoop has base radius 2 cm and height 6 cm, what is the volume of a single scoop?

c. How many scoops of ice cream are in a tub? Show how you found your answer.

d. What is the surface area of a tub?

e. What is the total surface area of the ice cream from one tub *after* it has all been scooped out?

f. Which has more surface area: the ice cream in the tub or the scooped ice cream? Which would melt faster?

Mixed Review

Evaluate each expression.

21. $0.6 \cdot 0.6$ **22.** $0.3 \cdot 0.3$ **23.** $0.02 \cdot 0.02$

24. $0.8 + 0.01$ **25.** $0.8 - 0.01$ **26.** $2 - 0.8$

27. Complete the table.

n	0	1	2	3	5	47	
$2n + 1$							201

Write each expression without using multiplication signs.

28. $b \times b \times b \times b$ **29.** $y \times y \times y \times 3$ **30.** $a \times a \times a \times a \times 4$

Find the value for $2k^2$ for each value of k.

31. $k = 0$ **32.** $k = \frac{1}{2}$ **33.** $k = 1.2$

Use the given expression to complete each table. Then write another expression that gives the same values.

34.

t	0	1	2	3	4	100
$2(t + 10)$						

35.

r	2	3	4	5	6	100
$6(r - 2)$						

36. Ecology The table lists the number of endangered species for five groups of animals.

Group	Number of Endangered Species
mammals	316
birds	253
fishes	82
reptiles	78
snails	22

a. Make a circle graph of these data. On each section, write the percentage each category is of the total number of species listed. Round to the nearest tenth.

b. About what percentage of the endangered species in the five groups are mammals or birds?

c. Write three statements comparing the number of endangered fish species to the number of endangered snail species.

Source: The World Almanac and Book of Facts 2003

VOCABULARY
base
cylinder
net
prism
surface area
volume

Chapter Summary

In this chapter, you have explored four representations of three-dimensional figures:

- top-count views
- engineering views
- regular views (front, top, and right)
- nets

You learned that volume is measured in cubic units, and that area and surface area are measured in square units. You found volume and surface area for block structures by counting cubes and counting exposed faces. You also found a formula for the volume of prisms, and a method for finding surface area by using nets.

Finally, you investigated a scientific application of surface area and volume: the relationship between surface area and volume in human beings, and what it reveals about cooling and dehydration.

MATERIALS

cubes

Strategies and Applications

The questions in this section will help you review and apply the important ideas and strategies developed in this chapter.

Working with block patterns

1. Create your own block pattern.

    a. Draw top-count views for the first three stages of your pattern.

    b. Describe how you arrange blocks as you build from one stage of your pattern to the next. Your description should be detailed enough that someone could use it to build any stage from the previous one.

    c. Describe the number of blocks added from one stage of your pattern to the next. Write an expression for the number of blocks you add to Stage s to make Stage $s + 1$.

    d. Write an expression for the total number of blocks in Stage s.

Representing three-dimensional structures

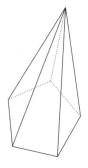

2. Use 10 blocks to create a structure.

    a. Draw the top-count view for your structure.

    b. Draw regular front, top, and right views of your structure.

    c. Draw engineering views for your structure.

3. Draw a net for the solid shown at left.

Finding the volume of a solid

4. A block prism has a base 4 cm long and 3 cm wide. The prism is 5 cm tall.

    a. What is the prism's volume?

    b. Imagine a different block prism with the same height and volume as the original prism. What are its dimensions?

    c. Imagine a cylinder with height 5 cm and the same volume as the original prism. What is the area of its base? Estimate its radius.

Finding the surface area of a solid

5. Find all the rectangular prisms you can make with 24 blocks. Try to do this without building all of them.

    a. Give the dimensions of each prism.

    b. Which of your prisms has the greatest surface area? What is its surface area?

    c. Which of your prisms has the least surface area? What is its surface area?

Find the surface area of each solid.

6.

7.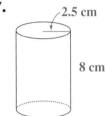

8. The square in the figure at the right has sides 3 cm long. The triangles are isosceles, and the height of each triangle is 4 cm. Is the figure a net? In other words, does it fold up to form a closed solid? Explain how you know.

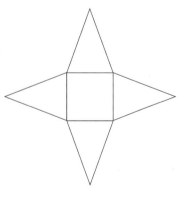

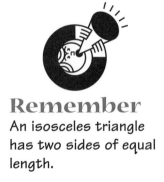

Remember
An isosceles triangle has two sides of equal length.

Demonstrating Skills

In Questions 9–16, find the volume and surface area of the solid.

9.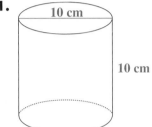

10.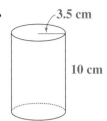
3.5 cm

10 cm

11.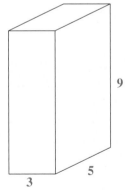
10 cm

10 cm

12.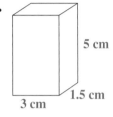
5 cm

1.5 cm

3 cm

13. a cylindrical storage tank with radius 5 ft and height 12 ft

14. a cardboard box with length 2 m, width 2.5 m, and height 3 m

15. a can with height 10 cm and circumference 18 cm

16. a cube with side length 1.5 m

17. Which of these three solids has the greatest surface area? Which has the least surface area?

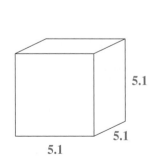

9

3

5

0.3

7.5

5.1

5.1

5.1

Remember
The circumference of a circle is $2\pi r$, where r is the radius.

CHAPTER 3

Exploring Exponents

Real-Life Math

Astronomical Figures The distances from the sun to each of the nine planets in our solar system varies from about 35,980,000 miles to 3,666,000,000 miles! These distances are easier to write in shorthand: 3.598×10^7 miles and 3.666×10^9 miles. The distance from the sun to the star nearest to it, Proxima Centauri, is about 25,000,000,000,000 miles. It would be much easier for an astronomer to write this distance as 2.5×10^{13} miles.

Mars, the fourth planet in our solar system, is 1.41×10^8 miles from the sun. On July 4, 1997, the *Mars Pathfinder* from the National Aeronautics and Space Administration landed on Mars. The spacecraft put the surface rover *Sojourner* on the planet. *Sojourner* sent detailed photos back to Earth, giving us our first up-close views of our sister planet!

Think About It The Mars Pathfinder Mission was part of a NASA program with a spending limit of $150,000,000. Can you write this dollar amount without listing all of the zeros?

Family Letter

Dear Student and Family Members,

Our class is about to begin Chapter 3 about exponents and extremely large or extremely small numbers. Exponents can be thought of as a shortcut method of expressing repeated multiplication. For example, $4 \times 4 \times 4$ is the same as 4^3. The base is 4—the number to be multiplied; the exponent is 3—the number of 4s you multiply together.

We will use a machine model to help us learn about exponents. *Stretching machines* are a model of multiplication. They stretch any input by the number on the machine. This machine will stretch something 4 times. Suppose you put a 1-inch piece of gum into the machine. How long will it be when it comes out?

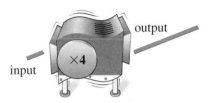

A *repeater machine* is a special type of stretching machine that models exponents. Look at the repeater machine at the right. It will stretch an input 4 times, then 4 times again, and then 4 times again. A 1-inch piece of gum goes through the $\times 4$ machine 3 times, for a total of 64 stretches, and comes out 64 inches long!

Once we are comfortable with the idea of exponents, we will learn what it means to add, subtract, multiply, and divide numbers with exponents.

Vocabulary Along the way, we'll be learning about these new vocabulary terms:

base	**exponential growth**
exponent	**exponential increase**
exponential decay	**power**
exponential decrease	**scientific notation**

What can you do at home?

During the next few weeks, your student may show interest in different ways exponents are used in the world outside of school. You might help them think about one common use of exponents—compound interest in savings accounts. Let's say you have $100 in an account that earns 5% interest a year. Without adding money to the account, after the first year you will have 100×1.05, or $105. After 2 years, you will have $100 \times 1.05 \times 1.05$ or $110.25. After 3 years, you will have $100 \times 1.05 \times 1.05 \times 1.05$ or $115.76. After 20 years, the account total will be 100×1.05^{20}, or $265.33—all from your original investment of $100!

Stretching and Shrinking Machines

Congratulations! You are now the proud owner of a resizing factory. Your factory houses a magnificent set of machines that will stretch almost anything. Imagine the possibilities: you could stretch a 5-meter flagpole into a 10-meter flagpole, a 10-foot ladder into a 30-foot ladder, or a 10-inch gold chain into a 100-inch gold chain!

With the $\times 4$ machine, for example, you can put a regular stick of gum into the input slot . . .

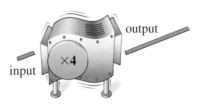

. . . and, in no time at all, a stick of gum four times as long emerges from the output slot!

Think & Discuss

If you put a 2-inch carrot into a $\times 4$ machine, how long will it be when it comes out?

If you put a 3-inch crayon into a $\times 4$ machine, how long will it be when it comes out?

Your factory has other stretching machines as well. If you put a 5-inch piece of wire into a $\times 7$ machine, how long will it be when it comes out?

A pen went through a ×6 machine and emerged 30 inches long.
How long was it when it entered?

? ×6   30 in.

What happens if a pen of length *g* goes through a ×8 machine?

Investigation 1 Machine Hookups

As with any machine, the stretching machines at your factory occasionally break down. Fortunately, with a little ingenuity, your employees usually find a way to work around this problem.

Problem Set A

1. One day, one of the ×10 machines broke down. Caroline figured out she could hook two machines together to do the same work as a ×10 machine. When something is sent through this hookup, the output from the first machine becomes the input for the second.

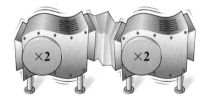

What two machines hooked together do the same work as a ×10 machine? Is there more than one arrangement of two machines that will work?

2. What stretching machine does the same work as two ×2 machines hooked together?

3. It was a bad day at the factory: one of the ×36 machines also broke down. Describe four ways Caroline could hook together other machines to replace the ×36 machine. Assume she can use any number of machines (not just two) and that several machines of each type are available.

4. It was a *very* bad day at the resizing factory: in the afternoon, a ×20 machine broke down. Caroline considered hooking together a ×2 machine and a ×10 machine, but her ×10 machine was still broken. What machines might Caroline hook together to replace the ×20 machine?

5. You have seen that some of the machines in your factory can be replaced by "hookups" of other machines. Not every machine can be replaced, however. After all, you need some machines to build the hookups.

a. Work with your partner to figure out which of the machines— from the ×1 machine to the ×36 machine—are essential, and which are replaceable.

To do this, start with a chart of all the machines, like the one below. If a machine can be replaced by two other machines, cross it out and write a hookup that replaces it. Then see if one of the machines in your hookup can be replaced. If so, cross out that hookup, and write a new hookup with more machines. (See the ×10 and ×20 machines in the table for examples.) Continue this process until the hookups can't be "broken down" any further.

Hookups

×1	×2	×3	×4	×5	×6
×7	×8	×9	~~×10~~ ×2×5	×11	×12
×13	×14	×15	×16	×17	×18
×19	~~×20~~ ~~×4×5~~ ×2×2×5	×21	×22	×23	×24
×25	×26	×27	×28	×29	×30
×31	×32	×33	×34	×35	×36

b. How would you describe the machines that are essential? That is, what do the numbers on the machines have in common?

Investigation 2 ▶ Repeater Machines

If you send a 1-inch piece of taffy through a ×2 machine four times, its length becomes $1 \times 2 \times 2 \times 2 \times 2 = 16$ inches. In Chapter 1, you learned that $2 \times 2 \times 2 \times 2$ can be written 2^4. The expression 2^4 is read "two to the fourth **power.**" The 2 is called the **base,** and the raised 4 is called the **exponent.**

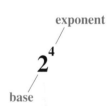

Imagine that you rig up some of the machines in your factory to automatically feed a piece of taffy through several times. You add an exponent to the machine labels to indicate the number of times each machine is applied. For example, sending a piece of taffy through a $\times 2^4$ machine is the same as putting it through a ×2 machine four times.

You call these adjusted machines *repeater* machines because they repeatedly stretch the input by the same factor. In the example above, the original machine, or *base* machine, is a ×2 machine.

Problem Set B

For each repeater machine, tell how many times the base machine is applied and how much the total stretch is.

1.

2.

3.

4. Find three repeater machines that will do the same work as a ×64 machine. Draw them, or describe them using exponents.

5. Surinam played a joke on Jeff by giving him this machine to run. What will it do to a 2-inch-long piece of chalk?

6. In a repeater machine with 0 as an exponent, the base machine is applied 0 times.

a. What do these machines do to a piece of chalk?

b. What do you think the value of 6^0 is?

Evaluate each expression without using a calculator.

7. 7^0 **8.** 2^5 **9.** 4^2 **10.** 3^3

Peter found a $\times\frac{1}{2}$ machine in a corner of the factory. He wasn't sure what it would do, so he experimented with some licorice. This is what he found.

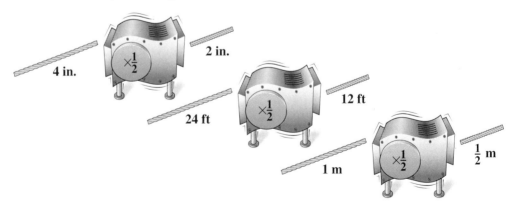

Each piece of licorice was compressed to half its original length. Peter decided this was a *shrinking* machine.

Problem Set C

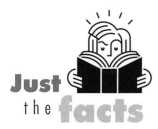
Try a shrinking machine for yourself.

1. If a foot-long sandwich is put into the machine below, how many inches long will it be when it emerges?

Like stretching machines, shrinking machines can be used in hookups and repeater machines.

2. What happens when 1-inch gummy worms are sent through these hookups?

a. 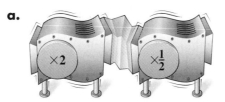 **b.**

3. Evan put a 1-inch stick of gum through a $\times\left(\frac{1}{3}\right)^2$ machine. How long was the stick when it came out?

4. This stick of gum came out of a $\times\left(\frac{1}{2}\right)^2$ machine. Without measuring, estimate the length of the input stick in inches. Explain how you found your answer. It may be helpful if you copy this stick onto your paper.

5. Antonio had a 1-inch piece of gum. He put it through this repeater machine, and it came out $\frac{1}{100,000}$ in. long. What is the missing exponent?

Find a single machine that will do the same job as the given hookup.

6. a $\times 2^3$ machine followed by a $\times\left(\frac{1}{2}\right)^2$ machine

7. a $\times 2^4$ machine followed by a $\times\left(\frac{1}{2}\right)^2$ machine

8. a $\times 5^{99}$ machine followed by a $\times\left(\frac{1}{5}\right)^{100}$ machine

Evaluate each expression without using a calculator.

9. $\left(\frac{1}{2}\right)^5$ **10.** $\left(\frac{1}{4}\right)^2$ **11.** $\left(\frac{1}{3}\right)^3$ **12.** $\left(\frac{1}{7}\right)^0$

Share & Summarize

1. If you put a 2-inch toothpick through this machine, how long will it be when it comes out?

2. A 1-inch-long beetle crawled into a repeater machine and emerged $\frac{1}{16}$ in. long. What repeater machine did it go through? Is there more than one possibility?

3. What single repeater machine does the same work as a $\times 3^5$ machine followed by a $\times \left(\frac{1}{3}\right)^3$ machine?

Investigation ▶3 Strings of Machines

Evan's supervisor asked him to stretch a 1-inch noodle into a 32-inch noodle. Evan intended to send the noodle through a $\times 2^5$ machine, but he accidentally put it through a $\times 2^3$ machine. His coworkers had several suggestions for how he could fix his mistake.

Think & Discuss

Will all the suggestions work? Why or why not?

Evan's mistake wasn't so terrible after all, because he figured out how to find single repeater machines that do the same work as some hookups.

EXAMPLE

Evan found a single machine for this hookup. He reasoned like this: "The first machine tripled the noodle's length 3 times. The second machine took that output and tripled *its* length 5 times. So, the original noodle's length was tripled $3 + 5$ times, or 8 times in all. That means a $\times 3^8$ machine would do the same thing."

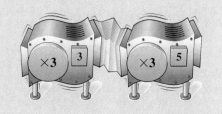

Problem Set D

Follow Evan's reasoning to find a single repeater machine that will do the same work as each hookup.

1.

2.

3.

4.

You have seen that a hookup of repeater machines with the same base can be replaced by a single repeater machine. Similarly, when you multiply exponential expressions with the same base, you can replace them with a single expression.

EXAMPLE

Malik thought about how he could rewrite the expression $2^{20} \times 2^5$.

Malik's idea is one of the *product laws of exponents,* which can be expressed like this:

> **Multiplying Expressions with the Same Base**
>
> $$a^b \times a^c = a^{b + c}$$

Actually, this law can be used with more than two expressions. As long as the bases are the same, to find the product you can add the exponents and use the same base. For example:

$$3^2 \times 3^3 \times 3^{10} = 3^{2 + 3 + 10}$$
$$= 3^{15}$$

Problem Set E

Rewrite each expression as a power of 2. It may help to think of the expressions as hookups of ×2 machines.

1. 2×2 **2.** $2^5 \times 2^4$ **3.** $2^{10} \times 2$

4. $2^{10} \times 2^{10}$ **5.** $2^0 \times 2^3$ **6.** $2^m \times 2^n$

For each hookup, determine whether there is a single repeater machine that will do the same work. If so, describe or draw it.

7.

8.

9.

10.

11.

Tell what number should replace the question mark to make each statement true.

12. $5 \times 5 \times 5 = ?^3$ **13.** $2.7^{10} \times 2.7^? = 2.7^{12}$

14. $\left(\frac{1}{5}\right)^4 \times \left(\frac{1}{5}\right)^6 = \left(\frac{1}{5}\right)^?$ **15.** $1.2^? \times 1.2^2 = 1.2^2$

16. $\frac{2}{3} \times ? = \left(\frac{2}{3}\right)^2$ **17.** $a^{20} \times a^5 = a^?$

If possible, rewrite each expression using a single base. For example, $6^2 \times 6^3$ can be rewritten 6^5.

18. $3^3 \times 3^3 \times 3^4$ **19.** $2^5 \times 5^2$ **20.** $4 \times 4 \times 4$

21. $4x^3 \times y^2$ **22.** $4^4 \times 2^2$ **23.** $a^n \times a^m$

When you simplify algebraic expressions involving exponents, it's important to keep the *order of the operations* in mind.

Remember

By the order of operations, $2b^3$ means $2 \times b^3$, not $(2 \times b)^3$.

> **EXAMPLE**
>
> $2b^3 \times 5b^2 = 2 \times b^3 \times 5 \times b^2 = 2 \times 5 \times b^3 \times b^2 = 10b^5$

Problem Set F

Simplify each expression.

1. $2b \times 2b$

2. $5z^2 \times 6z^2$

3. $5a^2 \times 3a^4$

4. $z^2 \times 2z^3 \times 6z^2$

Share & Summarize

1. Find a single repeater machine to do the same work as this hookup.

2. In your own words, write the rule that lets you figure out the missing exponent in equations like $3^3 \times 3^{16} = 3^?$ and $r^{15} \times r^? = r^{23}$.

Investigation 4 ▶ More Strings of Machines

Combining expressions that have the same base can make products much easier to understand. Do you think you can simplify products that *don't* have the same base?

Problem Set G

1. Caroline has an order from a golf course designer to put palm trees through a $\times 2^3$ machine and then through a $\times 3^3$ machine. She thinks she can do the job with a single repeater machine. What single repeater machine should she use?

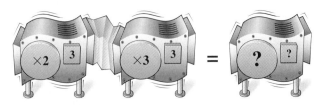

2. Caroline also needs to stretch some pine trees to 10^2 times their original lengths, but her $\times 10$ machine is still broken and someone is using the $\times 10^2$ machine. Find a hookup of two repeater machines that will do the same work as a $\times 10^2$ machine. To get started, think about the hookup you could use to replace the $\times 10$ machine.

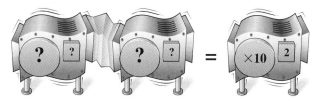

For each hookup, find a single repeater machine to do the same work.

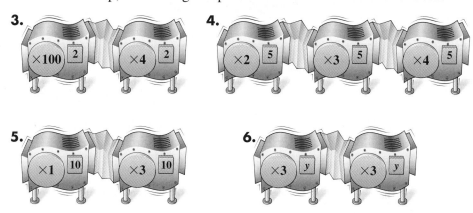

3. $\times 100 \;^2$ $\times 4 \;^2$

4. $\times 2 \;^5$ $\times 3 \;^5$ $\times 4 \;^5$

5. $\times 1 \;^{10}$ $\times 3 \;^{10}$

6. $\times 3 \;^y$ $\times 3 \;^y$

You can use the same kind of thinking you used in the problems above to multiply expressions with the same exponent.

EXAMPLE

Maya multiplied $4^2 \times 3^2$ by thinking about stretching machines.

I can think of $4^2 \times 3^2$ as a $\times 4^2$ machine hooked to a $\times 3^2$ machine.

I can replace the $\times 4^2$ machine with two $\times 4$ machines and the $\times 3^2$ machine with two $\times 3$ machines. In symbols, this is $4 \times 4 \times 3 \times 3$.

I can rearrange the machines into two groups, each with one $\times 4$ and one $\times 3$ machine. This is $(4 \times 3) \times (4 \times 3)$.

I can replace the machines in each group with a $\times 12$ machine and then replace two $\times 12$ machines with a $\times 12^2$ machine. So, $4^2 \times 3^2 = 12^2$. That means I can just multiply the bases and keep the same exponent.

Use Maya's idea to multiply $5^3 \times 2^3$. Use your calculator to check your answer.

Maya's idea is another *product law of exponents*.

> **Multiplying Expressions with the Same Exponents**
>
> $$a^c \times b^c = (a \times b)^c$$

You can use this law with more than two expressions. If the exponents are the same, multiply the expressions by multiplying the bases and using the same exponent. For example, $2^8 \times 3^8 \times 7^8 = (2 \times 3 \times 7)^8 = 42^8$.

Problem Set H

Rewrite each expression using a single exponent. For example, $2^3 \times 3^3$ can be rewritten 6^3.

1. $100^2 \times 3^2$

2. $10^{20} \times 25^{20}$

3. $5^{100} \times 5^{100} \times 2^{100}$

4. $\left(\frac{1}{3}\right)^a \times \left(\frac{1}{5}\right)^a$

5. $1{,}000^5 \times 3^5$

6. $x^2 \times y^2$

For each hookup, determine whether there is a single repeater machine that will do the same work. If so, describe it.

7.

8.

9.

10.

11.

Tell what number should replace the question mark to make each statement true.

12. $100^? \times 3^2 = 300^2$

13. $4^{20} \times 25^{20} = ?^{20}$

14. $5^{100} \times \left(\frac{1}{2}\right)^{100} = ?^{100}$

15. $\left(\frac{1}{3}\right)^3 \times \left(\frac{1}{5}\right)^3 = \left(\frac{1}{15}\right)^?$

You have worked with the product laws of exponents. Do you think there are similar laws for sums of exponential expressions? You will explore this question in the next problem set.

Problem Set I

Determine whether each statement is true or false, and explain how you decided.

1. $2^3 + 2^4 = 2^7$ **2.** $3^2 + 5^2 = 8^2$

3. From your answers to Problems 1 and 2, do you think $2^a + 2^b = 2^{a + b}$ for any numbers a and b?

4. From your answers to Problems 1 and 2, do you think $a^2 + b^2 = (a + b)^2$ for any numbers a and b?

Share & Summarize

1. In your own words, write the rule that lets you find the missing base in equations like $3^{30} \times 2^{30} = ?^{30}$ and $5^{15} \times ?^{15} = 100^{15}$.

Find the missing numbers.

2. $15^3 \times 4^3 = ?^3$ **3.** $3^7 \times 3^? = 3^{12}$

4. $\left(\frac{2}{5}\right)^{12} \times \left(\frac{2}{5}\right)^3 = \left(\frac{2}{5}\right)^?$ **5.** $2^2 \times ?^2 = 18^2$

6. You have seen two product laws for working with exponents. Explain how you know when to use each rule.

Product Laws

$$a^b \times a^c = a^{b + c}$$
$$a^c \times b^c = (a \times b)^c$$

Practice Apply

Supply the missing information for each diagram.

1.

5 cm   **?**   5 cm

2.

3 in.   **?**   15 in.

3.

1.25 ft   ×4   **?**

4.

?   ×4   ×3   36 cm

If possible, find a hookup that will do the same work as the given stretching machine. Do not use ×1 machines.

5.  ×100

6. ×99

7. ×37

8. ×1,111

9. Find two repeater machines that will do the same work as a ×81 machine.

10. Find a repeater machine that will do the same work as a ×$\frac{1}{8}$ machine.

11. This stick of gum exited a ×3^2 machine. Without measuring, estimate the length of the input stick in centimeters. Explain how you found your answer.

Evaluate each expression without using your calculator.

12. 6^2

13. 9^2

14. $\left(\frac{1}{3}\right)^3$

15. 3×2^3

16. 8^1

17. x^1

Do each calculation in your head. (Hint: Think about stretching and shrinking machines.)

18. $\left(\frac{1}{3}\right)^5(3)^5$

19. $(4)^6\left(\frac{1}{4}\right)^6$

20. $\left(\frac{1}{10}\right)^7(10)^8$

If possible, find a repeater machine to do the same work as each hookup.

21.

22.

23.

24.

If possible, rewrite each expression using a single base.

25. $6^4 \times 6^2$ **26.** $4^{10} \times 5^3$ **27.** $1.1^{100} \times 1.2^{101}$

28. $7^{10} \times 7^k$ **29.** $4^n \times 4^m$ **30.** $a^{20} \times a^5$

Simplify each expression.

31. $3c \times 5c$ **32.** $10z^2 \times z^2$ **33.** $2a^3 \times 10a$

If possible, find a repeater machine to do the same work as each hookup.

34.

35.

36.

37.

If possible, rewrite each expression using a single base and a single exponent.

38. $2^3 \times \left(\frac{1}{3}\right)^3$ **39.** $4^0 \times 5^2$ **40.** $1^{100} \times 5^2$

41. $3^9 \times \left(\frac{1}{3}\right)^7$ **42.** $\left(\frac{1}{2}\right)^2 \times \left(\frac{1}{5}\right)^2$ **43.** $6^4 \times 2^4$

44. $4^{10} + 1^{10}$ **45.** $x^{10} \times x^8$ **46.** $4^n \times 100^n$

47. $a^{20} + a^5$ **48.** $a^4 \times b^4$ **49.** $n^a \times m^a$

50. A ×4 machine can be replaced by a hookup of two ×2 machines.

A ×8 machine can be replaced by a hookup of three ×2 machines.

a. Find three other machines that can be replaced by hookups of ×2 machines.

b. Can a ×100 machine be replaced by a hookup of ×2 machines? Explain.

c. Find three machines that can be replaced with hookups of ×5 machines.

In your
own
words

Choose one of the product laws of exponents and explain it so a fifth grader could understand. Use stretching and shrinking machines in your explanation if you wish.

51. The left column of the chart lists the lengths of input pieces of ribbon. Stretching machines are listed across the top. The other entries are the outputs for sending the input ribbon from that row through the machine from that column. Copy and complete the chart.

Input Length	Machine			
	×2			
	1	5		
3				15
	14		7	

52. The left column of the chart lists the lengths of input chains of gold. Repeater machines are listed across the top. The other entries are the outputs you get when you send the input chain from that row through the repeater machine from that column. Copy and complete the chart.

Input Length	Repeater Machine			
	$\times 2^3$			
	40		125	
2				
		162		24

Evaluate each expression.

53. 12^2 **54.** 4^3 **55.** 10^4 **56.** $\left(\frac{1}{3}\right)^2$ **57.** 0.2^5

58. Evaluate each expression without using a calculator.

  a. 3^0 **b.** 100^0 **c.** $\left(\frac{1}{2}\right)^0$

  d. Try to explain your answers from Parts a–c using the idea of repeater machines.

59. You can use exponents to write prime factorizations in a shorter form. For example, $36 = 2 \times 2 \times 3 \times 3 = 2^2 \times 3^2$. Write the prime factorization of each number using exponents.

  a. 27   **b.** 12   **c.** 100   **d.** 999

Remember

The only factors of a prime number are the number itself and 1. The *prime factorization* of a number expresses the number as a product of prime numbers: $12 = 2 \times 2 \times 3$.

60. Peter has a 1-inch sewing needle. He says that putting the needle through a $\times 3$ machine five times will have the same effect as putting it through a $\times 5$ machine three times. Caroline says the needles will turn out different lengths.

  a. Who is right? Explain how you know.

  b. In Part a you compared 3^5 and 5^3. Are there *any* two different numbers, a and b, for which $a^b = b^a$? Your calculator may help you.

Mixed Review

61. This is an input/output table from a game of *What's My Rule?* Find two ways of writing the rule in symbols.

Input	0	1	2	3	4
Output	2	6	10	14	18

Evaluate each expression.

62. $\frac{1}{2} + \frac{1}{9} + \frac{2}{3}$ **63.** $\frac{3}{5} \times 2\frac{1}{3}$ **64.** $3\frac{2}{7} - \frac{1}{12}$

65. $7\frac{1}{8} \div \frac{2}{7}$ **66.** $0.217 - 0.0104$ **67.** 0.27×0.004

68. $\frac{1}{9} + 3\frac{4}{7}$ **69.** $0.982 - 0.444$ **70.** 0.5×0.001

Tell whether each equation is true. If it is not true, rewrite the expression to the right of the equal sign to make it true.

71. $2(x + 3) = 2x + 3$ **72.** $2(x + 9) = 2x + 18$

73. $3(2x + 1) = 2x + 3$ **74.** $15x - 95 = 5(3x - 19)$

Shrinking and Super Machines

Peter arrived at the resizing factory one day to find a whole new set of machines. Instead of × symbols, the machines had ÷ symbols. Peter's supervisor explained that the factory was getting more orders for shrinking things. So, the factory purchased a new type of shrinking machine that uses whole numbers instead of fractions. For example, a ÷2 machine divides the length of whatever enters the machine by 2.

Think & Discuss

How long will a meterstick be after traveling through each machine?

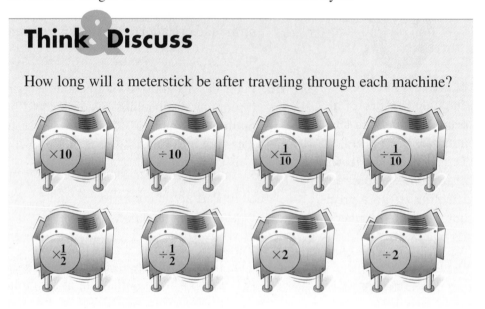

Investigation 1 A New Shrinking Machine

The new shrinking machines can be hooked to other shrinking machines and to stretching machines.

Problem Set A

If a yardstick is put into each machine, how many inches long will it be when it comes out?

1.

2.

3.

4.

5.

Some of the new shrinking machines are rigged up to make repeater machines. If a yardstick is put into each repeater machine, how many inches long will it be when it exits?

6.

7.

8.

9. If a 1-inch paper clip is put through this hookup, what will its final length be?

Find a single repeater machine to do the same work as each hookup.

10.

11.

12.

Share & Summarize

1. Find two repeater machines that will shrink a 1-inch gummy worm to $\frac{1}{16}$ of an inch.

2. A 1-inch gummy worm was sent through this hookup and emerged 25 inches long. What is the first machine in the hookup?

Investigation **2** Dividing and Exponents

The machine model can help you divide exponential expressions.

EXAMPLE

Here's how Shaunda thought about $3^5 \div 3^3$.

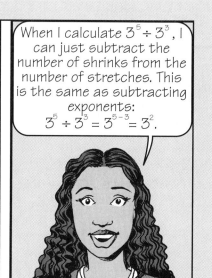

I think about a hookup of a x3^5 and a ÷3^3 machine. When something goes through the hookup, it is stretched 5 times and shrunk 3 times.

Each ÷3 shrink cancels a x3 stretch. So the three shrinks cancel three of the stretches...leaving two stretches. The hookup does the same work as a x3^2 machine.

When I calculate $3^5 \div 3^3$, I can just subtract the number of shrinks from the number of stretches. This is the same as subtracting exponents: $3^5 \div 3^3 = 3^{5-3} = 3^2$.

Shaunda had discovered one of the *quotient laws of exponents*.

Dividing Expressions with the Same Base

$$a^b \div a^c = a^{b-c}$$
$$\frac{a^b}{a^c} = a^{b-c}$$

Problem Set **B**

Write each expression as a power of 2.9.

1. $2.9^5 \div 2.9^3$ **2.** $2.9^{10} \div 2.9^6$ **3.** $2.9^{13} \div 2.9^{12}$

4. $2.9^{13} \div 2.9^{11}$ **5.** $2.9^{25} \div 2.9^{25}$ **6.** $2.9^8 \div 2.9^3$

Evaluate each expression without using your calculator.

7. $2^3 \div 2^3$ **8.** $2^{10} \div 2^{10}$ **9.** $2^{15} \div 2^{15}$

If possible, rewrite each expression using a single base.

10. $5^5 \div 5^2$ **11.** $3^4 \div 3^2$ **12.** $2^5 \div 3^4$ **13.** $3^2 \div 3^0$

14. $2^8 \div 3^2$ **15.** $a^{10} \div a$ **16.** $a^n \div a^m$ **17.** $a^n \div b^m$

Though you may not be able to rewrite an expression using a single base, there may sometimes still be a way to simplify the expression.

EXAMPLE

Kate wondered if it would be possible to simplify an expression like $2^2 \div 3^2$. Simon and Jin Lee tried to help her decide.

This led the class to discuss another quotient law of exponents.

Dividing Expressions with the Same Exponent

$$a^c \div b^c = (a \div b)^c$$

$$\frac{a^c}{b^c} = \left(\frac{a}{b}\right)^c$$

Problem Set C

Write each expression in the form c^m. The variable c can be a fraction or a whole number.

1. $2^2 \div 5^2$ **2.** $5^6 \div 5^2$ **3.** $a^2 \div 2^2$

4. $a^9 \div b^9$ **5.** $3^7 \div 3^4$ **6.** $x^n \div y^n$

7. Prove It! Show that $\frac{a^3}{b^3} = \left(\frac{a}{b}\right)^3$, for $b \neq 0$. Hint: Write a^3 and b^3 as products of a's and b's, and regroup the quotient into a string of $\left(\frac{a}{b}\right)$'s.

8. Do you think $\frac{a^c}{b^c} = \left(\frac{a}{b}\right)^c$ for $b \neq 0$ and any value of c greater than or equal to 0? Explain.

Share & Summarize

In your own words, write the rule that lets you find the missing exponent in equations like $5^{16} \div 5^8 = 5^?$ and $r^? \div r^{10} = r^5$.

Investigation ▶3▶ Super Machines

Your resizing factory is such a success, you decide to order a shipment of *super machines*. Here's one of the new machines.

Peter was excited to try the machine, but he wasn't sure what it would do. He found this diagram in the manufacturer's instructions:

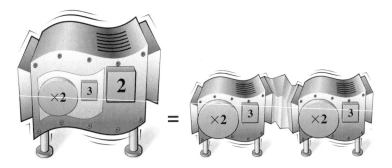

Just the facts

Improperly discarded chewing gum can be a litter nightmare. Schools and businesses spend thousands of dollars every year cleaning chewing gum from floors, carpets, and furniture.

Think & Discuss

If Peter puts a 1-inch stick of gum through the super machine above, how long will it be when it comes out?

How many times would Peter need to apply a $\times 2$ machine to do the same work as this new super machine?

Peter used the shorthand notation $\times (2^3)^2$ to describe this super machine. Why is this shorthand better than $\times 2^{3\ 2}$?

Describe or draw a hookup of three repeater machines that will do the same work as a $\times (4^2)^3$ machine.

Problem Set D

Careta has an order from a kite maker who wants to stretch different types of kite string. If a 1-inch piece of kite string is sent through each super machine, how long will it be when it exits?

1.

2.

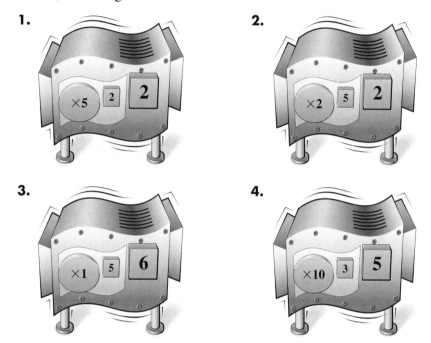

3.

4.

5. For each super machine in Problems 1–4, describe or draw a hookup of two or more repeater machines that would do the same work.

6. Describe a super machine that will stretch a 1-inch strand of kite string to 81 inches. Do not use 1 for either exponent.

7. Consider the expression $(4^3)^2$.

   a. Draw a super machine that represents this expression.

   b. How many times would you need to apply a $\times 4^3$ machine to do the same work as your super machine?

   c. How many times would you need to apply a $\times 4$ machine to do the same work as your super machine?

The expressions represented by the super machines can be simplified.

In Problem 7, Zoe added the exponents and decided it would take five ×4 machines to do the same work as the $\times(4^3)^2$ super machine. Maya thought the answer was six. She drew a diagram to explain her thinking to Zoe.

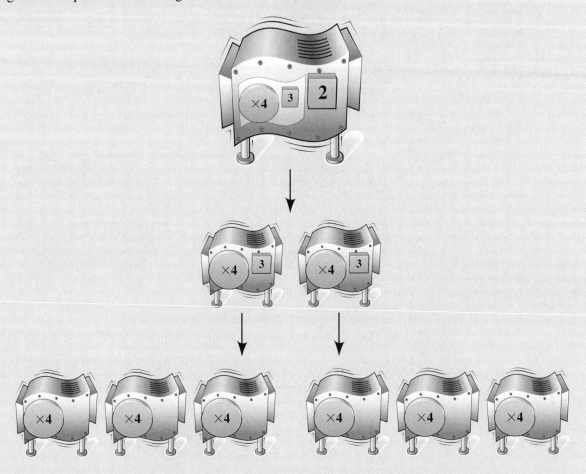

Maya's diagram convinced Zoe she should have multiplied the exponents rather than adding them. The diagram shows that $(4^3)^2 = 4^{3 \times 2} = 4^6$.

The diagram Maya used demonstrates an example of the *power of a power law of exponents*.

Raising a Power to a Power

$$(a^b)^c = a^{b \times c}$$

TI-34 II

Problem Set E

Write each expression as a power of 2.

1. $(2^4)^2$
2. $(2^2)^3$
3. $(2^4)^3$
4. $(2^2)^2$
5. $(2^1)^7$
6. $(2^0)^5$

Tell what number should replace the question mark to make each statement true.

7. $4 = 2^?$
8. $4^? = (2^2)^3$
9. $4^4 = (?^2)^4$
10. $2^3 \times 4^? = 2^{13}$
11. $2^7 \div 4^3 = 2^?$
12. $?^3 \div 2^3 = 2^3$
13. $8^2 = 2^?$
14. $2^7 + 2^7 = 2^?$ (Be careful!)

15. Use a calculator to check your answers for Problems 1–14. To evaluate an expression like $(2^4)^2$, push the exponent key twice. To find $(2^4)^2$, you could use the following keystrokes:

2 ⌃ 4 ⌃ 2

or

2 y^x 4 y^x 2

Share & Summarize

Tell whether each equation is true or false. If an equation is false, rewrite the expression to the right of the equal sign to make it true.

1. $3^5 \times 3^3 = 3^{15}$
2. $(2^5)^0 = 1$
3. $(a^3)^4 = a^7$
4. $a^{30} \div a^2 = a^{15}$
5. $(a^5)^3 = a^{15}$
6. $a^4 \times a^3 = a^7$

7. Explain the difference between $(a^b)^c$ and $a^b \times a^c$. Refer to stretching and shrinking machines if they help you explain.

On Your Own Exercises

Practice & Apply

A 12-inch ruler is put through each machine. Find its exit length.

1.

2.

3.

4.

5.

Find a single repeater machine that will do the same work as the given hookup.

6.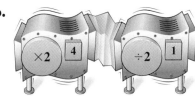

7.

8. This pencil emerged from a $\div 3$ machine. Estimate the length of the output in inches. Then find the length of the input pencil.

9. This pencil came out of a $\div \frac{1}{2}$ machine. Estimate the length of the output in inches. Then find the length of the input pencil.

Write each expression using a single base and a single exponent.

10. $2.7^{10} \div 2.7$

11. $0.2^{10} \div 0.2^{10}$

12. $a^{20} \div a^5$

13. $\dfrac{s^{100}}{s}$

14. $\dfrac{a^n}{a^m}$

15. $r^2 \div 5^2$

Write each expression as simply as you can.

16. $5a^4 \div 5a^4$

17. $\dfrac{12z^2}{6z^2}$

18. $\dfrac{15a^3}{5a^3}$

19. Caroline was working with this super machine.

   a. How many times would Caroline need to apply a $\times 3^3$ machine to do the same work as this super machine?

   b. How many times would she need to apply a $\times 3$ machine to do the same work as this machine?

   c. If Caroline puts a 1-inch bar of silver through this machine, how long will it be when it exits?

20. Here is Evan's favorite super machine.

   a. How many times would Evan need to apply a $\times 10^3$ machine to do the same work as this super machine?

   b. How many times would he need to apply a $\times 10$ machine to do the same work?

   c. If Evan puts a 1-inch bar of gold in this machine, how long will it be when it comes out?

Write each expression as a power of 2.

21. $(4^2)^3$ **22.** $2^4 \div 4^2$ **23.** $2^3 \div 2^3$

Write each expression using a single base and a single exponent.

24. $(0.5^2)^2$ **25.** $(2.5^9)^0$ **26.** $(1.9^5)^3$ **27.** $(p^{25})^4$

28. $(p^4)^{25}$ **29.** $(5^n)^2$ **30.** $(5^2)^n$ **31.** $(4n)^3$

32. $(3^2)^3$ **33.** $(5^2)^2$ **34.** $(99^1)^5$ **35.** $(100^0)^5$

Connect & Extend

36. Describe a hookup of repeater machines that will shrink a traffic ticket to $\frac{16}{27}$ of its original length.

37. Which one of these is not equal to the others?

$\left(\frac{1}{4}\right)^2$ $\left(\frac{1}{2}\right)^4$ $\frac{1}{8}$ 6.25% $\frac{625}{10,000}$

Preview Write each expression in the form $\frac{1}{5^n}$.

38. $5^4 \div 5^{10}$ **39.** $5^4 \div 5^{14}$ **40.** $5^{10} \div 5^{20}$

Your factory has received a special order to stretch a 1-inch bar of gold to *over* 1,000 inches. In Exercises 41–45, give the exponent of the repeater machine that would get the job done. If it is not possible to do the work with the given base machine, say so.

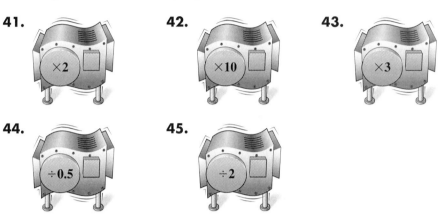

41. ×2

42. ×10

43. ×3

44. ÷0.5

45. ÷2

46. Preview Describe two super machines that would stretch a bar of gold to one million times its original length.

Preview A 1-inch bar of titanium is put through each super machine. How long will it emerge? Give your answers in inches, feet, and miles.

Remember
5,280 feet = 1 mile

47. ×10 | 3 | 1

48. ×10 | 3 | 2

49. ×10 | 3 | 3

50. ×10 | 3 | 4

51. How long will the bar of copper be when it exits this hookup?

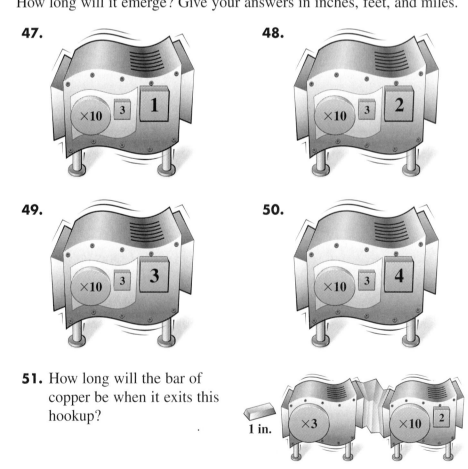

1 in. → ×3 → ×10 | 2 → ?

Rewrite each expression using a single base or a single exponent, if possible. For example, $3^7 \div 5^7$ can be rewritten $\left(\frac{3}{5}\right)^7$. You will have to decide which exponent laws to apply. Here are the laws you have seen so far.

In your
own words

Choose one of the quotient laws of exponents and explain the law so a fifth grader could understand it. You can refer to stretching and shrinking machines if they help you explain.

Product Laws	**Quotient Laws**	**Power of a Power Law**
$a^b \times a^c = a^{b+c}$	$a^b \div a^c = a^{b-c}$	$(a^b)^c = a^{b \times c}$
$a^c \times b^c = (a \times b)^c$	$a^c \div b^c = (a \div b)^c$	

52. $7^5 \div 3^5$

53. $7^5 \times 5^7$

54. $s^{100} \times s$

55. $1{,}000^{14} \div 2^{14}$

56. $6^5 \div 6^2$

57. $100^{100} \div 25^{25}$

58. $(2^0)^3$

59. $10^2 \times \left(\frac{1}{2}\right)^2$

60. $6^b \times 6^b$

61. $(x^2)^x$

62. $d^2 \times d^0$

63. $10^p \div 5^p$

64. $a^n \times a^m$

65. $(4x)^2 \div x^2$

66. $(10^2)^3 \div 10^2$

Challenge Use what you know about exponents to write each expression in a simpler form.

67. $m^3 n^5 \times m^5 n^2$

68. $a^3 b^4 c \times 2ab^3 c^2$

69. $14^p \div 7^{2p}$

70. $2y^3 z \times (6yz^4)^2$

71. $\dfrac{x^2 y z^3}{x^3 y^2 z}$

72. $\dfrac{15m^3 n^2}{60k^3 m^4 n^2}$

Mixed Review

73. Geometry Consider this cylinder.

   a. Find the volume of this cylinder.

   b. Find the volume of a cylinder with the same height as this cylinder and twice the radius.

   c. Find the volume of a cylinder with the same radius as this cylinder and twice the height.

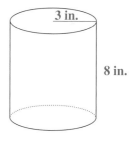

3 in.

8 in.

Order each set of numbers from least to greatest.

74. $4, 2, -\frac{1}{2}, 0, ^-4$

75. $^-30, 30, 3, ^-3, 0$

76. $200, ^-500, ^-50, 20, 2, ^-5$

77. $0.001, 0.1, ^-0.1, ^-1, ^-0.001$

Algebra Solve each equation for a.

78. $5a - 3a = 19$

79. $7a - 3a = a$

80. $a - 5 = 1.5a - 6$

81. Use a factor tree to find the prime factors of 75.

82. Use a factor tree to find the prime factors of 72.

83. Use a factor tree to find the prime factors of 100.

3.3 Growing Exponentially

Have you ever heard someone say, "It's growing exponentially"? People use this expression to describe things that grow very rapidly.

In this lesson, you will develop a more precise meaning of *exponential growth*. You'll also investigate what it means to decrease, or decay, exponentially.

Explore

Suppose you make up a funny joke one evening. The next day, you tell it to two classmates. The day after that, your two classmates each tell it to two people. Everyone who hears the joke tells two new people the next day.

How many new people will hear the joke on the fourth day after you made it up?

On which day will more than 1,000 new people hear the joke?

If the joke was told only to people in your school, how many days would it take for everyone in your school to hear it?

The number of students who hear the joke for the first time doubles each day. *Repeated doubling* is one type of exponential growth. In the next two investigations, you will look at other situations in which things grow exponentially.

Investigation Telling the Difference

The expressions 2^x, $2x$, and x^2 look similar—they all include the number 2 and the variable x—but they have very different meanings. The same is true of 3^x, $3x$, and x^3. Looking at tables and graphs for these expressions will help you discover just how different they are.

Problem Set A

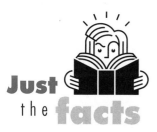

Mr. Brooks brought two fruit flies to his biology class. He told his students that the flies would reproduce, and that the population of flies would double every week. Luis and Zoe wanted to figure out how many flies there would be each week.

Luis made this table:

Week	1	2	3	4	5
Flies	2	4	6	8	10

Zoe made this table:

Week	1	2	3	4	5
Flies	2	4	8	16	32

1. Whose table is correct? That is, whose table shows the fruit fly population doubling every week?

2. How do the fly-population values in the incorrect table change each week?

3. Complete this table to compare the expressions $2x$, 2^x, and x^2.

x	0	1	2	3	4	5
$2x$	0	2				
2^x	1	2				
x^2	0	1				

4. As x increases from 0 to 5, which expression's values increase most quickly?

5. Which expression describes Luis's table? Which describes Zoe's table?

6. How many flies does the expression for Luis's table predict there will be in Week 10? How many flies will there actually be (assuming none of the flies die)?

7. Use your table from Problem 3 to plot the points for each set of ordered pairs— $(x, 2x)$, $(x, 2^x)$, and (x, x^2)—on one set of axes. If you can, use a different color for each set of points to help you remember which is which. How are the graphs similar? How are they different?

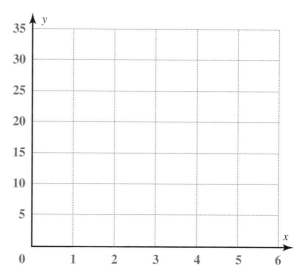

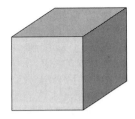

Problem Set B

Remember

The formula for the volume of a cube with edge length x is $V = x^3$.

Kate and Darnell made tables to show the relationship between the edge length and the volume of a cube.

Kate made this table:

Edge Length	1	2	3	4	5
Volume	3	9	27	81	243

Darnell made this table:

Edge Length	1	2	3	4	5
Volume	1	8	27	64	125

1. Whose table shows the correct volumes?

2. Complete this table to compare the expressions $3x$, 3^x, and x^3.

x	0	1	2	3	4	5
$3x$	0					
3^x	1					
x^3	0					

3. As x increases from 0 to 5, which expression's values increase most quickly?

4. Which expression describes Kate's table? Which describes Darnell's table?

5. Look at the values in the table for 3^x. Describe in words how to get from one value to the next.

6. Look at the values in the table for $3x$. Describe in words how to get from one value to the next.

Share & Summarize

1. Compare the way the values of $2x$ change as x increases to the way the values of 2^x change as x increases.

2. Consider the expressions $4x$, 4^x, and x^4. Without making a table, predict which expression's values grow the fastest as x increases.

Investigation 2 ▶ Exponential Increase and Decrease

Suppose you buy four comic books each week. The table shows how your comic book collection would grow.

Weeks	1	2	3	4	5	6
Comics	4	8	12	16	20	24

Since you add the same number of comics each week, your collection grows at a *constant rate*. You add 4 comics each week, so the number of comics in your collection after x weeks can be expressed as $4x$.

Now consider the fruit fly population from Problem Set A.

Weeks	1	2	3	4	5	6
Flies	2	4	8	16	32	64

The fly population grows at an *increasing rate*. That is, the number of flies added is greater each week. In fact, the total number of flies is multiplied by 2 each week.

Quantities that are repeatedly multiplied by a number greater than 1 are said to grow exponentially, or to show **exponential increase**, or **exponential growth.** The number of flies after x weeks can be expressed as 2^x.

Think & Discuss

Look back at your tables from Investigation 1.

The values of $2x$ and $3x$ grow at a constant rate. How do the values of $2x$ and $3x$ change each time x increases by 1? How is the change related to the expression?

The values of the expression 3^x grow exponentially. How do the values of 3^x change each time x increases by 1? How is the change related to the expression?

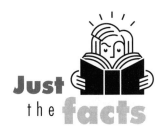

Just the facts

Rice, the staple food of millions of people in Asia and eaten around the world, is actually a grain harvested from a plant in the grass family.

Problem Set C

In an ancient legend, a ruler offers one of his subjects a reward in return for a favor. For the reward, the subject requests that the ruler place 2 grains of rice on the first square of a chessboard, 4 grains on the second square, 8 on the third, and so on, doubling the number of grains of rice on each square.

The ruler thinks this wouldn't take much rice, so he agrees to the request.

1. Copy and complete the table to show the number of grains of rice on the first six squares of the chessboard.

Square	1	2	3	4	5	6
Grains of Rice	2	4	8			

2. Which expression describes the number of grains on Square x?

$$x^2 \qquad\qquad 2^x \qquad\qquad 2x$$

3. Use the expression to find the number of grains on Square 7, Square 10, and Square 20.

4. For one type of uncooked, long-grain rice, 250 grains of rice have a volume of about 5 cm^3. What would be the volume of rice on Square 5? On Square 7? On Square 10?

5. When the ruler saw there were about 20 cm^3 of rice on Square 10, he assumed there would be 40 cm^2 on Square 20. Is he correct? Is he even close? How many cubic centimeters of rice will actually be on Square 20?

6. The number of grains of rice on the last square is 2^{64}. Use your calculator to find this value. What do you think your calculator's answer means?

The value of 2^{64} is too large for calculators to display normally, so they use a special notation, which you will learn about in Lesson 3.4.

In Problem Set C, the number of grains of rice on each square grows exponentially—it is multiplied by 2 each time. In Problem Set D, you will look at a different kind of exponential change.

Problem Set D

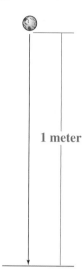

Imagine that a superball is dropped onto concrete from a height of 1 meter and is allowed to bounce repeatedly. Suppose the height of each bounce is 0.8 times the height of the previous bounce. (To understand how the height changes, you can imagine sending a stick of gum through a $\times 0.8$ machine repeatedly.)

1 meter

Remember
1 m = 100 cm

1. How high does the ball rise on the first bounce?

2. How high does the ball rise on the second bounce?

3. How high does the ball rise on the third bounce?

4. Complete the table to show how the ball's height in centimeters changes with each bounce.

Bouncing a Superball

Initial Height	100	100
1st Bounce Height	100×0.8	100×0.8^1
2nd Bounce Height	$100 \times 0.8 \times 0.8$	100×0.8^2
3rd Bounce Height		
4th Bounce Height		
10th Bounce Height		

5. How high does the ball rise on the 10th bounce?

6. How many times do you think the ball will bounce before coming to rest?

7. Use the pattern in the right column of your table to write an expression for the ball's height on the nth bounce.

8. Assuming the ball continues to bounce, use your expression to find the height of the ball on the 35th bounce. Does your answer surprise you?

exponential
decrease
exponential decay

In both the bouncing-ball problem and the rice problem, a quantity is repeatedly multiplied by the same factor. For the ball problem, however, the factor is 0.8, which is between 0 and 1. This means the bounce height gets smaller with each bounce. This type of exponential change is called **exponential decrease,** or **exponential decay.**

Share & Summarize

1. Briefly explain why the amount of rice on each square of the chessboard grows so rapidly.

2. As n increases, what happens to the value of $\left(\frac{1}{5}\right)^n$?

3. As n increases, what happens to the value of 5^n?

On Your Own Exercises

Practice & Apply

1. The rows of this table represent the expressions 4^x, x^4, and $4x$. Copy the table. Fill in the first column with the correct expressions, and then fill in the missing entries.

x	0	1	2	3	4	5	6
	1	4		64			
	0	4				20	
	0	1	16				

2. For a science experiment, Jinny put a single bacterium in a dish and placed it in a warm environment. Each day at noon, she counted the number of bacteria in the dish. She made this table from her data:

Day	0	1	2	3
Bacteria	1	4	16	64

a. How do the bacteria population values change each day?

Jinny repeated her experiment, beginning with 1 bacterium. She wanted to see if she could get the number of bacteria to triple each day by lowering the temperature of the environment. She predicted the number of bacteria in the dish with this table:

Day, d	0	1	2	3
Bacteria, b	1	3	6	9

b. Which expression describes her table for $d = 1$, 2, and 3: $b = 3d$, $b = d^3$, or $b = 3^d$?

c. Is the table correct? If not, which expression should Jinny have used?

3. It's the first day of camp, and 242 campers are outside the gates waiting to enter. The head counselor wants to welcome each camper personally, but she doesn't have enough time. At 9:00 A.M., she welcomes two campers and leads them into camp. This takes 10 minutes.

Now the camp has three people: the head counselor and the two campers. The three return to the gate, and each welcomes two more campers into the camp. This round of welcomes—for six new campers this time—also takes 10 minutes.

a. How many people are in the camp after this second round of welcomes?

b. At what time will all of these people be ready to welcome another round of campers?

LESSON 3.3 Growing Exponentially **183**

c. With every round, each person in the camp welcomes two new campers into camp. After the third round of welcomes, how many people will be in the camp? What time will it be then?

d. Copy and complete the table.

Welcome Rounds, n	0	1	2	3	4	5	6	7	8
Time at End of Round	9:00	9:10							
People in Camp, p	1	3							

e. At what time will all 242 campers have been welcomed into camp? (Don't forget about the head counselor!)

f. Look at the last row in your table. By what number do you multiply to get from one entry to the next?

g. Which expression describes the total number of people in the camp after n rounds: 3^n, $3n$, or n^3? Explain your choice.

h. Use your expression from Part g to find the number of welcome rounds that would be required to bring at least 1,000 campers into camp and to bring at least 10,000 campers into camp.

i. Lunch is served at noon. Could a million campers be welcomed before lunch? (Ignore the fact that it would be very chaotic!)

4. Match the given expression with the situation it describes.

$$3p \qquad\qquad p^2 \qquad\qquad 2^p$$

a. area of the squares in a series, with side length p for the square at Stage p

b. a population of bacteria that doubles every hour

c. a stamp collection to which 3 stamps are added each week

5. A botanist recorded the number of duckweed plants in a pond at the same time each week.

Week, w	0	1	2	3	4
Plants, d	32	48	72	108	162

a. Look at the second row of the table. By what number do you multiply each value to get the next value?

b. Predict the number of plants in the fifth week.

c. Which expression describes the number of plants in the pond after w weeks?

$$32 \times 1.5^w \qquad\qquad 1.5w + 32 \qquad\qquad w^2 + 32$$

6. Start with a scrap sheet of paper. Tear it in half and throw half away. Then tear the piece you have left in half and throw half away. Once again, tear the piece you have left in half and throw half away.

   a. What fraction of the original piece is left after the first tear? After the second tear?

   b. Make a table showing the number of tears and the fraction of the paper left.

   c. Look at the second row of your table. By what number do you multiply each value to get the next value?

   d. Write an expression for the fraction of the paper left after t tears.

   e. If you continue this process, will the paper ever disappear? Why or why not?

7. Preview You can evaluate the expression 2^x for negative values of x. This table is filled in for x values from 0 to 5.

x	$^-5$	$^-4$	$^-3$	$^-2$	$^-1$	0	1	2	3	4	5
2^x						1	2	4	8	16	32

   a. How do the 2^x values change as you move from *right to left*?

   b. Use the pattern you described in Part a to complete the table.

Tell whether each expression represents exponential growth, exponential decay, or neither.

8. 3.2^x **9.** $3x^2$ **10.** $32x$ **11.** $\left(\frac{3}{2}\right)^x$ **12.** $\left(\frac{2}{3}\right)^x$

Tell whether each sentence describes exponential growth, exponential decay, or neither. Explain how you decided.

13. Each time a tennis ball is used, its pressure is $\frac{999}{1,000}$ what it was after the previous use.

14. Each year the quantity of detergent in a liter of groundwater near an industrial area is expected to increase by 0.05 gram.

15. A star becomes twice as bright every 10 years.

16. It is predicted that the membership in a club will increase by 20 people each month.

17. The number of tickets sold in a national lottery is expected to increase 1.2 times each year.

18. **Ecology** "Whale Numbers up 12% a Year" was a headline in a 1993 Australian newspaper. A 13-year study had found that the humpback whale population off the coast of Australia was increasing significantly. The actual data suggested the increase was closer to 14%! This means that the population each year was 1.14 times the previous year's population.

a. When the study began in 1981, the humpback whale population was 350. If the population grew to 1.14 times this number in the next year, what was the humpback whale population in 1982?

b. What would you expect the whale population was in 1983?

c. Complete the table to show how the population grew each year.

Humpback Whale Population

1981: Year 0	350	350
1982: Year 1	350×1.14	350×1.14^1
1983: Year 2	$350 \times 1.14 \times 1.14$	350×1.14^2
1984: Year 3		
1985: Year 4		
1993: Year 12		

d. Write an expression for the number of whales x years after the study began in 1981.

e. How long did it take the whale population to double from 350?

In y o u r
own
words

Describe the difference in the way 4^x and $4x$ grow. In your description, tell which expression grows exponentially, and explain how you could tell from a table of values.

Remember
An *equilateral triangle* has three sides of the same length.

19. Challenge A ball of Malaysian rubber, displayed in a museum of science and technology, rolled off the 1.25-meter-high display table and onto the floor. It bounced, rose to a height of 0.75 meter, and was caught at the top of its bounce.

  a. What fraction of the original height was the bounce height of the ball?

  b. Assume each bounce height is a fixed fraction of the previous bounce height. Suppose the ball falls off the table again and is not caught. Write an expression for the height of the bounce, in centimeters, after n bounces.

  c. Darnell wondered whether he would have been able to slide his hand between the ball and the floor on the 10th bounce. Would he have been able to? Explain.

  d. Suppose the same ball fell from a 2-meter-high shelf. Would Darnell have been able to fit his hand between the ball and the floor on the 10th bounce? Explain.

20. This drawing was created by first drawing a large equilateral triangle. A second equilateral triangle was drawn inside the first by connecting the midpoints of its sides. The same process was used to draw a third triangle inside the second, and so on. The sides of each triangle are half as long as the sides of the previous triangle.

  a. If the sides of the smallest triangle are 1 cm long, what are the side lengths of the other triangles?

  b. What is the length of the purple spiral in the drawing?

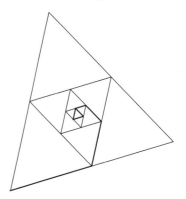

**Mixed
  Review**

21. Each expression below is equal to one of the other expressions. Find the matching pairs.

  a. 2^5 **b.** $c \times c \times c$ **c.** 2×5

  d. $2 \cdot 2 \cdot 2 \cdot 2 \cdot 2$ **e.** $c + c + c$ **f.** 5^2

  g. 5×5 **h.** c^3 **i.** $5 + 5$

  j. $3c$

22. Suppose you flip a coin two times.

a. What is the probability you will get two heads? Two tails?

b. What is the probability you will get one head and one tail?

23. Earth Science A middle school class in Florida conjectured that the temperature in coastal areas varies less dramatically than it does inland. They gathered the data shown in the table.

Average Temperature on First Day of the Month

Month	Coastal Temperature, °F (St. Augustine)	Inland Temperature, °F (Lakeland)
January	59	63
February	58	62
March	60	65
April	66	70
May	72	75
June	77	79
July	81	82
August	81	82
September	80	81
October	76	79
November	69	70
December	61	63

Source: Southeast Regional Climate Center Home Page, *water.dnr.state.sc.us/climate/sercc/.*

a. On a graph like the one shown, plot the points for coastal temperatures in one color and those for inland temperatures in another.

Average Temperature on First Day of the Month

b. What are the highest and lowest temperatures shown for Saint Augustine? For Lakeland?

c. Does the information for these two cities support the idea that coastal temperatures vary less dramatically than inland temperatures? Explain.

24. Consider this pattern of squares.

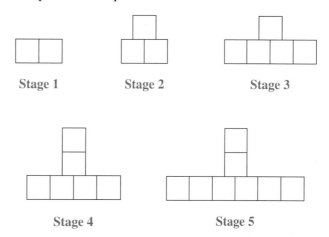

Stage 1   Stage 2   Stage 3

Stage 4   Stage 5

a. For the part of the pattern shown, how many squares are added from each stage to the next?

b. How many squares are added to go from Stage *s* to Stage *s* + 1? Where are the squares placed? (Hint: Consider two cases, when *s* is even and when *s* is odd.)

c. How many squares in all are in each stage shown?

d. How many squares will be needed to make Stage 7? Stage 10?

e. What stage has 20 squares?

Simplify each expression.

25. $\frac{1}{3} \div \frac{2}{3}$

26. $\frac{7}{5} \div \frac{5}{7}$

27. $\frac{7}{5} \div \frac{7}{5}$

28. $\frac{1}{8} \div \frac{1}{4}$

Fill in each blank to make a true statement.

29. $0.25 + \underline{\hspace{1cm}} = 1$

30. $0.6 + 0.15 + \underline{\hspace{1cm}} = 1$

31. $0.05 + \underline{\hspace{1cm}} = 1$

32. $\frac{3}{10} + \frac{9}{10} - \underline{\hspace{1cm}} = 1$

Can you imagine what a million quarters look like? If you stacked a million quarters, how high would they reach? How long is a million seconds? How old is someone who has been alive a billion seconds?

In this lesson, you will explore these questions, and you will learn a new way of writing and working with very large numbers.

Think & Discuss

What is funny about each of these cartoons?

Investigation Millions and Billions

It's difficult to understand just how large 1 million and 1 billion are. It helps to think about these numbers in contexts that you can imagine. The problems in this investigation will help you get a better sense of the size of a million and a billion.

Think & Discuss

How many zeroes follow the 1 in 1 million?

How many zeroes follow the 1 in 1 billion?

How many millions are in 1 billion?

Problem Set A

1. How old is someone who has been alive 1 million seconds?

2. How old is someone who has been alive 1 billion seconds?

MATERIALS
a quarter

Problem Set B

1. Find the diameter of a quarter in inches. Give your answer to the nearest inch.

2. The distance around Earth's equator is about 24,830 mi.

   a. If you lined up 1 million quarters end to end, how far would they reach? Give your answer in miles.

   b. Would they reach around the equator? If not, how many quarters *would* you need to reach around the equator?

3. How many quarters do you need to make a stack 1 in. high?

4. The average distance from Earth to the moon is 238,855 mi.

   a. If you stacked 1 million quarters, how high would they reach?

   b. Would they reach the moon? If not, how many quarters *would* you need to reach the moon? Estimate your answer without using a calculator, and explain how you found your answer.

Remember
5,280 ft = 1 mi

5. A baseball diamond is a square, 90 ft on each side. Would 1 million quarters cover a baseball diamond if you spread them out? Explain how you found your answer.

Hint: Think of 1 million quarters forming a square with 1,000 quarters on each side. Since quarters are round, they don't fit together exactly. Don't worry about the extra space left uncovered.

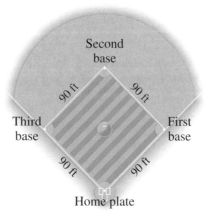

Second base

90 ft   90 ft

Third base   First base

90 ft   90 ft

Home plate

Share & Summarize

1. The average distance from Earth to the sun is about 93,000,000 miles. How many quarters would you need to stack for the pile to reach the sun?

2. Why do you think many people don't understand the difference between a million and a billion?

Investigation ▶ 2 ▶ Powers of 10

In Investigation 1, you investigated 1 million and 1 billion. Both of these numbers can be expressed as powers of 10.

$$1 \text{ million} = 1,000,000 = 10^6$$

$$1 \text{ billion} = 1,000,000,000 = 10^9$$

It is often useful to talk about large numbers in terms of powers of 10. Imagining ×10 repeater machines can help you get used to working with such powers.

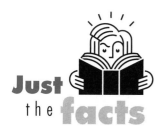
Think & Discuss

Your resizing factory is so successful, you plan a celebration for all your employees. To make decorations, you send pieces of colored party streamers through stretching machines.

Find the length of the streamer that exits each machine. Give your answers in centimeters.

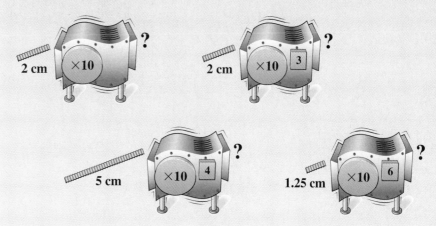

If you know the length of the input streamer and the exponent on the ×10 repeater machine, how can you find the length of the output streamer?

Find the exponent of each repeater machine. (Note: The output streamer is not drawn to scale.)

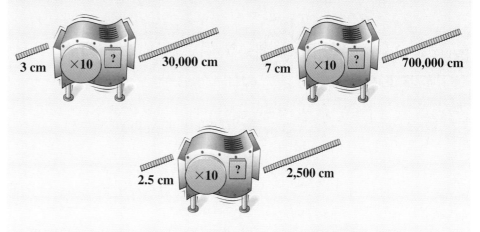

If you know the length of the input and output streamers, how can you find the exponent of the ×10 repeater machine?

Problem Set C

1. This party straw exited a $\times 10^2$ machine. Without measuring, estimate the length of the input straw in centimeters. It may be helpful if you copy the straw onto your paper. Explain how you found your answer.

2. This straw went into a $\times 10$ machine. Without measuring, estimate the length of the output straw in centimeters.

Find the length of each output straw.

3.

1.5 cm

4.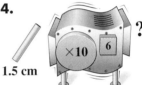

1.5 cm

5.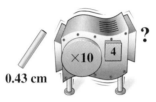

0.43 cm

Find each exponent.

6.

2 m   2 km

7.

17.95 ft   1,795 ft

8.

0.25 in.   25,000 in.

Find the length of each input straw.

9.

2 m

10.

70 cm

11.

? $\times 10$ | 4 | 45,000 mi

In Problem Set D, you will work with expressions involving powers of 10. As you work, it may help to think of repeater machines. For example, you could use this machine to think about 5×10^3.

5 | $\times 10$ | 3 |

Problem Set D

Find each product.

1. 8×10^5 **2.** 18.6×10^8 **3.** $9,258 \times 10^4$

Find the value of N in each equation.

4. $6 \times 10^N = 600$ **5.** $8.6 \times 10^N = 86,000$

6. $54 \times 10^N = 5,400,000$ **7.** $2,854 \times 10^N = 285,400$

8. $N \times 10^2 = 800$ **9.** $N \times 10^3 = 47,000$

10. $N \times 10^6 = 54,800,000$ **11.** $N \times 10^4 = 3,958$

12. $85 \times 10^4 = N$ **13.** $9.8 \times N = 9,800,000$

14. $0.00427 \times 10^N = 42.7$ **15.** $N \times 10^{11} = 24,985,000,000$

Just the facts

The longest gum-wrapper chain on record was 18,721 feet long and took more than 30 years to complete.
Source: Guinness Book of World Records. New York: Bantam Books, 1998.

Share & Summarize

1. What repeater machine would stretch a 6-inch gum-wrapper chain to 6 million inches?

2. What repeater machine would stretch a 3-inch gum-wrapper chain to 3 billion inches?

3. What repeater machine would stretch a 2.3-million-inch gum-wrapper chain to 2.3 billion inches?

Investigation ▶3 Scientific Notation

Powers of 10 give us an easy way to express very large and very small numbers without writing a lot of zeros. This is useful in many fields, such as economics, engineering, computer science, and other areas of science. For example, astronomers need to describe great distances—like the distances between planets, stars, and galaxies. Chemists often need to describe small measurements—like the sizes of molecules, atoms, and quarks.

In this investigation, you will focus on a method for expressing large numbers. In Chapter 4, you will see how this method can be used to express small numbers, as well.

Just the facts

The Centaurus system includes Alpha Centauri, the star—other than our own sun—that is closest to Earth.

EXAMPLE

Astronomical distances are often expressed in light-years. For example, the Centaurus star system is 4.3 light-years from Earth. A *light-year* is the distance light travels in one year. One light-year is approximately 5,878,000,000,000 miles.

There are lots of ways to express the number of miles in a light-year as the product of a number and a power of 10. Here are three.

$$5{,}878 \times 10^9 \qquad 58.78 \times 10^{11} \qquad 5.878 \times 10^{12}$$

Can you explain why each of these three expressions equals 5,878,000,000,000?

The third expression above is written in scientific notation. A number is in **scientific notation** when it is expressed as the product of a number greater than or equal to 1 but less than 10, and a power of 10.

at least 1 but
less than 10   a power of 10

$$\overbrace{5.878} \times \overbrace{10^{12}}$$

The chart lists the outputs of some repeater machines in scientific and standard notation.

Machine	Scientific Notation	Standard Notation	Example of Number
3 ×10 2	3×10^2	300	length of a football field in feet
6.5 ×10 7	6.5×10^7	65,000,000	years since dinosaurs became extinct
2.528 ×10 13	2.528×10^{13}	25,280,000,000,000	distance from Earth to Centaurus star system in miles

Scientific notation can help you compare the sizes of numbers.

Think & Discuss

Here are three numbers written in standard notation.

74,580,000,000,000,000,000

8,395,000,000,000,000,000

242,000,000,000,000,000,000

• Which number is greatest? How do you know?

• Which number is least? How do you know?

Here are three numbers written in scientific notation.

3.723×10^{15} 9.259×10^{25} 4.2×10^{19}

• Which number is greatest? How do you know?

• Which number is least? How do you know?

Problem Set E

Tell whether each number is written correctly in scientific notation. For those that are not written correctly, describe what is incorrect.

1. 6.4535×10^{52}

2. 41×10^{3}

3. 0.4×10^{6}

4. 1×10^{1}

Write each number in standard notation.

5. 1.28×10^{6}

6. 9.03×10^{5}

7. 6.02×10^{23}

8. 5.7×10^{8}

Write each number in scientific notation.

9. 850

10. 7 thousand

11. 10,400,000

12. 659,000

13. 83 million

14. 27 billion

15. The table shows the average distance from each planet in our solar system to the sun.

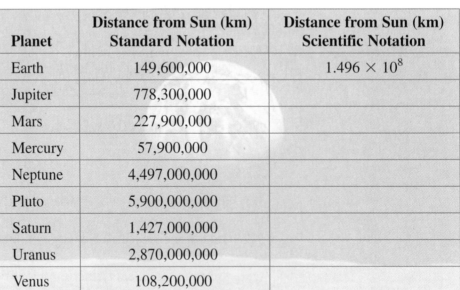

Planet	Distance from Sun (km) Standard Notation	Distance from Sun (km) Scientific Notation
Earth	149,600,000	1.496×10^{8}
Jupiter	778,300,000	
Mars	227,900,000	
Mercury	57,900,000	
Neptune	4,497,000,000	
Pluto	5,900,000,000	
Saturn	1,427,000,000	
Uranus	2,870,000,000	
Venus	108,200,000	

a. Complete the table by expressing the distance from each planet to the sun in scientific notation.

b. Order the planets from closest to the sun to farthest from the sun.

Just the facts

Pluto's orbit is unusual. Pluto is sometimes closer to the sun than its neighbor Neptune is, as it was from January 1979 to February 1999.

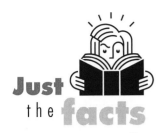
Scientific notation is useful for ordering numbers only if the convention is followed properly—that is, if the first part of the number is between 1 and 10. Otherwise, the power of 10 may not tell you which number is greatest.

The expressions below show the same number written in four ways as the product of a number and a power of 10. However, none of these expressions is in scientific notation. If you compare these numbers by looking at the powers of 10, you might not realize that they are all equal.

$$0.0357 \times 10^8 \quad 35.7 \times 10^5 \quad 3{,}570 \times 10^3 \quad 357{,}000 \times 10^1$$

You can change numbers like these into scientific notation without changing them into standard form first.

EXAMPLE

Write 357×10^4 in scientific notation.

You can change 357 to a number between 1 and 10 by dividing it by 100. To compensate, you need to multiply by 100, or 10^2.

$$357 \times 10^4 = \frac{357}{100} \times 100 \times 10^4$$
$$= 3.57 \times 10^2 \times 10^4$$
$$= 3.57 \times 10^6$$

Problem Set F

Write each number in scientific notation.

1. 13×10^2 **2.** 0.932×10^3

3. 461×10^4 **4.** 5.9×10^5

5. 98.6×10^9 **6.** 197×10^6

Share & Summarize

How can you tell whether a number greater than 1 is written in scientific notation?

Investigation 4 Scientific Notation on Your Calculator

In Lesson 3.3 you worked on a problem about a ruler who had to place rice on the squares of a chessboard as a reward for one of his subjects. The first square had 2 grains of rice, and every square after that had twice as many as the previous square.

The table describes the number of grains of rice on each square.

Square	Number of Grains (as a product)	Number of Grains (as a power of 2)
1	1×2	2^1
2	$1 \times 2 \times 2$	2^2
3	$1 \times 2 \times 2 \times 2$	2^3
4	$1 \times 2 \times 2 \times 2 \times 2$	2^4
5	$1 \times 2 \times 2 \times 2 \times 2 \times 2$	2^5
6	$1 \times 2 \times 2 \times 2 \times 2 \times 2 \times 2$	2^6
7	$1 \times 2 \times 2 \times 2 \times 2 \times 2 \times 2 \times 2$	2^7

Think & Discuss

There are 2^{40} grains of rice on Square 40. Evaluate this number on your calculator. What does the calculator display? What do you think this means?

Numbers too large to fit in a calculator's display are expressed in scientific notation. Different calculators show scientific notation in different ways. When you entered 2^{40}, you may have seen one of these displays:

Both of these represent $1.099511628 \times 10^{12}$. When you read the display, you may need to mentally insert the "$\times 10$."

If you continued the table, what power of 2 would represent the number of grains of rice on Square 64, the last square on the chessboard?

Use your calculator to find the number of grains of rice on Square 64. Give your answer in scientific notation.

Problem Set G

Just the facts

A day is the time it takes for one rotation of a planet; a year is the time it takes for a planet to make one revolution around the sun. On Jupiter, a day is about 10 Earth hours long, and a year is about 12 Earth years.

Use your calculator to evaluate each expression. Give your answers in scientific notation.

1. 3^{28} **2.** 4.05^{21} **3.** 7.95^{12} **4.** 12^{12}

5. It would be difficult to order the numbers in Problems 1–4 as they are given. It should be much easier now that you've written them in scientific notation. List the four numbers from least to greatest.

6. Light travels at a speed of about 186,000 miles per second.

    a. How many miles does light travel in a day?

    b. How many miles does light travel in a year?

When the result of a calculation is a very large number, your calculator automatically displays it in scientific notation. You can also enter numbers into your calculator in scientific notation. Different calculators use different keys, but a common one is $\boxed{\text{EE}}$.

For example, to enter 2.4×10^{12}, press

$$2.4 \boxed{\text{EE}} \; 12 \; \boxed{\overset{\text{ENTER}}{=}}.$$

Problem Set H

Estimate the value of each expression, and then use your calculator to evaluate it. Give your answers in scientific notation.

1. $5.2 \times 10^{15} + 3.5 \times 10^{15}$

2. $(6.5 \times 10^{18}) \times (1.8 \times 10^{15})$

3. $(8.443 \times 10^{18}) \div 2$

4. Estimate the value of each expression, and then use your calculator to evaluate it. Give your answers in scientific notation.

    a. $6 \times 10^{12} + 4 \times 10^{2}$

    b. $8.5 \times 10^{20} + 1.43 \times 10^{45}$

    c. $4.92 \times 10^{22} - 9.3 \times 10^{5}$

    d. How do your results in Parts a–c compare to the numbers that were added or subtracted?

    e. Write the numbers in Part a in standard notation and then add them. How does your result compare to the result you found in Part a? If the answers are different, try to explain why.

Problem Set I

In 1945 a librarian at Harvard University estimated that there were 10 million books of "worthwhile" printed material. Suppose you wanted to enter all this printed material into a computer for storage.

Use your calculator to solve the following problems. When appropriate, record your answers in scientific notation. Make these assumptions as you work:

- The average book is 500 pages long.
- The average page holds 40 lines of text.
- The average line holds 80 characters.

1. How many characters are on one page of an average book?

2. How many characters are in the average book?

3. How many characters were in all the "worthwhile" books in 1945?

4. One *byte* of computer disk space stores about one character of text. A *gigabyte* of computer disk space is 1×10^9 bytes, so a gigabyte can store about 1×10^9 characters. How many gigabytes are needed to store all the "worthwhile" books from 1945?

Share & Summarize

1. Explain, in a way that a fifth grader could understand, why a calculator sometimes displays numbers in scientific notation.

2. Explain why sometimes when you add two numbers in scientific notation on your calculator, the result is just the greater of the two numbers.

Lab Investigation ▶ The Tower of Hanoi

MATERIALS

6 blocks, labeled 1, 2, 3, 4, 5, and 6

The Tower of Hanoi is a famous puzzle invented in 1883 by Edouard Lucas, a French mathematician. Lucas based the puzzle on this legend:

At the beginning of time, the priests in a temple were given three golden spikes. On one of the spikes, 64 golden disks were stacked, each one slightly smaller than the one below it.

The priests were assigned the task of moving all the disks to one of the other spikes while being careful to follow these rules:

- *Move only one disk at a time.*

- *Never put a larger disk on top of a smaller disk.*

When they completed the task, the temple would crumble and the world would vanish.

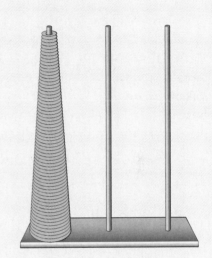

In this lab, you will figure out how long it would take to move all the disks from one spike to another. We will start by assuming the disks are so large and heavy that the priests can move only one disk per minute.

Make a Prediction

Imagine that the spikes are labeled A, B, and C and that the disks start out on Spike A. Since it can be overwhelming to think about moving all 64 disks, it may help to first consider a much simpler puzzle.

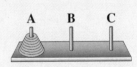

Remember

A larger disk cannot be placed on top of a smaller disk.

1. Suppose the puzzle started with only 1 disk on Spike A. How long would it take to move the disk to Spike B?

2. Suppose the puzzle started with 2 disks on Spike A. How long would it take to move both disks to Spike B? What would the moves be?

3. Try again with 3 disks. How long would it take? What would the moves be?

4. Predict how the total time required to solve the puzzle will change each time you increase the number of disks by 1.

5. Predict how long it would take to move all 64 disks. Write down your prediction so you can refer to it later.

Try It Out

Luckily, you don't need 64 golden disks to try the Tower of Hanoi puzzle. You can model it with some simple equipment. Your puzzle will have 5 "disks," rather than 64. You'll need a blank sheet of paper and five blocks, labeled 1, 2, 3, 4, and 5.

Label your paper with the letters A, B, and C as shown. Stack the blocks in numerical order, with 5 on the bottom, next to the A.

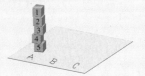

To solve the puzzle, you need to move all the blocks to another position—either B or C—following these rules:

• Move only one block at a time.

• Never put a larger number on a smaller number.

This is not an easy puzzle. To solve it, you might want to start with a puzzle using only 2 or 3 blocks. As you explore, look for a *systematic* way to move all the blocks to a new position.

Try It Again

6. Solve the puzzle again for towers of 1, 2, 3, 4, and 5 blocks. This time, count the number of moves it takes you to solve each puzzle. Record your data in a table.

Tower Height	1	2	3	4	5
Number of Moves					

7. Describe any patterns you see that might help you make predictions about how many moves it would take for larger towers.

8. Use your pattern to fill in a table like the one below. Then use a sixth cube to test your prediction for a tower of height 6.

Tower Height	6	7	8	9	10
Number of Moves					

9. Write an expression for the number of moves it would take to solve the puzzle for a tower of height *t*. (Hint: Add 1 to each entry in the second row of your table for Question 6, and then look at the pattern again.)

Back to the Legend

10. Assume that one disk is moved per minute. Figure out how long it would take to solve the puzzle for the heights shown in the table below. Report the times in appropriate units. (After a while, minutes are not very useful.)

Tower Height	1	2	3	4	5	6	7	8	9	10
Number of Moves										
Time										

11. How long would it take to move all 64 disks? Give your answer in years. How does your answer compare to your prediction in Question 5?

12. You are able to move blocks at a much faster pace than one per minute. What if the disks the priests used were smaller and lighter, so they could also work faster?

a. If one disk is moved per second, how long would it take to finish the puzzle?

b. If 10 disks are moved per second, how long would it take to finish?

What Did You Learn?

13. When you move a piece in the Tower of Hanoi puzzle, you often have two choices of where to place it. Explain how you decide which move to make.

14. Suppose the legend is true and the priests can move pieces at the incredible rate of 10 per second. Do you think they are likely to finish the puzzle in your lifetime? Explain.

15. Write a newspaper article about the Tower of Hanoi puzzle. You might mention the legend and the time it takes to move the disks for towers of different heights.

Practice  **Apply**

1. How many millimeters are in a kilometer?

2. In this exercise, you will think about the volume of 1 million quarters.

   a. Imagine a box that has dimensions of 1 inch on each edge. What is the volume of the box?

   b. Now imagine putting a stack of quarters in the box. How many quarters would fit in the box? (Notice that there will be some empty space because the quarters are round instead of square. For this exercise, don't worry about the extra space.)

   c. How many quarters would fit in a cubic foot of space? Explain. (Hint: Think carefully about the number of cubic inches in a cubic foot. It is not 12.)

   d. Would 1 million quarters fit in your refrigerator? (If you can't measure your refrigerator, use 20 ft^3 as an estimate of the volume.)

3. In this exercise, you will figure out how far you would walk if you took 1 million steps and if you took 1 billion steps.

   a. Measure or estimate the length of a single step you take.

   b. If you took 1 million steps, about how far would you walk? Give your answer in miles.

   c. If you took 1 billion steps, about how far would you walk? Give your answer in miles.

   d. Do you think you will walk 1 million steps in your lifetime? What about 1 billion steps? Explain your answers.

Supply each missing value. For Exercises 6 and 7, give the answer in kilometers.

4.

5 in.   ×10   2   ?

5.
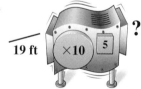
19 ft   ×10   5   ?

6.
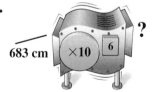
683 cm   ×10   6   ?

7.
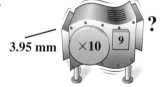
3.95 mm   ×10   9   ?

impactmath.com/self_check_quiz

Find each product.

8. 4×10^3 **9.** 62×10^4 **10.** 15.8×10^2

Find each exponent.

11.

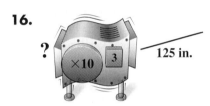

6 in. ×10 ?   600 in.

12.

15 m ×10 ?   150 km

Find the value of N in each equation.

13. $8 \times 10^N = 80$ **14.** $53 \times 10^N = 5{,}300$

15. $9.9 \times 10^N = 99{,}000$

Find the length of each input.

16.

? ×10 3   125 in.

17.

? ×10 5   280,000 mi

18.

? ×10 4   75.45 km

Find the value of N in each equation.

19. $N \times 10^2 = 900$ **20.** $523 \times N = 52{,}300$

21. $0.0614 \times N = 6.14$ **22.** $N \times 10^6 = 39{,}650{,}000$

Write each number in standard notation.

23. 8×10^1 **24.** 5×10^3

25. 7.5×10^2 **26.** 3.54×10^8

Write each number in scientific notation.

27. 300

28. ten thousand

29. 158,000

30. 8,350,000,000

31. 183 billion

32. 421,938,000,000,000

33. Astronomy The table shows the mass of the planets in our solar system and of the sun and the moon.

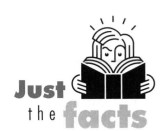

Celestial Body	Mass (kg) Standard Notation	Mass (kg) Scientific Notation
Sun	1,990,000,000,000,000,000,000,000,000,000	1.99×10^{30}
Mercury	330,000,000,000,000,000,000,000	
Venus	4,870,000,000,000,000,000,000,000	
Earth	5,970,000,000,000,000,000,000,000	
Mars	642,000,000,000,000,000,000,000	
Jupiter	1,900,000,000,000,000,000,000,000,000	
Saturn	568,000,000,000,000,000,000,000,000	
Uranus	86,800,000,000,000,000,000,000,000	
Neptune	102,000,000,000,000,000,000,000,000	
Pluto	12,700,000,000,000,000,000,000	
Moon	73,500,000,000,000,000,000,000	

    a. Write the mass of each planet and the moon in scientific notation.

    b. Order the planets and the moon by mass, from least to greatest.

    c. Which planet has about the same mass as Earth?

34. Refer to the table of masses in Exercise 33. Find how many times greater the mass of Earth is than the mass of each body.

    a. Mercury   **b.** Venus   **c.** Mars

    d. Pluto   **e.** the moon

35. Refer to the table of masses in Exercise 33. Find how many times greater the mass of each body is than the mass of Earth.

    a. the sun   **b.** Jupiter   **c.** Saturn

    d. Uranus   **e.** Neptune

Estimate the value of each expression. Then use your calculator to evaluate it. Give your answers in scientific notation.

36. $7.35 \times 10^{22} - 1.33 \times 10^{21}$

37. $6.42 \times 10^{45} + 4.55 \times 10^{45}$

38. $6.02 \times 10^{23} \times 15$

Connect & Extend

39. Economics In this exercise, you will investigate what you could buy if you had $1 million. To answer these questions, it might help to look at advertisements in the newspaper.

a. How many cars could you buy? Tell how much you are assuming each car costs.

b. How many houses could you buy? Tell how much you are assuming each house costs.

c. Make a shopping list of several items that total about $1 million. Give the price of each item.

40. Astronomy The average distance from Earth to the sun is 93,000,000 miles. How many of you, stacked on top of yourself, would it take to reach to the sun? Explain how you found your answer.

Remember

12 in. = 1 ft
5,280 ft = 1 mi

41. How many dollars is 1 million quarters?

42. Count the number of times your heart beats in 1 minute.

a. At your current heart rate, approximately how many times has your heart beaten since your birth?

b. How many times will your heart have beaten when you reach age 20?

c. How many times will your heart have beaten when you reach age 80?

43. A dollar bill is approximately 15.7 cm long.

a. How long would one thousand dollar bills laid end to end be?

b. How long would one million dollar bills laid end to end be?

c. The distance from Earth to the sun is about 1.5×10^8 km. How many dollar bills, taped together end to end, would it take to reach the sun?

Find the length of each output.

44.  **?**

2 in.

45.  **?**

21 ft

46.  **?**

16 mm

47.  **?**

276 km

48. 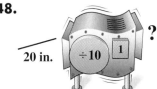 **?**

20 in.

49. 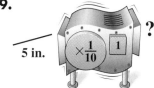 **?**

5 in.

50.  **?**

300 in.

51. 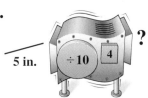 **?**

5 in.

Find the value of *N* in each equation.

52. $N \times \frac{1}{10^2} = 8$ **53.** $N \times \frac{1}{10^4} = 12$ **54.** $N \times \frac{1}{10^6} = 4.927$

55. $17 \div 10^1 = N$ **56.** $128.4 \div 10^3 = N$ **57.** $714 \div 10^8 = N$

58. List these numbers from least to greatest.

a quarter of a billion 2.5×10^7 two thousand million

10^9 half a million 10^5

59. Geometry Suppose you arranged 10^3 1-cm cubes into a large cube.

   a. What is the length of an edge of the cube, in centimeters?

   b. What is the volume of the cube, in cubic centimeters?

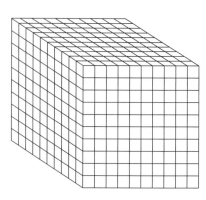

In your
own
words

Explain why scien-
tific notation is a
useful way to write
very large numbers.

Now suppose you arranged 10^6 1-cm cubes into a large cube.

c. What is the edge length of the new cube, in centimeters? What is the volume of the new cube, in cubic centimeters?

d. What is the volume of the new cube, in cubic meters? What is the volume of the first large cube, in cubic meters?

e. What is the area of a face of the new cube, in square centimeters?

60. Geometry Suppose you stacked 10^6 2-cm cubes into a tower. Part of the tower is shown here. What would be the height of the tower in meters?

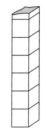

61. Preview Fill in the exponents below. (Hint: Fill in the exponents you know first. Then fill in the others by continuing the pattern.)

$$5{,}000 = 5 \times 10^?$$
$$500 = 5 \times 10^?$$
$$50 = 5 \times 10^?$$
$$5 = 5 \times 10^?$$
$$0.5 = 5 \times 10^?$$
$$0.05 = 5 \times 10^?$$

62. On May 29, 2003, the U.S. federal deficit was $6.5452 trillion, or 6.5452×10^{12}.

a. If the government could pay back $1 million each year, and the deficit did not increase, how long would it take to pay off the entire deficit?

b. A typical railway freight car measures 3 m $\times$ 4 m $\times$ 12 m. A dollar bill is approximately 15.7 cm long and 6.6 cm wide. A stack of 100 dollar bills is 1 cm high. How many freight cars would it take to carry the 2002 U.S. federal deficit in $1 bills? In $100 bills (a $100 bill is the same size as a $1 bill)?

c. On May 29, 2003, the estimated population of the United States was 291,095,000. Imagine that every person in the country gave the government $1 each minute. How many times would you pay $1 during your math class? During a day (24 hours)? How long would it take to pay off the deficit?

d. Invent another example to illustrate the size of the federal deficit.

63. A block structure and its top engineering view are shown. What are the volume and surface area of this structure?

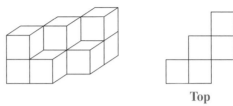

Top

64. Here is the top-count view of a block structure. What are the structure's volume and surface area?

2	2	2	2
2	1	1	2
	1	1	

Decide whether each equation is true. If an equation is not true, rewrite the right side to make it true.

65. $(y + 1) \times 7 = 7y + 7$

66. $(y - 1) \times 8 = y - 8$

67. $(3 + y) \times 7 = 3 \times 7 + y \times 7$

68. $2(4b + 1) = 8b + 2$

69. $2(4b + 4) = 8b + 4$

70. $1.5(A + 9) = 1.5A + 9.5$

Order each set of numbers from least to greatest.

71. $^-25, \, ^-42, \, ^-12, \, ^-10$

72. $0, 5, \, ^-5, \, ^-2, \, ^-2.5$

73. If a point on a graph has 0 as its *x*-coordinate, what do you know about the point's location?

74. What does the graph of all points with 3 as their *y*-coordinate look like?

75. Name a point that lies to the right and above point (1, 3).

Evaluate each expression.

76. 0.25×0.4

77. 0.25×4

78. 0.25×0.1

79. 0.04×0.04

80. 0.02×4

81. 0.2×0.7

Chapter Summary

V O C A B U L A R Y
base
exponent
**exponential
decay**
**exponential
decrease**
**exponential
growth**
**exponential
increase**
power
scientific notation

In this chapter, stretching and shrinking machines helped you understand the laws of exponents.

Product Laws	**Quotient Laws**	**Power of a Power Law**
$a^b \times a^c = a^{b+c}$	$a^b \div a^c = a^{b-c}$	$(a^b)^c = a^{b \times c}$
$a^c \times b^c = (a \times b)^c$	$a^c \div b^c = (a \div b)^c$	

You looked at quantities that grow exponentially, and investigated how exponential and constant growth differ. You explored situations in which quantities decrease exponentially. You learned how to recognize exponential growth and decay in tables, expressions, and written descriptions.

You worked on problems that helped you develop a sense of large numbers like a million and a billion. You learned about powers of 10, and you found that sometimes it is useful to express large numbers in scientific notation.

Strategies and Applications

The questions in this section will help you review and apply the important ideas and strategies developed in this chapter.

Working with stretching and shrinking machines

Supply the missing information for each diagram.

1.
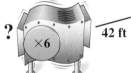

2.

3.
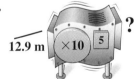

4.

If a 1-cm stick of gum is sent through each super machine, how long will it be when it exits?

5.

6.

Find a single repeater machine to do the same work as each hookup.

7.  **8.**

9. How long will the gummy worm be when it exits this hookup? Give your answer in meters.

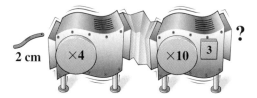

Understanding the laws of exponents

10. Explain why $a^b \div a^c = a^{b-c}$. You can refer to stretching and shrinking machines if they help you explain.

11. Explain why $2^0 = 1$. You can refer to stretching and shrinking machines if they help you explain.

Identifying and working with exponential growth

12. The table shows the values of several expressions containing x.

x	0	1	2	3	4	5
$3x$	0					
2^x	1					
x^2	0					

a. Complete the table.

b. Which expression grows at a constant rate? Explain.

c. Which expression grows exponentially? Explain.

13. It is said that some Swiss banks are still holding the funds of people who died in the collapse of the Hapsburg Empire in Europe. The accounts keep growing, but no one has come forward to claim the fortunes. Suppose someone invested $100 in a Swiss bank in 1804, and that the account earned 7% interest each year. This means that the amount each year is 1.07 times the amount the previous year.

a. How much money is in the account at the end of the first year?

b. How much money is in the account at the end of the second year?

c. Copy and complete the table to show how the account grows for the first four years after it is opened.

Year after 1798	Amount
0 (1798)	100
1	100×1.07^1
2	$100 \times 1.07 \times 1.07 = 100 \times 1.07^2$
3	
4	

d. Write an expression for the amount in the account after n years.

e. How much money was in the account in 1998, 200 years after it was opened?

Developing a sense of large numbers

14. What repeater machine will stretch a stick 1 million miles long into a stick 1 billion miles long?

15. Which is worth more: a billion $1 bills, or a 100 million $1 bills?

Expressing large numbers in scientific notation

16. Explain how to write a large number in scientific notation. Use an example to demonstrate your method.

17. Explain how to write $1,234 \times 10^9$ in scientific notation.

Demonstrating Skills

Rewrite each expression in exponential notation.

18. $3^2 \times 27$

19. $3 \times d \times d \times d \times 2 \times d$

Write each expression in the form 4^c.

20. $\dfrac{4^5}{4^3}$
  21. $4^5 \times 4^7$
  22. $(4^3)^5$
  23. 4×4^7
  24. $\dfrac{4^6}{4^6}$

Find the value of N in each equation.

25. $3.56 \times 10^N = 356$

26. $N \times 10^5 = 6,541$

27. $0.23 \times 10^3 = N$

28. $N \times \frac{1}{10^5} = 0.34$

Write each number in scientific notation.

29. $123,000$
  30. $57,700,000$
  31. 45×10^6

CHAPTER 4

Working with Signed Numbers

Real-Life Math

Soaring to New Heights How high is the highest mountain? How deep is the deepest part of the ocean? How are height and depth measured?

We call height and depth *elevation*, and we measure elevation from sea level, the average level of the ocean. Parts of the world that are at sea level have an elevation of 0. Denver, the "mile-high" city, is approximately 1 mile above sea level; its elevation is approximately 1 mile. Death Valley is 282 feet below sea level; we say that its elevation is $^-282$ feet.

Think About It If the Dead Sea is 30,324 feet less than Mt. Everest's elevation, is it below sea level?

Family Letter

Dear Student and Family Members,

In Chapter 4, our class will be working with negative numbers. We will learn how to work with numbers that have negative signs that appear in sums, differences, products, quotients, exponents, and graphs.

We will use the number line to calculate sums and differences involving negative numbers. For example, to add 3 + (⁻2), you begin by placing a pointer, point up, on 3. Because the number to be added is negative, you point the pointer in the negative direction. Since the operation is addition, you move the pointer forward (the same direction the tip is pointing) 2 places to get the result, 1.

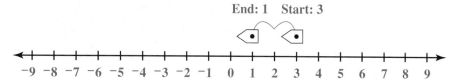

End: 1   Start: 3

Once we are comfortable adding and subtracting negative numbers, we will learn how to multiply and divide negative numbers. We will then explore graphing in four quadrants and find distances between points on a coordinate grid. We will also review exponents and use the stretching and shrinking machines as a way to think about negative exponents.

Vocabulary
Along the way, we'll be learning about two new vocabulary terms:

distance formula   **quadrant**

What can you do at home?

During the next few weeks, your student may show interest in different ways that negative numbers are used in the world outside of school. You might help them think about common occurrences of negative numbers—elevations below sea level, the national debt, temperatures below zero, or balances of over-drawn bank accounts.

Adding and Subtracting with Negative Numbers

You have probably seen negative numbers used in many situations. Golf scores below par, elevations below sea level, temperatures below 0°, and balances of overdrawn bank accounts can all be described with negative numbers.

Zoe and Darnell have different ways of thinking about negative numbers.

Zoe and Darnell both used positive numbers when they thought about negative numbers. Zoe actually used the *absolute value* of ⁻4. Since ⁻4 and 4 are both four units from 0, both have absolute value 4.

Remember

The *absolute value* of a number is its distance from 0 on the number line.

To show the absolute value of a number, draw a vertical segment on each side of the number.

$$|4| = 4 \qquad\qquad |^-4| = 4$$

Think & Discuss

These questions will help you review some important ideas about negative numbers.

- Order these numbers from least to greatest:

  $$4 \qquad ^-5 \qquad 0 \qquad ^-3.5 \qquad 4.2 \qquad ^-0.25 \qquad 1.75$$

- Express each number using the fewest negative signs possible.

  $$^-(^-2) \qquad\qquad ^-(^-(^-(^-(7)))) \qquad\qquad ^-(^-(^-5.5))$$

- How can you tell whether a number is positive or negative by counting how many negative signs it has?

- How can you tell whether ^-x is positive or negative?

- Explain why the following is true:

  If x is positive, then $|x| = x$.

- Tell for what numbers the following is true:

  $$|x| = ^-x$$

In this lesson, you will expand your understanding of negative numbers by learning how to add and subtract negative and positive numbers. Throughout the lesson, do your calculations without a calculator. This will help you better understand why the operations work the way they do.

Lab Investigation ▶ Walking the Plank

Imagine that you are on the crew of a pirate ship commanded by a very generous captain. He has decided to give victims who must walk the plank a chance to survive. He has formulated a game he can play with his victims, but he's asked your help in testing and revising the game.

The Captain's Game

The captain has set up an 8-foot plank that extends from the ship toward the water. He has drawn lines on the plank at intervals of 1 foot.

MATERIALS

- a copy of the plank
- several blank sheets of paper
- a paper arrow

- a direction cube with 3 faces labeled *Ship* and 3 labeled *Shark*
- a walking cube with faces labeled *F1, F2, F3, B1, B2,* and *B3*

The victim starts by standing in the center of the plank, with 4 feet in front of him or her and 4 feet behind him or her. To decide where the victim should move, the captain rolls two cubes, a *direction* cube and a *walking* cube.

Direction Cube   If the captain rolls *Ship*, the victim faces the ship. If he rolls *Shark*, the victim faces the sharks in the water.

Walking Cube   If the captain rolls *F1, F2,* or *F3,* the victim must walk forward 1, 2, or 3 steps. If he rolls *B1, B2,* or *B3,* the victim must walk backward 1, 2, or 3 steps.

The captain continues to roll the cubes until the victim walks off the edge of the plank—either into the shark-infested water or onto the safety of the ship's deck.

Try It Out

Play the game with a partner using a drawing of the plank and a paper arrow. Take turns playing the roles of the captain and the victim. For each game, keep track of the number of times the cubes are rolled before the victim falls to the sharks or reaches safety. Then answer the questions.

1. Did the victim in each game make it to the ship, or was he or she forced to jump into the water?

2. How many moves did it take to complete each game?

Make a Prediction

The captain is thinking about changing the length of the plank and would like some advice on how long it should be. If the plank is too short, games won't last long enough to entertain him. If the plank is too long, he will become bored with the game. Your challenge is to decide on a plank length that results in an interesting game that is not too short or too long.

3. To keep the game interesting, what do you think the ideal number of moves should be?

4. How long do you think the plank should be so that the average number of moves is about the right number?

Check Your Prediction

One way to check your prediction in Question 4 is to collect data. Play at least three games with the plank length you specified. Record the number of moves for each game, and then find the average number of moves for all the games you played.

5. Is the average number of moves close to the number you specified in Question 3?

6. If the plank length you predicted is too long or too short, adjust it. Play the game at least three times with the new length. Continue adjusting the plank length until the average number of moves for a particular length is about what you recommended in Question 3.

You might put the results of your games in a table like this:

Plank Length	Number of Moves			Average Number of Moves
	Game 1	Game 2	Game 3	

What Did You Learn?

7. In one round of the game, the victim moved two steps in the direction of the ship. What might the captain have rolled when he rolled the two cubes? List as many possibilities as you can.

8. Write a letter to the captain giving him your recommendation. Include the following in your letter:

- The number of moves you think is ideal

- How long the plank should be

- How you know that this plank length will give your ideal number of moves

Investigation 1 ▶ Walking the Number Line

In earlier grades, you may have learned to add and subtract positive numbers on the number line. For example, to find 2 + 3, you start at 2 and move three spaces to the right. The number on which you end, 5, is the sum.

To find 6 − 4, start at 6 and move four spaces to the left.

In a similar way, you can "walk the number line" to calculate sums and differences involving negative numbers.

Before you can walk the number line, you need to create a large number line that goes from ⁻9 to 9, and a pointer that looks like the one below.

Once you have created your number line and pointer, you are ready to walk the number line. Here's how to find the solution to 3 + ⁻2 by using your number line.

- Place your pointer, point up, on 3.

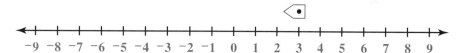

- Look at the number being added. Since it is *negative,* point your pointer in the *negative direction* (to the left).

- Now look at the operation. Since it is *addition,* move the pointer *forward* (in the direction the pointer is facing) two spaces. The pointer ends on 1, which means the answer is 1. The complete addition equation for this sum is 3 + ⁻2 = 1.

End: 1   Start: 3

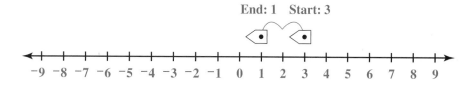

Here's how you would find the solution to $^-2 - 6$.

- Place your pointer, point up, on $^-2$.

- Look at the number being subtracted. Since it is *positive,* point your pointer in the *positive direction* (to the right).

- Now look at the operation. Since it is *subtraction,* move the pointer *backward* (opposite the direction the pointer is facing) six spaces. The answer is $^-8$. The complete subtraction equation for this difference is $^-2 - 6 = {}^-8$.

End: $^-8$ **Start: $^-2$**

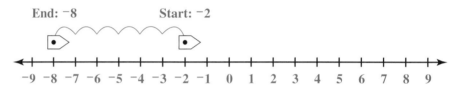

When you walk the number line, the sign of the number being added or subtracted determines the direction the pointer faces.

- If the number is positive, point the pointer in the positive direction.

- If the number is negative, point the pointer in the negative direction.

The operation determines the direction to move the pointer.

- If the operation is addition, move the pointer forward, in the direction it is pointing.

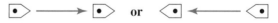

- If the operation is subtraction, move the pointer backward, opposite the direction it is pointing.

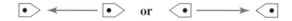

Think & Discuss

The number lines below show the steps for finding $5 + {}^-3$. Explain each step.

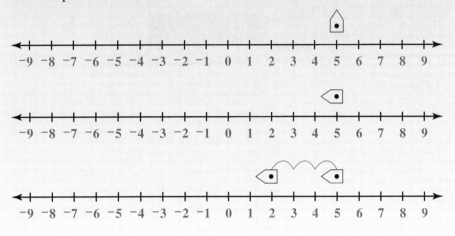

Now you will have an opportunity to compute some sums and differences with the number line.

MATERIALS

number line
and pointer

Problem Set A

For Problems 1–4, walk the number line to find each sum or difference. Describe your steps in words. Be sure to include the following in your descriptions:

- Where on the number line you start

- The direction the pointer is facing

- Whether you move forward or backward and how far you move

- The point on the number line at which you end (the answer)

- The complete equation that represents your moves on the number line

1. $5 + {}^-2$ **2.** ${}^-4 + {}^-3$ **3.** $3 - {}^-2$ **4.** ${}^-8 - {}^-2$

Find each sum or difference by walking the number line.

5. $3 + {}^-5.25$ **6.** $5 - 6\frac{1}{3}$ **7.** ${}^-6 - 2.23$ **8.** ${}^-\frac{2}{3} - {}^-1\frac{1}{3}$

9. When Brynn woke up one morning, the temperature was ${}^-15°F$. By noon the temperature had risen $7°F$. What was the temperature at noon?

10. Colleen was hiking in Death Valley, parts of which are at elevations below sea level. She began her hike at an elevation of 300 feet and hiked down 450 feet. At what elevation did she end her hike?

You can make drawings to record your moves on the number line. Once you get comfortable walking the number line, you may be able to add and subtract just by making a drawing. Your drawings can show very easily the steps you take to find a sum or a difference.

EXAMPLE

Make a drawing to record the steps for finding ⁻3 − ⁻7.

First, draw a pointer over ⁻3. Since the number being subtracted is negative, draw the pointer so that it is facing to the left. Make sure the dot on the pointer is directly over ⁻3.

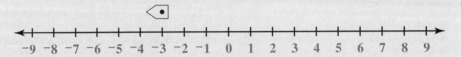

Since you are subtracting ⁻7, you need to move backward seven spaces. Draw an arrow from the pointer to the ending number.

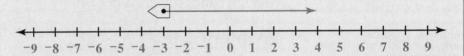

Finish your drawing by adding some labels. Write *Start* and the number where you started above the pointer. Write *End* and the number where you ended above the arrowhead. Adding labels is especially important when you find sums and differences involving fractions and decimals, because it is hard to show the numbers precisely on the number line.

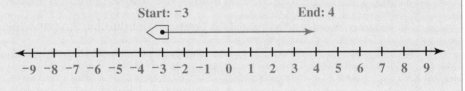

Notice that the final number-line drawing captures all your moves: It shows the number at which you began, the direction of your pointer, whether you moved forward or backward, how far you moved, and your ending number.

Problem Set B

Walk the number line to compute each sum or difference, and record your steps by making a number-line drawing. Write the complete addition or subtraction equation next to the drawing.

1. $^-3 + {}^-2$
2. $5 - 8$
3. $^-5 - {}^-2$
4. $^-5 + 7$
5. $2 + {}^-7.2$
6. $2\frac{2}{3} - {}^-6\frac{1}{3}$

Write the addition or subtraction equation represented by each drawing.

7.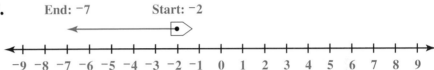

8.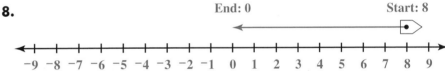

9.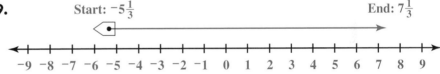

Share & Summarize

Write one addition equation and one subtraction equation. On a separate sheet of paper, make a number-line drawing for each.

Trade drawings with your partner, and write equations that match your partner's drawings. Check with your partner to make sure the equations match the drawings.

Investigation 2 Equivalent Operations

You have practiced writing addition and subtraction equations to match number-line drawings. For each complete number-line drawing, there is only one equation that matches.

But what if a number-line drawing shows only the starting and ending values, and not the pointer or the arrow?

Think & Discuss

What number-line drawings start at ⁻3 and end at 2?

Start: ⁻3　　　　End: 2

What equation does each number-line drawing represent?

In Problem Set C, you will use number-line drawings to explore the relationship between addition and subtraction.

Problem Set C

In Problems 1–6, make two copies of each number line. Complete one to represent an addition equation and the other to represent a subtraction equation. Write the equations your drawings represent.

1.　　　　　　　　　　　　　　　End: 2　　Start: 5

2.　　　Start: ⁻7　　End: ⁻3

3.   End: ⁻1   Start: 3

4.

End: ⁻9   Start: ⁻6

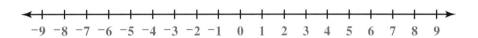

5.   Start: 5.25   End: 7.75

6.   Start: $-3\frac{1}{3}$ End: $2\frac{1}{2}$

Think & Discuss

Describe the two ways you can move from a start point to an end point when you walk the number line.

What other move on the number line accomplishes the same thing as facing the negative direction and moving forward?

What other move accomplishes the same thing as facing the positive direction and moving forward?

In Problem Set C, you wrote two equations that started and ended at the same place on the number line. Since you started with the same number and found the same result after adding or subtracting, you're really doing *equivalent* operations.

Problem Set D

Compute each difference, and write your answer in the form of a subtraction equation. Then write an addition equation that is equivalent to the subtraction equation.

1. $2 - 5$

2. $^-2 - {}^-3.2$

Compute each sum, and write your answer in the form of an addition equation. Then write an equivalent subtraction equation.

3. $^-6 + {}^-5$

4. $^-1 + 7\frac{1}{4}$

In Problems 5 and 6, write an addition and a subtraction equation that gives the answer to the question.

5. When Simon went to sleep, the temperature outside was $^-2°$F. The temperature dropped 12°F overnight. What was the temperature when Simon woke up?

6. Challenge Rosa was hiking near the Dead Sea. She began her hike at an elevation of $^-600$ feet and ended the hike at 92 feet. What was the change in her elevation?

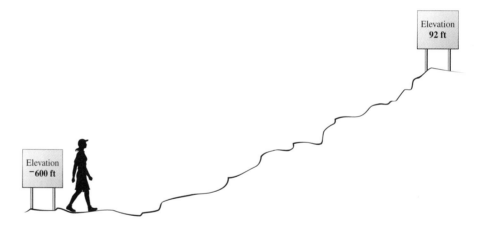

Elevation 92 ft

Elevation $^-600$ ft

Share & Summarize

Complete each sentence. The first is done for you.

1. Adding a negative is equivalent to _subtracting a positive_.

2. Subtracting a negative is equivalent to _____.

3. Adding a positive is equivalent to _____.

4. Subtracting a positive is equivalent to _____.

5. Using the number line or anything else you want, explain why the statements above are true.

Investigation 3 ▶ Taking It Further

Several occupations require adding and subtracting positive and negative numbers. In this investigation, you will apply what you've learned to solve some chemistry and astronomy problems. Then you will look at inequalities involving positive and negative numbers.

Problem Set E

Temperature is usually recorded in degrees Celsius (°C) or degrees Fahrenheit (°F), but scientists often use a third scale, the Kelvin scale. The Kelvin unit is simply "kelvins," and Kelvin temperatures are abbreviated with a K. So, a temperature of 5 kelvins is written 5 K.

You can use this formula to convert a Celsius temperature, *C,* to a Kelvin temperature, *K:*

$$K = C + 273$$

1. Helium is usually a gas, but if it gets cold enough, it will become a liquid. The temperature at which a substance changes from gas to liquid (or from liquid to gas) is called its *boiling point.* The boiling point of helium is ⁻269°C. What is the boiling point of helium on the Kelvin scale?

2. If it gets even colder, helium will actually freeze into a solid. The freezing point of helium—the temperature at which it changes from liquid to solid—is 1 K. What is the freezing point of helium in degrees Celsius?

3. Earth and its moon are about the same distance from the sun, so they receive about the same amount of heat-generating sunlight. You might expect their temperatures to be about the same, but this is not the case. Earth's temperature ranges from ⁻130°F to 136°F, while the moon's temperature ranges from ⁻280°F to 212°F.

    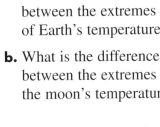

    a. What is the difference between the extremes of Earth's temperature?

    b. What is the difference between the extremes of the moon's temperature?

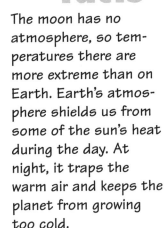

Just the facts

The moon has no atmosphere, so temperatures there are more extreme than on Earth. Earth's atmosphere shields us from some of the sun's heat during the day. At night, it traps the warm air and keeps the planet from growing too cold.

In the next two problem sets, you will add and subtract with negative numbers as you solve inequalities.

Problem Set F

Tell what number you could put in each blank to make a sum of 0.

1. 3 + ___

2. ⁻7.2 + ___

3. Look at your answers to Problems 1 and 2.

    a. If you have a positive number, what number could you add to it to get a sum of 0?

    b. If you have a negative number, what number could you add to it to get a sum of 0?

Give one number you could put in each blank to make a sum less than 0.

4. ⁻6 + ___

5. ⁻4.3 + ___

6. $\frac{1}{2}$ + ___

7. 3 + ___

8. In Problems 4–7, how did you figure out what number to put in the blank?

9. In Problem 1, you found a number that gives a sum of 0 when it is added to 3. This is the same as solving the *equation* $3 + x = 0$. In Problem 7, you found a number that gives a sum less than 0 when it is added to 3. To solve the *inequality* $3 + x < 0$, though, it isn't enough to find a single number.

    a. Find three more solutions of the inequality $3 + x < 0$. Draw a number-line picture for each sum.

    b. Describe *all* the x values that are solutions of $3 + x < 0$. It may help to think about walking the number line.

10. Now think about what you might do if you wanted to solve an inequality related to each of Problems 4–6.

    a. What number could you add to a given positive number to get a sum less than 0?

    b. What number could you add to a given negative number to get a sum less than 0?

Remember

An *inequality* is a mathematical statement that one quantity is greater than or less than another. Inequalities use these symbols:

> greater than
< less than
≥ greater than or equal to
≤ less than or equal to

Problem Set G

Tell what number you could put in each blank to make a difference of 0.

1. $\frac{2}{3} -$ ___

2. $^-3 -$ ___

3. Look at your answers to Problems 1 and 2.

   a. If you have a positive number, what number could you subtract from it to get a difference of 0?

   b. If you have a negative number, what number could you subtract from it to get a difference of 0?

Give one number you could put in each blank to make a difference greater than 0.

4. $7 -$ ___   **5.** $^-2.4 -$ ___   **6.** $\frac{3}{4} -$ ___   **7.** $5 -$ ___

8. In Problems 4–7, how did you figure out what number to put in the blank?

9. In Problem 7, you found a number that, when subtracted from 5, gives a difference greater than 0. That is, you found one solution of the inequality $5 - x > 0$.

   a. Find three more solutions of the inequality $5 - x > 0$. Draw a number-line picture for each difference.

   b. Describe all the solutions of the inequality $5 - x > 0$. It might help to use a number-line picture.

10. Now think about what you might do if you wanted to solve an inequality related to each of Problems 4–6.

   a. What number could you subtract from a given positive number to get a difference greater than 0?

   b. What number could you subtract from a given negative number to get a difference greater than 0?

Share & Summarize

Without doing the calculation, tell whether each sum or difference is positive or negative, and explain how you know.

1. $^-25 + 36$ **2.** $^-53 - {}^-14$ **3.** $45 - 87$ **4.** $123 - {}^-220$

5. Choose two of Questions 1 through 4. Find the sum or difference, and explain your steps so that someone who is not in your class could understand them. You may refer to the number-line model or another method that makes sense to you.

Investigation 4 ▶ Predicting Signs of Sums and Differences

Look back over your answers to the Share & Summarize on the previous page. Can you always use your strategy to predict whether a sum will be positive or negative? If you can, you have a way to check that sums and differences you calculate have the correct sign.

Problem Set H

1. If possible, give an example in which the sum of a positive number and a negative number is negative. If it is not possible, explain why.

2. If possible, give an example in which the difference between two negative numbers is positive. If it is not possible, explain why.

3. If possible, give an example in which the sum of two negative numbers is positive. If it is not possible, explain why.

Positive and negative numbers can be combined in sums and differences in eight ways:

- positive + positive
- positive + negative
- negative + negative
- negative + positive

- positive − positive
- positive − negative
- negative − negative
- negative − positive

You will now explore each of these combinations.

Problem Set I

Figure out whether the sums or differences in each category are

- always negative
- always positive
- sometimes positive, sometimes negative, and sometimes 0

Explain how you know you are right.

1. positive + positive
2. positive + negative
3. negative + negative
4. negative + positive
5. positive − positive
6. negative − positive
7. positive − negative
8. negative − negative

9. Look at your work from Problems 1–8.

    a. Which combinations produce sums or differences that are sometimes negative, sometimes positive, and sometimes 0?

    b. Choose one of the combinations from Part a. Figure out how you can tell—without doing the calculation—whether a given sum or difference involving that combination will produce a negative number, a positive number, or 0.

Share & Summarize

1. The sum of two numbers is less than 0.

    a. Could both numbers be negative? If so, give an example. If not, explain why not.

    b. Could both numbers be positive? If so, give an example. If not, explain why not.

    c. Could one number be negative and one positive? If so, give an example. If not, explain why not.

2. The difference between two numbers is greater than 0.

    a. Could both numbers be positive? If so, give an example. If not, explain why not.

    b. Could both numbers be negative? If so, give an example. If not, explain why not.

    c. Could one number be negative and one positive? If so, give an example. If not, explain why not.

On Your Own Exercises

In Exercises 1–3, walk the number line to find each sum or difference. Describe your steps in words. Include the following in your descriptions:

- Where on the number line you start
- The direction the pointer is facing
- Whether you move forward or backward and how far you move
- The point on the number line at which you end (the answer)
- The complete addition or subtraction equation

1. $^-8 + 5$ **2.** $8 + {}^-4$ **3.** $3 - {}^-5$

Walk the number line to compute each sum or difference.

4. $^-6 - 3$ **5.** $^-2 + {}^-4$

6. $2 + {}^-11$ **7.** $^-9 - {}^-5$

8. $4 - 9$ **9.** $^-6 + 14$

10. $^-5 - {}^-11$ **11.** $^-3.5 - 2.2$

12. $2\frac{1}{5} - {}^-3\frac{2}{5}$ **13.** $^-8.25 + 1.75$

14. $^-\frac{13}{2} - {}^-\frac{13}{4}$ **15.** $4.9 - {}^-3.2$

Walk the number line to compute each sum or difference, and record your steps by making a number-line drawing. Write the complete addition or subtraction equation next to the drawing.

16. $^-2 + {}^-7$ **17.** $6 - {}^-2$

18. $5 - 7$ **19.** $^-3 + 11$

20. $2.2 - 1.4$ **21.** $^-5\frac{1}{3} + 2\frac{5}{6}$

In Exercises 22–25, write the addition or subtraction equation represented by the drawing.

22.

End: $^-6$ Start: $^-3$

```
<-+--+--+--+--+--+--+--+--+--+--+--+--+--+--+--+--+--+->
 -9 -8 -7 -6 -5 -4 -3 -2 -1  0  1  2  3  4  5  6  7  8  9
```

23.

Start: $^-5$ End: 7

```
<-+--+--+--+--+--+--+--+--+--+--+--+--+--+--+--+--+--+->
 -9 -8 -7 -6 -5 -4 -3 -2 -1  0  1  2  3  4  5  6  7  8  9
```

 impactmath.com/self_check_quiz

24.

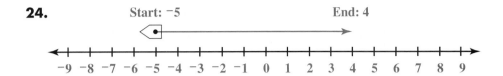

25.

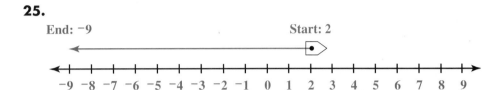

26. The sum of two numbers is ⁻8. What could the numbers be? Give three possibilities.

Write an addition equation and a subtraction equation that each number-line picture could describe. You might want to make a drawing.

27.   Start: ⁻7   End: 9

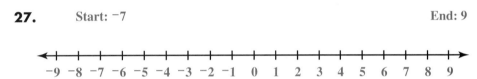

28.

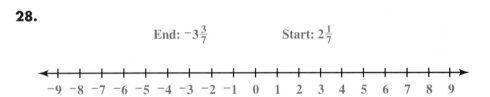

Compute each difference, and write your answer in the form of a subtraction equation. Then write an equivalent addition equation.

29. $25.2 - {}^-3.4$

30. ${}^-43 - 18$

Compute each sum, and write your answer in the form of an addition equation. Then write an equivalent subtraction equation.

31. ${}^-6 + {}^-9$

32. ${}^-21.8 + 17.4$

Remember

The formula for converting Celsius temperatures to Kelvin temperatures is $K = C + 273$.

33. Physical Science At room temperature, mercury is a liquid. The freezing point of mercury—the point at which it turns from liquid to solid—is $^-39°C$. What is the freezing point of mercury on the Kelvin scale?

34. Algebra In this problem, you will solve this inequality: $^-5 - y < 0$.

   a. Find three positive values of y that make the inequality true.

   b. Find three negative values of y that make the inequality true.

   c. Describe *all* the values of y that make the inequality true.

35. If possible, give an example in which a negative number is subtracted from a positive number and the result is negative. If it is not possible, explain why.

36. The sum of two numbers is greater than 0.

   a. Could both numbers be negative? If so, give an example. If not, explain why not.

   b. Could both numbers be positive? If so, give an example. If not, explain why not.

   c. Could one number be negative and one positive? If so, give an example. If not, explain why not.

37. The difference between two numbers is less than 0.

   a. Could both numbers be positive? If so, give an example. If not, explain why not.

   b. Could both numbers be negative? If so, give an example. If not, explain why not.

   c. Could one number be negative and one positive? If so, give an example. If not, explain why not.

Do each computation by walking the number line. Explain each step you take.

38. $3 + {^-4} + {^-2} - 5$ **39.** $^-2 - 9 + 2 - 4 - {^-15}$

40. Statistics The chart shows the high temperature, in degrees Fahrenheit, for the first 10 days of the year in a small Alaskan town. What is the average of these temperatures?

Daily High Temperatures

Date	1/1	1/2	1/3	1/4	1/5	1/6	1/7	1/8	1/9	1/10
Temp (°F)	3	1	13	5	2	2	$^-4$	$^-7$	$^-2$	$^-2$

41. Luis said, "If you add $^-3$ to my sister's age, you will get my age." Luis is 12. How old is his sister?

Without calculating each sum or difference, figure out whether it is less than 0, greater than 0, or equal to 0.

42. $^-8 - 9$ **43.** $^-3 - ^-\pi$ **44.** $^-78 + 2^6$

45. Copy and complete the flowchart.

$$\boxed{9} \xrightarrow{\times 3} \bigcirc \xrightarrow{-20} \bigcirc$$

46. *Integers* are whole numbers and their opposites: $\ldots, ^-3, ^-2, ^-1, 0, 1, 2, 3, \ldots$. The sum of two negative integers is $^-6$.

    a. What could the integers be? List all the possibilities.

    b. Find two more pairs of negative numbers (not integers) whose sum is $^-6$.

Algebra Use the distributive property to rewrite each expression as simply as you can.

47. $6x - ^-2x$ **48.** $3z - 5z$ **49.** $7x + ^-2x$

50. **Prove It!** Maya thinks that for all values of x and y, this equation is true:

$$|x + y| = |x| + |y|$$

For example, if x is $^-5$ and y is $^-6$, the two sides of the equation become

$$|x + y| = |^-5 + ^-6| = |^-11| = 11$$

and

$$|x| + |y| = |^-5| + |^-6| = 5 + 6 = 11$$

Is Maya right? If so, explain why. If not, give an example of values for x and y such that $|x + y| \neq |x| + |y|$.

51. Choose a number, and write down both the number and its opposite.

    a. What happens when you add the two numbers?

    b. Will the same thing happen when you add *any* number to its opposite? Explain why or why not.

Remember

The *absolute value* of a number—its distance from 0—is represented with vertical lines: $|5| = 5$ because 5 is 5 units from 0, and $|^-7| = 7$ because $^-7$ is 7 units from 0.

Algebra Solve each equation by using backtracking or another method.

52. $x + {}^-9 = 1.5$ **53.** $y - 3 = {}^-0.5$ **54.** $3 - x = 8.4$

55. What values of y make each inequality true?

  a. $2 + y < 0$

  b. ${}^-10 < 2 + y$

  c. ${}^-10 < 2 + y$ and $2 + y < 0$

56. Copy each problem, and fill in the blank with a number that gives a sum between $-1\frac{1}{2}$ and $-\frac{1}{2}$.

  a. $10 +$ ____   **b.** ${}^-1.2 +$ ____

  c. ${}^-2 +$ ____   **d.** ${}^-84 +$ ____

  e. In Parts a–d, what did you need to do to make the answer be between $-1\frac{1}{2}$ and $-\frac{1}{2}$?

Algebra Without solving each equation, determine whether the value of x is less than 0, equal to 0, or greater than 0.

57. $x - 3 = {}^-5$ **58.** ${}^-x + 3 = 5$

59. $4 - x = 9$ **60.** $x - 3 = 12$

61. In this exercise, you will investigate the possible results for the sum of two positive numbers and a negative number.

  a. Choose two positive numbers and one negative number. Is their sum positive, negative, or 0?

  b. Suppose you repeat Part a with different numbers. Will the sum be always positive; always negative, or sometimes positive, sometimes negative, and sometimes 0? How do you know?

Mixed Review

If possible, rewrite each expression using the laws of exponents.

62. $y^2 \times y^6$ **63.** $a^n \times a^n$

64. $2b \times 3b^2$ **65.** $4^5 \times 9^5$

66. $5a^4 \times 5a^4$ **67.** $a^{100} \times b^{100}$

68. $(1.1^2)^4$ **69.** $(0.9^4)^3$

70. $(a^3)^4$ **71.** $(a^4)^0$

72. $(x^2)^3$ **73.** $(3^n)^m$

Find each product.

74. $6 \times \frac{1}{10^2}$

75. $35,900 \times \frac{1}{10^3}$

76. $564,890 \times \frac{1}{10^2}$

77. $90,500 \times 10^4$

Copy and complete each flowchart.

78.

$\boxed{12} \xrightarrow{+3} \bigcirc \xrightarrow{\times 2} \bigcirc$

79.

$\boxed{2.5} \xrightarrow{+5} \bigcirc \xrightarrow{\div 1.5} \bigcirc$

80. On the first Saturday of her new weekend exercise plan, Jen bikes 2 miles to the park, 5 miles around the park on Perimeter Path, and then 2 miles back home. She plans to gradually build her workout by adding one loop on Perimeter Path each Saturday.

   a. On which Saturday will Jen ride more than 25 miles?

   b. After the first month on her new plan (four Saturdays), what average distance will Jen have ridden each Saturday? Show how you found your answer.

   c. How many more miles will Jen ride on the sixth Saturday than on the first Saturday?

   d. Jen doesn't want to ride more than 50 miles in a day. After which Saturday will she stop increasing the distance she rides? What will be the distance she rides on that Saturday?

Complete each table.

81.

m	0	1	2	3	4	40	
$3m + 10$							109

82.

t	1	2	4	8	11		
$4t - 4$						96	120

4.2 Multiplying and Dividing with Negative Numbers

In this lesson, you will use what you have learned about addition and subtraction of signed numbers to help you understand how to multiply and divide them.

Explore

The table lists the lowest temperatures in degrees Celsius for each day during the month of November at a weather station in Alaska.

The first row lists the different low temperatures, and the second row tells the number of days in November with each daily low temperature. For example, the first column indicates that the low temperature was $^-4°C$ two days during the month.

November Temperatures in Alaska

Daily Low Temperature (°C)	$^-4$	$^-3$	$^-2$	$^-1$	0	1	2	3	4
Days at This Temperature	2	1	2	4	2	3	6	6	4

Use what you learned in Lesson 4.1 to find the average low temperature during the month of November.

How did you compute the average?

If you calculated the total of the temperatures by using only addition, you probably found the problem rather tedious. In this lesson, you will learn how to multiply and divide with negative numbers so you can solve problems like this one quickly and efficiently. Again, try to do the calculations without a calculator. This will help you understand why the operations work the way they do.

Investigation 1 ▶ The Product of a Positive and a Negative

When you first learned to multiply positive numbers, you thought of multiplication as repeated addition. For example, the product of 3 and 5 can be thought of as three 5s, or $5 + 5 + 5$. It can also be thought of as five 3s, or $3 + 3 + 3 + 3 + 3$.

Problem Set A

Use the way of thinking described above to calculate each product.

1. $2 \times {}^-8$ **2.** ${}^-3 \times 2$ **3.** ${}^-5 \times 4$

4. $3 \times {}^-7.5$ **5.** ${}^-5\frac{1}{3} \times 3$ **6.** ${}^-1 \times 6$

Think & Discuss

Look for patterns in your computations in Problem Set A.

Can you see a shortcut you could use to compute the product of a positive and a negative number?

Use your shortcut to find each product.

$2 \times {}^-45$ ${}^-32 \times 11$

Problem Set B

Now you will investigate products in which the positive number is a fraction or a decimal.

1. $\frac{1}{2} \times {}^-10$ **2.** $\frac{1}{3} \times {}^-9$

3. $\frac{2}{3} \times {}^-9$ **4.** $\frac{4}{5} \times {}^-15$

5. ${}^-12 \times 0.25$ **6.** $1\frac{1}{2} \times {}^-20$

7. ${}^-5 \times 1.2$ **8.** $3.2 \times {}^-1.1$

9. ${}^-\frac{2}{3} \times \frac{3}{8}$ **10.** $\frac{1}{7} \times {}^-\frac{14}{3}$

Multiplying with negative numbers can help you solve some interesting problems. Use what you learned in Problem Sets A and B to complete Problem Set C.

Problem Set C

You might want to use a calculator to help you answer Problems 1 and 2.

1. A diver jumps into the ocean from a boat. She starts at an elevation of 0 feet, and her elevation decreases 50 feet every minute.

   a. What is her elevation after 1 min?

   b. What is her elevation after 5 min?

   c. What is her elevation after n min?

In Lesson 4.1, you learned about the Kelvin temperature scale. Kelvin and Celsius temperatures are related according to the formula $K = C + 273$.

You also know that you can convert Celsius temperatures to Fahrenheit temperatures with the formula $F = \frac{9}{5}C + 32$, where C is the temperature in degrees Celsius and F is the temperature in degrees Fahrenheit.

2. Temperatures on Mercury—the planet closest to our sun—range from $^-173°C$ to $427°C$. Convert these temperatures to find the range of temperatures on Mercury in degrees Fahrenheit.

3. *Absolute zero,* 0 K, is theoretically the coldest anything in our universe can ever be. How many degrees Fahrenheit is absolute zero?

4. Make up a word problem that requires calculating $3 \times {}^-8$.

5. The product of two integers is $^-14$. (Remember: The *integers* are the whole numbers and their opposites: . . . , $^-3, ^-2, ^-1, 0, 1, 2, 3, ...$)

   a. What could the integers be? List all the possibilities.

   b. Find three more pairs of numbers—not necessarily integers—that have a product of $^-14$.

Solve each equation.

6. $3x = {}^-6$ 7. $^-2y = {}^-12$ 8. $4x + 15 = 3$

Share & Summarize

Is the product of a positive number and a negative number

- always negative?

- always positive?

- sometimes positive, sometimes negative, and sometimes 0?

Investigation ▶2 The Product of Two Negatives

In Investigation 1, you used addition to figure out how to multiply a negative number by a positive number. You can't use that strategy with two negative numbers, because you can't add a negative number of times! However, the pattern in products of a positive number and a negative number can help you figure out how to multiply two negative numbers.

Problem Set D

1. Find each product.

    a. $^-3 \times 4$

    b. $^-3 \times 3$

    c. $^-3 \times 2$

    d. $^-3 \times 1$

    e. $^-3 \times 0$

2. In Problem 1, what happens to the product from one part to the next? Why?

3. Now use your calculator to compute these products.

    a. $^-3 \times ^-1$

    b. $^-3 \times ^-2$

    c. $^-3 \times ^-3$

    d. $^-3 \times ^-4$

    e. $^-3 \times ^-5$

4. Did the pattern you observed in Problem 2 continue?

Think & Discuss

Look for patterns in your computations in Problem Set D. What do you think the rule is for finding the product of *any* two negative numbers?

Use your rule to find each product. Check your results with your calculator.

$^-5 \times ^-7$ $^-10 \times ^-90$ $^-2 \times ^-4.4$ $^-1.2 \times ^-7$

There are four ways to combine positive and negative numbers in products:

- positive × positive
- positive × negative
- negative × positive
- negative × negative

You now know how to multiply each combination, and what kind of results to expect.

Problem Set E

If you know the result of a multiplication, can you figure out what the factors might have been? Try these problems.

1. The product of two integers is 12. What could the integers be? List all the possibilities.

2. The product of *three* integers is 12. What could the integers be? List all the possibilities.

Solve each equation.

3. $^-4x = 12$

4. $^-3x + 5 = 11$

5. $3 - 4x = {}^-17$

When you use exponents to indicate repeated multiplication of a negative number, you need to be careful about notation. Put the negative number *inside* the parentheses, and put the exponent *outside* the parentheses.

EXAMPLE

Calculate $(^-2)^4$ and $^-2^4$.

Notation	Meaning	Calculation
$(^-2)^4$	$^-2$ to the fourth power	$^-2 \cdot {}^-2 \cdot {}^-2 \cdot {}^-2 = 16$
$^-2^4$	the opposite of 2^4	$^-(2 \cdot 2 \cdot 2 \cdot 2) = {}^-16$

Problem Set F

Evaluate each expression.

1. $(^-3)^2$

2. $^-3^2$

3. $^-4^2$

4. $(^-4)^2$

5. $(^-2)^1$

6. $(^-2)^2$

7. $(^-2)^3$

8. $(^-2)^4$

9. $(^-2)^5$

10. $(^-2)^6$

11. Look for patterns in your answers to Problems 5–10.

    a. For what values of n is $(^-2)^n$ positive?

    b. For what values of n is $(^-2)^n$ negative?

12. Simon said the solution of $x^2 = 16$ is 4. Luis said there is another solution. Is Luis correct? If so, find the other solution.

Solve each equation. Be careful: Each equation has two solutions.

13. $x^2 = 36$

14. $x^2 = \frac{1}{36}$

Share & Summarize

1. Is the product of two negative numbers

   • always negative?

   • always positive?

   • sometimes positive, sometimes negative, and sometimes 0?

2. Consider the expression $(^-3)^m$.

    a. If m is an even number, is $(^-3)^m$ positive or negative? How can you tell?

    b. If m is an odd number, is $(^-3)^m$ positive or negative? How can you tell?

Investigation Dividing with Negative Numbers

You can solve any division problem by thinking about a corresponding multiplication problem. Look at how Kate solves $30 \div 5$.

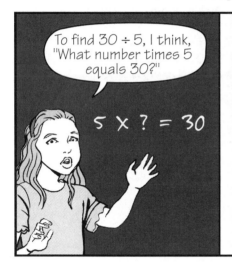

Problem Set G

Use Kate's method to solve each division problem.

1. $21 \div {}^-3$

2. ${}^-64 \div {}^-2$

3. ${}^-24 \div 48$

4. ${}^-2 \div {}^-32$

5. ${}^-\dfrac{6}{5}$

6. $\dfrac{9}{{}^-27}$

7. ${}^-2.16 \div {}^-54$

8. $\dfrac{{}^-3}{{}^-0.3}$

Think & Discuss

Look for patterns in your computations in Problem Set G.

• Can you see a shortcut for computing a quotient when exactly one of the numbers is negative?

• Can you see a shortcut for computing the quotient of two negative numbers?

Use your shortcut to find each quotient.

$44 \div {}^-2.2$ $\qquad {}^-\dfrac{7}{3} \qquad\qquad {}^-56 \div {}^-2$

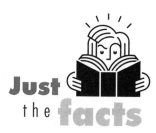

Just the facts

Water exerts pressure on a scuba diver—the deeper the diver, the greater the pressure. Divers are trained to make the pressure in the body's air spaces—the lungs, sinuses, and ears—equal to the outside water pressure.

Problem Set H

1. Suppose a scuba diver begins his dive at an elevation of 0 feet. During the dive, his elevation changes at a constant rate of $^-2$ feet per second. How long will it take for him to reach an elevation of $^-300$ feet?

2. The Marianas Trench, south of Guam, contains the deepest known spot in the world. This spot, called Challenger Deep, has an elevation of $^-36,198$ feet. Suppose a deep-sea diver entered the ocean above this spot. She started at an elevation of 0 feet and moved at a constant rate of $^-50$ feet per minute. If the diver were able to go deep in the ocean without being affected by the pressure, and her tank contained enough air, how long would it take her to reach the bottom of Challenger Deep?

3. Write four division problems with a quotient of $^-4$.

Solve each equation.

4. $\frac{30}{x} = {}^-15$

5. $\frac{x}{^-4} = 20$

6. $^-\frac{x}{6} + 1 = \frac{4}{3}$

7. $\frac{6}{x} + 5 = 3$

Share & Summarize

1. When you divide a positive number by a negative number, is the result always negative; always positive; or sometimes positive, sometimes negative, and sometimes 0? Explain how you know.

2. When you divide a negative number by a positive number, is the result always negative; always positive; or sometimes positive, sometimes negative, and sometimes 0? Explain how you know.

3. When you divide a negative number by a negative number, is the result always negative; always positive; or sometimes positive, sometimes negative, and sometimes 0? Explain how you know.

On Your Own Exercises

Practice & Apply

Compute each product or sum.

1. $^-3 \times 52$

2. $6.2 \times {}^-5$

3. $^-0.62 \times 5$

4. $^-2 \times 4 + 3 \times {}^-4$

5. The product of two integers is $^-15$.

    a. What could the integers be? List all the possibilities.

    b. Find three more pairs of numbers—not necessarily integers—with a product of $^-15$.

6. Solve $^-3x + 4 = {}^-5$.

Compute each product.

7. $^-3 \cdot {}^-32$

8. $^-\frac{1}{2} \cdot {}^-35$

9. $^-5 \cdot {}^-0.7$

10. $^-2.5 \cdot {}^-7$

11. $^-2^5 \cdot {}^-3$

12. $^-3 \cdot {}^-5^3$

13. $^-3 \cdot ({}^-5)^3$

14. $^-3^2 \cdot ({}^-3)^2$

Without calculating each product, determine whether it is less than 0 or greater than 0.

15. $3 \times {}^-2$

16. 3×2

17. $^-3 \times 2$

18. $^-3 \times {}^-2$

19. The product of two integers is 9. What could the integers be? List all the possibilities.

20. Solve $^-5y + 13 = 3$.

Find each quotient.

21. $^-3 \div 2.5$

22. $^-3 \div {}^-25$

23. $45 \div {}^-360$

24. $\frac{^-45}{^-90}$

25. One number divided by a *negative* number is 0.467. What could the two numbers be? Think of three possibilities and write them as division equations.

Solve each equation.

26. $\frac{x}{^-3} + 4 = 6$

27. $\frac{x}{8} - 2.5 = {}^-3$

28. $^-\frac{12}{x} + 45 = 49$

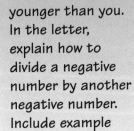

In your **own words**

Write a letter to a student a year younger than you. In the letter, explain how to divide a negative number by another negative number. Include example problems in your letter.

impactmath.com/self_check_quiz

29. As a parachutist descends into Death Valley, her elevation decreases 15 feet every second. Five seconds before she lands, she is at an elevation of $^-127$ feet. What is the elevation of the place she lands?

30. Physical Science When it is cold outside and the wind is blowing, it feels colder than it would at the same temperature without any wind. Scientists call the temperature that it feels outside the *windchill*.

For example, if the temperature outside is 20°F and the wind speed is 20 mph, the equivalent windchill temperature is about $^-10$°F. That means that even though the thermometer reads 20°F, it *feels* as cold as it would if you were in no wind at $^-10$°F.

Here's how meteorologists calculate windchill. The variable s represents wind speed in mph, and t represents actual temperature in °F.

$$\text{windchill} = 0.0817(3.71\sqrt{s} + 5.81 - \tfrac{s}{4})(t - 91.4) + 91.4$$

This formula works when wind speeds are between 4 mph and 45 mph.

a. If the wind blows at 20 mph and the temperature is $^-5$°F, what is the windchill?

b. If the wind blows at 30 mph and the temperature is $^-10$°F, what is the windchill?

c. Challenge If the wind blows at 25 mph and the windchill is $^-60$°F, what is the actual temperature?

31. Pedro said, "I'm thinking of a number. When I multiply my number by $^-2$ and subtract 4, I get $^-16$." What is Pedro's number?

Without calculating, predict whether each product is less than 0, equal to 0, or greater than 0.

32. $(^-4)^3 \cdot (^-2)^2$

33. $(^-3)^5 \cdot (^-6)^3$

34. $(^-1)^5 \cdot 5^2$

35. $(^-2)^2 \cdot (^-972)^{90}$

36. $^-45^2 \cdot (^-35)^5$

37. $(^-2)^8 \cdot (^-5)^8$

Challenge If possible, solve each equation. If it is not possible, explain why.

38. $y^3 = ^-8$

39. $x^2 = ^-9$

Describe the values of b that make each inequality true.

40. $3b < 0$

41. $3(b - 2) < 0$

Without calculating, determine whether the quotient in Exercises 42–44 is less than ⁻1, between ⁻1 and 0, between 0 and 1, or greater than 1.

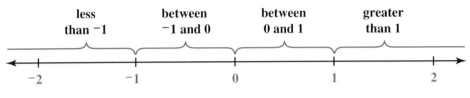

42. $1 \div -\frac{3}{5}$

43. $\frac{-555.233}{-5,552}$

44. $78.3636 \div {}^-25.33$

45. The average of 10 temperatures is ⁻21°F. What could the temperatures be?

46. The average of 5 temperatures is ⁻21°F. What could the temperatures be?

Solve each equation. Be careful: Each equation has two solutions.

47. $\left(\frac{1}{y}\right)^2 = \frac{1}{9}$

48. $y^2 = \frac{1}{9}$

Challenge Solve each equation.

49. $\left(\frac{1}{z}\right)^2 = 16$

50. $-\frac{5}{x} + 4 = 2$

Mixed Review

Evaluate each expression.

51. $0.125 \cdot 0.3$

52. $0.125 \cdot 0.5$

53. $0.125 \cdot 0.04$

54. $0.125 \cdot 0.004$

55. $0.125 \cdot 2$

56. $0.125 \cdot 200$

Tama drew the following flowcharts. For each flowchart, tell what equation Tama was trying to solve. Then, copy and complete the flowchart.

57. ⬭ —× 16→ ⬭ —− 4→ (28)

58. ⬭ —+ 6→ ⬭ —× 2→ (18)

59. ⬭ —− 5→ ⬭ —÷ 2→ (35)

60. Find the value for $\frac{1}{2}k^2$ for each value of k.

   a. 0   **b.** $\frac{1}{2}$ **c.** 1.2   **d.** 4

61. Find the value for $10m^2$ for each value of m.

   a. 0   **b.** $\frac{1}{2}$ **c.** 1.2   **d.** 4

62. Find the value for $3p^2 + 10$ for each value of p.

   a. 0   **b.** $\frac{1}{2}$ **c.** 1.2   **d.** 4

63. A calculator costs $14 and requires three batteries that are not included in the price of the calculator.

   a. Suppose each battery costs b dollars. Write an expression for the total cost of the calculator and batteries.

   b. What is the total cost if each battery is $.89?

64. Geometry Consider this cylinder.

   a. Find the volume of the cylinder.

   b. Find the volume of a cylinder with a radius twice the length of this cylinder's radius.

   c. Find the volume of a cylinder that is twice as high as this cylinder.

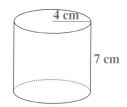

4 cm

7 cm

65. Probability Sherina bought two dozen superballs to give out at her birthday party, and she put them in a bag. A third of the superballs were blue, 25% were striped, an eighth were hot pink, and the rest were made to glow in the dark.

   a. Antonie arrived first, and Sherina asked him to reach into the bag without looking and take a ball. What is the probability that Antonie's ball will glow in the dark?

   b. Antonie picked a blue ball. Lucita arrived next and said she hoped to pick hot pink. What is the probability she will get her wish?

   c. Lucita didn't get what she'd wanted, but Emilio, who picked next, did get hot pink. What was the probability of this happening?

   d. Emilio traded with Lucita for her striped superball. Miki picked next, hoping for a glow-in-the-dark ball. What is the probability of her choosing one?

Plotting Points in Four Quadrants

Making a graph is a useful way to represent the relationship between two quantities. For example, during a two-week snorkeling vacation, Deane timed how long he could hold his breath under water. The graph shows his maximum breath-holding time each day.

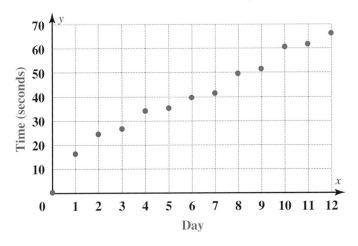

Deane's Breath-holding Time

Think & Discuss

Choose two points on the graph, and give their coordinates. Explain what the coordinates tell you about Deane's breath-holding time.

How long could Deane hold his breath at the end of his sixth day of practicing? How did you find your answer from the graph?

In the graph above, both quantities—day and time—are always positive. What if one or both of the quantities you want to graph are sometimes negative?

For example, suppose you wanted to investigate the relationship between month of the year and average temperature in a very cold place, like Antarctica. How would you plot points for the months that have average temperatures below 0?

Investigation Plotting Points with Negative Coordinates

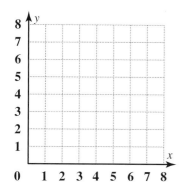

You know how to plot points on a coordinate grid that looks like the one at left. The *x*-axis of the grid is a horizontal number line, and the *y*-axis is a vertical number line.

In the graphs you have worked with so far, the number lines included only numbers greater than or equal to 0. But if they are extended to include negative numbers, the coordinate grid will look something like this:

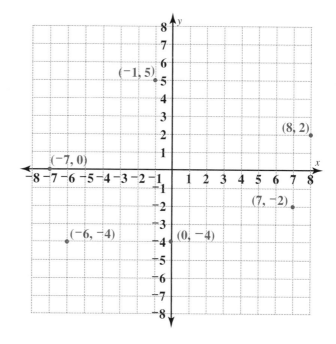

Using a grid like this, you can plot points with negative coordinates.

Think & Discuss

Shaunda plotted six points on the grid above. See if you can figure out the procedure she used to plot the points.

MATERIALS
graph paper

Problem Set A

1. Plot Points *A–F* on the same coordinate grid. Label each point with its letter.

   Point *A:* (6, ⁻1)   Point *B:* (⁻2, ⁻2)   Point *C:* (⁻1, ⁻3)

   Point *D:* (⁻1, 0)   Point *E:* (⁻2, 3.5)   Point *F:* (⁻$\frac{1}{3}$, 5)

2. Give the coordinates of each point plotted on this grid.

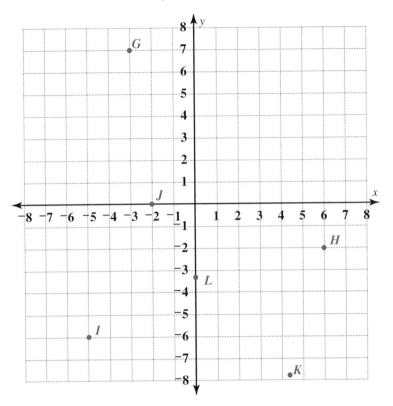

3. The graph shows daily average temperatures at the Gulkana Glacier basin in Alaska, from September 9 to September 18 in a recent year.

Daily Average Temperatures

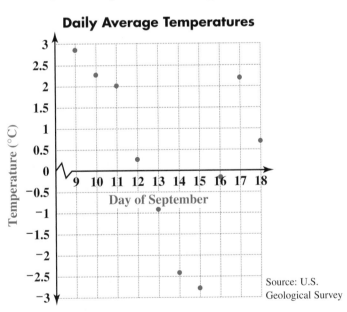

Source: U.S. Geological Survey

a. What was the lowest of these temperatures?

b. On which day was the temperature lowest?

c. What was the highest of these temperatures?

d. On which day was the temperature highest?

The game you will now play will give you practice locating points on a coordinate grid.

MATERIALS

coordinate grids with x-axis and y-axis from ⁻3 to 3

Problem Set B

In the *Undersea Search* game, you and your partner will hide items from each other on a coordinate grid. To win the game, you need to find your partner's hidden items before he or she finds yours.

Each player will need two coordinate grids with *x*- and *y*-axes that range from ⁻3 to 3. Think of each grid as a map of part of the ocean floor. During the game, you will be hiding a buried treasure and a coral reef on one of your grids.

The grid below shows one way you could hide the items.

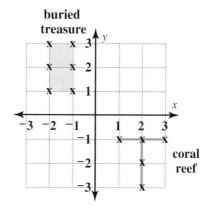

You can bury the items anywhere on the grid, but they must have the shapes shown above.

- The buried treasure must be a rectangle with two Xs along one side and three along the other.

- The coral reef must be a T-shape made from five Xs.

The Xs must all be placed where grid lines intersect. The buried treasure and the reef can't overlap, so they can't share points on your map.

Here's how you play the game.

- *Hide the buried treasure and the coral reef.* Start with one of your grids. Without showing your partner, use Xs to mark the places you want to hide the buried treasure and the coral reef. Make sure you put all your Xs where grid lines intersect.

- *Search the sea.* You and your partner take turns calling out the coordinates of points, trying to guess where the other has hidden the items.

  If your partner calls out a point where you have hidden something, say "X marks the spot." If your partner calls out any other point, say "Sorry, nothing there."

  Use your blank grid to keep track of your guesses. If you guess a point where your partner has hidden something, put an X on that point. If you guess a point where nothing is hidden, circle the point so you know not to guess it again.

- *Victory at sea.* The first person to guess all the points for both hidden items wins.

Play *Undersea Search* with your partner at least once, and then answer the questions.

1. Suppose your partner said "X marks the spot" when you guessed these points: ($^-$3, 1), ($^-$3, 2), and ($^-$2, 2). Can you tell whether you have found the buried treasure or the coral reef? Why or why not?

2. Suppose your partner said "X marks the spot" when you guessed these points: ($^-$2, $^-$2), ($^-$3, 0), and ($^-$1, 0). Can you tell whether you have found the buried treasure or the coral reef? Why or why not?

3. Suppose you have already found the coral reef, and you know that part of the buried treasure is at these points: (1, $^-$2), (0, $^-$2), and ($^-$1, $^-$2). What could be the coordinates of the other three points that make up the buried treasure? Name as many possibilities as you can.

Share & Summarize

Write a letter to a student a grade below you explaining how to plot points with negative coordinates on a coordinate grid.

Investigation 2 ▶ Parts of the Coordinate Plane

VOCABULARY
quadrant

The *x*- and *y*-axes divide the coordinate plane into four sections called **quadrants.** The quadrants are numbered with roman numerals as shown below. Points on the axes are not in any of the quadrants.

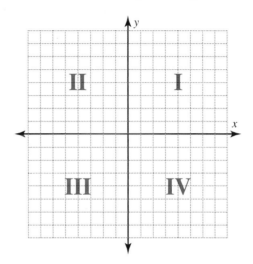

Problem Set C

Points *A* through *R* are plotted on the grid.

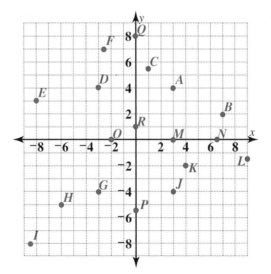

1. Look at the points in Quadrant I.

    a. Record the coordinates of each point in Quadrant I.

    b. What do you notice about the signs of the coordinates of each point?

    c. If someone gives you the coordinates of a point, how can you tell whether it is in Quadrant I without plotting the point?

2. Look at the points in Quadrant II.

   a. Record the coordinates of each point.

   b. What do you notice about the signs of the coordinates of each point?

   c. If someone gives you the coordinates of a point, how can you tell whether it is in Quadrant II without plotting the point?

3. Look at the points in Quadrant III.

   a. Record the coordinates of each point.

   b. What do you notice about the signs of these coordinates?

   c. If someone gives you the coordinates of a point, how can you tell whether it is in Quadrant III without plotting the point?

4. Look at the points in Quadrant IV.

   a. Record the coordinates of each point.

   b. What do you notice about the signs of these coordinates?

   c. If someone gives you the coordinates of a point, how can you tell whether it is in Quadrant IV without plotting the point?

5. Look at the points on the *x*-axis.

   a. Record the coordinates of each point.

   b. What do these coordinates have in common?

   c. If someone gives you the coordinates of a point, how can you tell whether it is on the *x*-axis without plotting the point?

6. Look at the points on the *y*-axis.

   a. Record the coordinates of each point.

   b. What do these coordinates have in common?

   c. If someone gives you the coordinates of a point, how can you tell whether it is on the *y*-axis without plotting the point?

Just the facts

The word *quadrant* comes from a Latin word meaning "four." *Quadrille*, a French dance for four couples; *quart*, one-fourth of a gallon; and *quadrilateral*, a geometrical figure with four sides, are all related to the Latin word.

You will now use what you've learned about the signs of the coordinates in each quadrant.

Problem Set D

Recall that the formula for converting between the Celsius and Fahrenheit temperature scales is

$$F = 1.8C + 32$$

where F is the temperature in degrees Fahrenheit and C is the temperature in degrees Celsius.

The graph of this equation is shown below. Celsius temperatures are on the horizontal axis, and Fahrenheit temperatures are on the vertical axis.

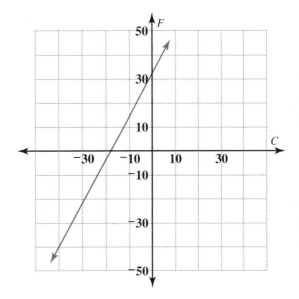

Use the graph to answer the questions.

1. Are Celsius and Fahrenheit temperatures ever both positive? If so, for which Celsius temperatures? If this never happens, explain how you know.

2. Are Celsius and Fahrenheit temperatures ever both negative? If so, for which Celsius temperatures? If this never happens, explain how you know.

3. Are Fahrenheit temperatures ever positive when Celsius temperatures are negative? If so, for which Celsius temperatures? If this never happens, explain how you know.

4. Are Fahrenheit temperatures ever negative when Celsius temperatures are positive? If so, for which Celsius temperatures? If this never happens, explain how you know.

5. **Challenge** Is the Celsius temperature ever equal to the Fahrenheit temperature? If so, for which temperature? Explain how you know.

Share & Summarize

1. Without plotting each point, determine in which quadrant or on which axis or axes it lies.

   a. $(^-5, ^-2)$

   b. $(0, 0)$

   c. $(3, ^-\frac{2}{7})$

   d. $(^-35, 0)$

2. In general, if you are given the coordinates of a point, how can you tell which part of the coordinate plane the point is in without plotting it? You might organize your ideas in a chart with these headings:

x-coordinate	*y*-coordinate	Part of Coordinate Plane

Investigation ▶3 Representing Operations on the Coordinate Plane

You can color points on the coordinate plane to create a representation of sums, differences, products, and quotients of signed numbers.

MATERIALS
- graph paper
- blue, black, and red pens or pencils

Problem Set E

Each member of your group will choose six points on the coordinate plane and color them according to these rules:

- *If the product of the coordinates is positive, color the point red.*
- *If the product of the coordinates is negative, color the point blue.*
- *If the product of the coordinates is 0, color the point black.*

When all the points are colored, you will look for a pattern.

1. Decide with your group which six points each member will color. Choose points in all four quadrants and on the axes, so you will get a good idea of the overall pattern. Each member of your group should plot his or her points, in the appropriate color, on the same coordinate grid.

2. What patterns do you notice in the colors of the points?

3. Imagine what the grid would look like if *every* point on it was colored. On a new coordinate grid, color *every* point red, blue, or black according to the rules above. Then compare your grid with the others in your group.

4. How do you know you colored the coordinate plane correctly?

In Problem Set E, you represented multiplication on a coordinate grid. In Problem Set F, you will look at addition.

MATERIALS
- graph paper
- blue, black, and red pens or pencils

Problem Set F

You will now color the points on the coordinate plane according to this set of rules:

- *If the sum of the coordinates is positive, color the point red.*
- *If the sum of the coordinates is negative, color the point blue.*
- *If the sum of the coordinates is 0, color the point black.*

1. Start with a new coordinate grid. Find at least 10 points that should be black, and color them. Be sure to check each quadrant.

2. Find some points that should be red, and color them. Again, check each quadrant.

3. Find some points that should be blue, and color them.

4. Now use the new coloring rules to plot all the points you plotted in Problem Set E, if you haven't already.

5. What patterns do you notice in the colors of the points?

6. Imagine coloring *every* point on the coordinate plane. On a new grid, color *every* point red, blue, or black according to the rules above.

Share & Summarize

In the grids you colored in Problem Sets E and F, the black lines form boundaries between red and blue sections.

1. For the grid in Problem Set E, explain why the black lines are located where they are.

2. For the grid in Problem Set F, explain why the black lines are located where they are.

On Your Own Exercises

Practice  **Apply**

1. Plot these points on a coordinate plane. Label each point with its letter.

$A\ (3\frac{2}{5}, 2)$ $B\ (^-2, 6)$ $C\ (0.4,\ ^-4.4)$ $D\ (^-5,\ ^-2)$

2. Find the coordinates of Points J through O.

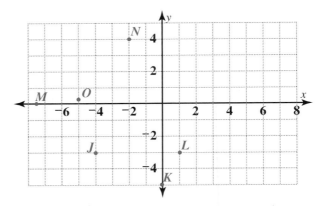

3. Physical Science The graph of the Celsius-Kelvin conversion formula, $K = C + 273$, is shown below.

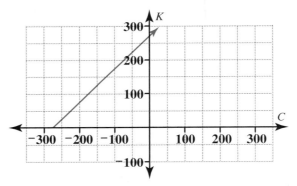

Just the facts

The Kelvin scale is named for the Scottish physicist who first proposed an absolute temperature scale, William Thomson—also known by his British title, Baron of Kelvin.

Use the graph or the equation to answer the questions.

a. Are Kelvin and Celsius temperatures ever both positive? If so, for which Celsius temperatures? If this never happens, explain how you know.

b. Are Kelvin and Celsius temperatures ever both negative? If so, for which Celsius temperatures? If this never happens, explain how you know.

c. Are Kelvin temperatures ever positive when Celsius temperatures are negative? If so, for which Celsius temperatures? If this never happens, explain how you know.

d. Are Kelvin temperatures ever negative when Celsius temperatures are positive? If so, for which Celsius temperatures? If this never happens, explain how you know.

     impactmath.com/self_check_quiz

4. Without plotting each point, tell in which quadrant or on which axis it lies.

a. (2, 0)   **b.** (0, ⁻24)   **c.** (35, ⁻23)

d. (3, 5)   **e.** (⁻2, ⁻2)   **f.** (⁻52, 5)

5. Challenge In this problem, you will color a coordinate plane to show the signs of differences.

a. Create a coordinate plane in which you color a point (c, d) red if $c - d$ is positive and blue if it is negative. If $c - d$ is 0, color the point black. Shade the plane to show what it would look like if you colored *every* point red, blue, or black.

b. Now create a coordinate plane in which you color a point (c, d) red if $d - c$ is positive and blue if it is negative. If $d - c$ is 0, color the point black. Shade the plane to show what it would look like if you colored *every* point red, blue, or black.

c. How are the two coordinate planes similar? How are they different? Why are they similar and different in these ways?

Connect & Extend

6. Astronomy The average surface temperatures of the planets in our solar system are related to their average distances from the sun.

Planets in the Solar System

Planet	Distance from Sun (millions of miles)	Surface Temperature (°F)
Mercury	36	662
Venus	67	860
Earth	93	68
Mars	142	⁻9
Jupiter	483	⁻184
Saturn	888	⁻292
Uranus	1,784	⁻346
Neptune	2,799	⁻364
Pluto	3,674	⁻382

a. Create a graph with distance from the sun on the *x*-axis and average surface temperature on the *y*-axis. Plot the nine points listed in the table.

b. Generally speaking, how does temperature change as you move farther from the sun?

c. Why do you think the relationship you noticed in Part b happens?

d. Which planet or planets don't fit the general pattern? Why might a planet not follow the pattern?

In your
own
words

Give one example
of a problem in
which you would
use a coordinate
plane with four
quadrants to help
you solve the prob-
lem. In what way
would the coordi-
nate plane help you
solve the problem?

7. Think about straight lines on the coordinate plane. You might draw a coordinate plane and experiment with lines to answer these questions.

    a. What is the greatest number of quadrants a straight line can go through? Explain.

    b. What is the least number of quadrants a straight line can go through?

8. If you plotted all points for which the y-coordinate is the square of the x-coordinate, in which quadrants or on which axes would the points lie? Explain how you know your answer is correct.

9. In this exercise, you will use the coordinate plane you colored for Problem Set F to help solve the inequality $x + 5 < 0$.

    a. Look at all the points on the coordinate plane you made in Problem Set F with a y-coordinate of 5. Of these points, which are red? Which are blue? Which are black?

    b. When is the sum of 5 and another number less than 0?

    c. For which values of x is $x + 5 < 0$?

10. Challenge In Problem Set E, you colored points on a coordinate plane to represent the sign of the products of coordinates. What about the quotients? Suppose you colored a coordinate plane so that point (c, d) is red if $c \div d$ is positive, blue if $c \div d$ is negative, black if $c \div d$ is 0, and green if it is impossible to calculate $c \div d$.

    a. Shade a coordinate plane to show what it would look like if you colored *every* point red, blue, black, or green.

    b. Compare your coordinate plane to the one you made in Problem 3 of Problem Set E. How are they similar and different? Why?

Mixed
Review

Evaluate each expression.

11. $4.5 \cdot 90.02$

12. $3.45 \div 0.5$

13. $0.034 \cdot 3.2 \cdot \frac{1}{2}$

14. Draw a flowchart for this rule: *output* $= 2n + 4$. Use it to find the output for the input 8.

15. Draw a flowchart for this rule: *output* $= 7n + 9$. Use it to find the output for the input 20.

16. Draw a flowchart for this rule: *output* $= \frac{n - 13}{12}$. Use it to find the output for the input 73.

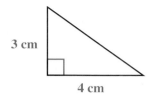

3 cm

4 cm

17. Geometry Consider the triangle at left.

    a. What is the area of the triangle?

    b. Suppose you use this triangle as the base of a prism 1 cm high. What would the prism's volume be?

    c. Draw another triangular base that would produce a 1-cm-high prism with this volume.

    d. Suppose you use the triangle to make a prism 4 cm high. What would the prism's volume be?

    e. Suppose you use the triangle to make a prism h cm high. What would the prism's volume be?

18. From the following list, find all the pairs of fractions with a sum of 1.

$$\frac{15}{21} \quad \frac{10}{25} \quad \frac{8}{14} \quad \frac{1}{6} \quad \frac{2}{7} \quad \frac{3}{7} \quad \frac{3}{6} \quad \frac{10}{12} \quad \frac{3}{5} \quad \frac{1}{2}$$

19. At top speed, Chet the cheetah can run about 60 mph for short distances. At this speed, how far can Chet run in 40 seconds?

20. Statistics Diego surveyed the students in his English class about how many siblings (brothers or sisters) they had. The pictograph shows his results.

    a. How many students did Diego survey?

    b. What are the mean, median, and mode of Diego's data?

    c. How many siblings do the students in Diego's English class have in all?

    d. If you choose a student at random from Diego's English class, what is the probability the student has exactly three siblings?

    e. If you choose a student at random from Diego's English class, what is the probability the student has more than two siblings?

English Class Survey

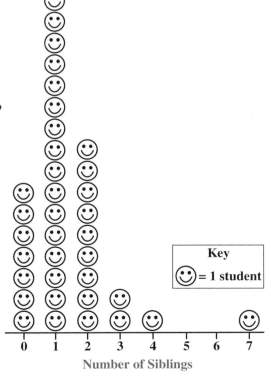

Key
☺ = 1 student

Number of Siblings

4.4 Finding Distances

When pilots chart courses, they use maps with grids similar to the coordinate grids you have been working with. When archaeologists set up a dig, they create a grid on the ground with string so they can record the locations of the objects they find. Pilots and archaeologists often use their grids to find the distance between two locations. In this lesson, you will learn how to find distances between points on a coordinate grid.

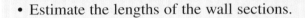

Explore

Dr. Davis is working on an archaeological dig. He has used string to lay out a grid on the ground. The lines of the grid are 1 foot apart.

So far, Dr. Davis has unearthed sections of two walls and an object he thinks might have been a toy. He drew this diagram to show the location of these objects on the grid. He labeled the corner of the walls (0, 0).

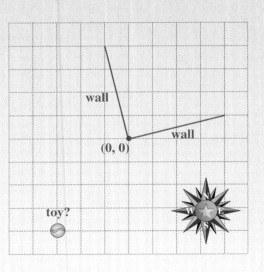

• Estimate the lengths of the wall sections.

• What are the coordinates of the toy?

• Estimate the distance from the toy to the corner where the walls meet. Save your estimate so you can refer to it later in this lesson.

Investigation ▶ 1 ▶ The Pythagorean Theorem

Sometimes estimates like the ones you made for the lengths of the wall segments are all you really need. But there are times when you need a more accurate—or even exact—measurement. In this investigation, you will learn a way to find measurements like this one by using two side lengths in a triangle to calculate the third.

Think & Discuss

Four line segments are drawn on the grid. Without measuring, determine which is the longest. Explain how you know it is the longest.

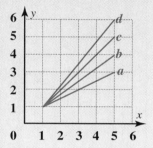

Without measuring, determine which segments below are the same length. Explain how you know.

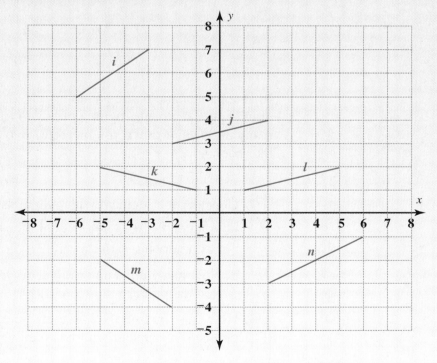

Problem Set A

The segments described in Problems 1–4 are drawn on the grid. For each segment, tell how many units left or right and how many units up or down you must move to get from the first endpoint to the second.

1. Segment *j*: (1, 6) to (⁻2, ⁻2)   **2.** Segment *k*: (3, ⁻4) to (⁻1, ⁻6)

3. Segment *l*: (⁻2, 5) to (1, ⁻3)   **4.** Segment *m*: (8, ⁻5) to (0, ⁻8)

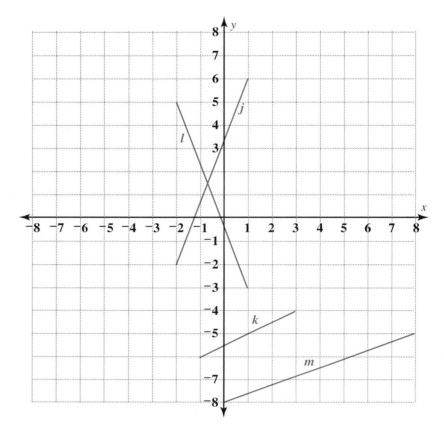

In Problems 5 and 6, the endpoints of a segment are given. Without drawing the segments, figure out how many units left or right and how many units up or down you must move to get from the first endpoint to the second.

5. (4, 7) to (⁻4, 11)   **6.** (9, 5) to (1, 2)

7. Consider the segment with endpoints (⁻3, ⁻4) and (3, ⁻2).

   a. How many units left or right do you need to move to get from the first endpoint to the second?

   b. How many units up or down do you need to move to get from the first endpoint to the second?

8. How can you find the horizontal and vertical distances between endpoints of a segment without counting?

Remember Dr. Davis and his archaeological dig? One day, he found pieces of broken pottery at (1, 2) and at (3, 5) on the grid, but he did not record the distance between them. As his assistant Luisa looks at Dr. Davis's notes, she has an idea. She draws dashed line segments to form a right triangle.

I've drawn a right triangle with the segment between the pieces of pottery as the hypotenuse.

To find the lengths of the horizontal and vertical sides I can just count the units.

Once I have these lengths, I can find the length of the hypotenuse using the Pythagorean Theorem. This length is the distance between the pottery fragments.

Remember

The *hypotenuse* of a right triangle is the side opposite the right angle. The other two sides are the *legs*.

The Pythagorean Theorem states that, in a right triangle with legs of lengths a and b and a hypotenuse of length c, $a^2 + b^2 = c^2$.

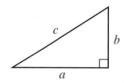

You can use the Pythagorean Theorem to find the distance between any two points on a coordinate grid.

EXAMPLE

The locations of the pieces of pottery Dr. Davis found are indicated on the grid. Find the distance between them.

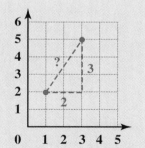

The legs of the right triangle have lengths 2 feet and 3 feet. Use the Pythagorean Theorem to find the length of the hypotenuse.

$$a^2 + b^2 = c^2$$
$$2^2 + 3^2 = c^2$$
$$4 + 9 = c^2$$
$$13 = c^2$$

The length of the hypotenuse is $\sqrt{13}$ feet, or about 3.6 feet. So, the pottery fragments are about 3.6 feet apart.

Problem Set B

Use the Pythagorean Theorem to find each missing side length.

1.

2.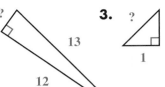

3.

4.

5. You will now explain the Pythagorean Theorem in your own words.

   a. How do you use the Pythagorean Theorem to find the length of the hypotenuse if you know the lengths of the other two sides?

   b. If you know the lengths of the hypotenuse and one leg, how can you use the Pythagorean Theorem to find the other leg's length?

6. Use the methods you described in Problem 5 to find the length of each segment on the grid shown at left.

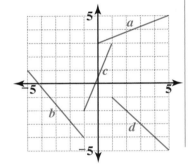

7. Dr. Davis's assistant Luisa uncovered one end of a farm tool at (4, ⁻1) and the other end at (9, ⁻4).

   a. How many units long is the farm tool?

   b. Remember that the grid lines are 1 foot apart. What is the actual length of the tool?

Share & Summarize

Explain to another student how you did Part a of Problem 7. Then work together to write steps for using the Pythagorean Theorem to find the distance between *any* two points on the coordinate plane. Your steps should be clear enough that someone else could follow them.

Investigation 2 The Distance Formula

You can use the Pythagorean Theorem to find lengths of segments on a coordinate grid, whether that grid is on a map, an archaeological site, or a construction site. You will now learn about a special formula that is related to the Pythagorean Theorem: the *distance formula.*

Problem Set C

1. In the Explore activity on page 268, you looked at a diagram of an archaeological site. You estimated the distance between the corner where the two walls meet and the toy. Now use the Pythagorean Theorem to find the distance. How close was your estimate?

2. Zach and Jin Lee calculated the distance in Problem 1 differently.

Zach's Steps		Jin Lee's Steps	
Step 1	How far left or right do I go? Find this length.	*Step 1*	How far left or right do I go? Find this length.
Step 2	How far up or down do I go? Find this length.	*Step 2*	How far up or down do I go? Find this length.
Step 3	Add the two lengths.	*Step 3*	Square both lengths.
Step 4	Square the sum from Step 3.	*Step 4*	Add the two squared lengths from Step 3.
Step 5	Find the square root of the answer from Step 4.	*Step 5*	Find the square root of the sum from Step 4.

a. Which student wrote the correct set of steps? Identify the mistakes the other student made.

b. How does your method for finding the distance (Problem 1) compare with the correct series of steps in the table? Do you have all the same steps? Are your steps in the same order?

3. The steps for finding the distance between two points can be written in symbols as well as with words. Let's use (x_1, y_1) to represent one point and (x_2, y_2) to represent the other.

Refer to the correct set of steps from Problem 2. Parts a–e each show one of the steps from that set in symbols. Tell which step the symbols represent.

a. $y_2 - y_1$

b. $(x_2 - x_1)^2$ and $(y_2 - y_1)^2$

c. $x_2 - x_1$

d. $\sqrt{(x_2 - x_1)^2 + (y_2 - y_1)^2}$

e. $(x_2 - x_1)^2 + (y_2 - y_1)^2$

VOCABULARY
distance formula

The **distance formula** gives the symbolic rule for calculating the distance between any two points in the coordinate plane.

Distance Formula

If (x_1, y_1) and (x_2, y_2) represent the points, then

$$distance = \sqrt{(x_2 - x_1)^2 + (y_2 - y_1)^2}$$

To find the distance between two given points, first decide which point will be (x_1, y_1) and which will be (x_2, y_2).

EXAMPLE

Find the distance between $(1, {}^-3)$ and $({}^-4, 5)$.

Let $(1, {}^-3)$ be (x_1, y_1), and let $({}^-4, 5)$ be (x_2, y_2). Substitute the coordinates into the distance formula.

$$distance = \sqrt{(x_2 - x_1)^2 + (y_2 - y_1)^2}$$
$$= \sqrt{({}^-4 - 1)^2 + (5 - {}^-3)^2}$$
$$= \sqrt{({}^-5)^2 + 8^2}$$
$$= \sqrt{25 + 64}$$
$$= \sqrt{89}$$
$$\approx 9.43 \text{ units}$$

Just the facts

The variables x_1, y_1, x_2, and y_2 are called *sub-scripted* variables. They are like any other variables, just with lowered numbers called *subscripts*.

Notice that the symbol $\approx$ is used in the last line above. This means that the distance between the points is *approximately equal* to 9.43 units.

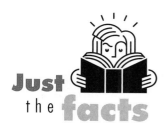

Problem Set D

1. What information do you need to use the distance formula?

2. Each part of the distance formula has a connection to applying the Pythagorean Theorem to a right triangle.

   a. When you subtract the second x-coordinate from the first x-coordinate, what does that difference correspond to in a right triangle?

   b. When you subtract the second y-coordinate from the first y-coordinate, what does that difference correspond to in a right triangle?

3. Tyrone and Luis were trying to find the distance between the two points on this grid.

   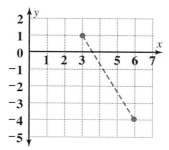

   • Tyrone said, "I'm going to let $(3, 1)$ be (x_2, y_2) and $(6, {}^-4)$ be (x_1, y_1). When I find $x_2 - x_1$, I get $3 - 6$, or $^-3$."

   • Luis said, "That's not right. You need to let $(6, {}^-4)$ be (x_2, y_2) and $(3, 1)$ be (x_1, y_1). When you find $x_2 - x_1$, you get $6 - 3$, or 3."

   a. Show how Tyrone would calculate the distance between the points.

   b. Show how Luis would calculate the distance between the points.

   c. Who is correct? Explain your answer.

Use the distance formula or the Pythagorean Theorem to find the distance between the given points.

4. $(^-3, {}^-5)$ and $(7, {}^-1)$

5. $(6, 4)$ and $(2, {}^-3)$

6. $(^-1, 5)$ and $(^-10, {}^-4)$

7. $(^-3, 1)$ and $(5, {}^-1)$

Share & Summarize

Explain how the distance formula and the Pythagorean Theorem are related. Use drawings or examples if they help you explain.

On Your Own Exercises

Practice & **Apply**

1. Four pirates are hunting for buried treasure on an island. They all start at the palm tree, which is at (2, 6) on their maps. All they know from their clues is that they must walk 10 paces either east or west, make a 90° turn, and walk another 12 paces. So each pirate walks 10 paces, two heading east and two heading west. Then one pirate from each pair heads north another 12 paces, while the other heads south for 12 paces.

   a. Where did each pirate end? Give the coordinates of each.

   b. How far is each pirate from the palm tree, as the crow flies? ("As the crow flies" means "in a straight line from where you started.") Explain how you found your answers.

For Exercises 2–6, use the Pythagorean Theorem to find the length of the segment.

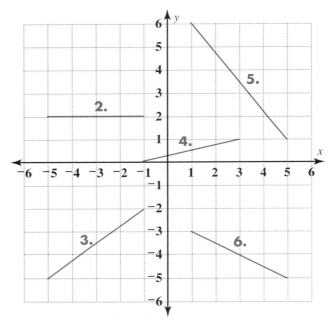

In y o u r
own
words

You have been asked to write about something you have learned for the parents' Math Newsletter. Describe to parents how using the distance formula is just like using the Pythagorean Theorem to find the length of a segment.

Use the distance formula to find the length of each segment.

7. (4, 2) to (⁻5, ⁻6)

8. (0, ⁻3) to (6, 2)

9. (⁻9, 2) to (⁻10, 11)

10. (6, 1) to (9, 5)

impactmath.com/self_check_quiz

11. Use the distance formula to order the segments below from shortest to longest. Give the length of each segment.

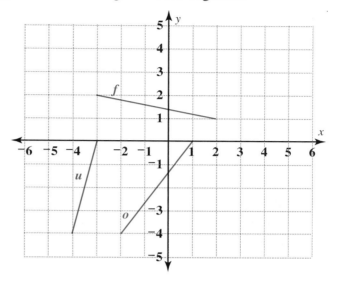

12. Life Science The grid shows the arrangement and length of mammal bones found embedded in a layer of the La Brea Tar Pits in California. (Assume the bones are lying flat in the tar pit, and are not tilted down into the pit.)

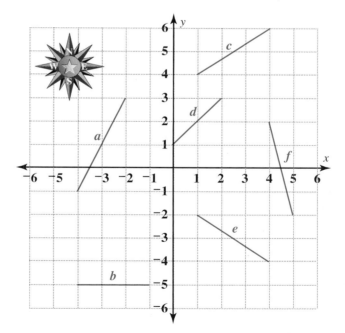

a. Which bone is the shortest? How long is it, if each unit on the grid represents 6 inches?

b. Which bone is the longest? How long is it?

13. Use graph paper to help you think about this exercise. Draw a Point *A* somewhere in the middle of the graph paper. Draw a new point labeled *B* by moving from Point *A*, left 5 units and up 2 units.

  a. From Point *A*, move left 2 units and up 5 units, and mark Point *C*. Is Segment *AC* the same length as Segment *AB*? Why or why not?

  b. From Point *A*, move up 2 units and left 5 units, and mark Point *D*. Is Segment *AD* the same length as Segment *AB*? Why or why not?

  c. Describe two ways you could move from Point *B* back to Point *A*.

14. Each grid unit on the map represents 15 miles. Suppose you are anchored at the point shown on White Sands Island.

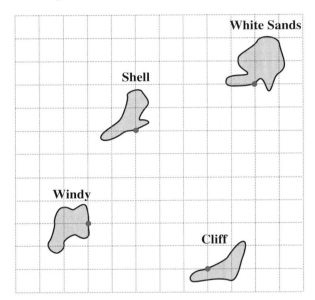

  a. If you can sail 100 miles per day, to which other islands could you travel in a day? You must travel to the harbors marked by the dots.

  b. Is there a pair of islands to which you could travel in the same day? If so, which are they? (For example, starting from White Sands to Shell to Windy?) If not, why not?

15. Make up a word problem to go with the graph at left. Design your problem so that you must use the distance formula to solve it.

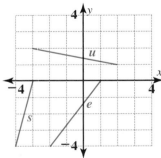

Mixed Review

Find the value of the variable in each equation.

16. $6b = 15$ **17.** $2k - 5 = 1.4$ **18.** $a = 25a$

19. $3.5m = 7$ **20.** $6 - 22p = 6$ **21.** $3n \div 7 = 3$

Rewrite each expression in exponential notation.

22. $c \cdot c \cdot c \cdot c$ **23.** $4 \cdot 4 \cdot 4$

Rewrite each number in standard notation.

24. 6×10^2 **25.** 9×10^5

Rewrite each number in scientific notation.

26. 654,000   **27.** 9,500,000,000

28. Shayla has made a sketch of the garden she would like to plant in the spring. Each grid square has an area of 1 square foot.

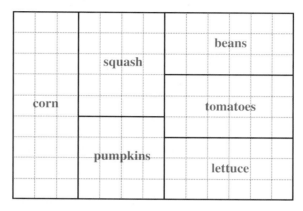

    a. How many more square feet will Shayla plant in corn than in pumpkins?

    b. How many more squware yards will Shayla plant in corn than in pumpkins? (Be careful!)

    c. What fraction of the total garden area will Shayla plant in squash?

    d. What percentage of the total garden area will Shayla plant in lettuce?

    e. Suppose Shayla adds more space to her garden in order to triple the size of the corn plot. What percentage of the total garden area will be planted in corn?

29. Match each point on the graph to the coordinates given.

    a. $(^-2, 0)$

    b. $(1, ^-3)$

    c. $(^-2, 3)$

    d. $(^-2, ^-4)$

    e. $(2, 1)$

    f. $(3, 4)$

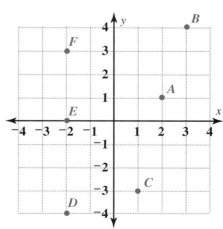

4.5 Negative Numbers as Exponents

You have seen that negative numbers can be used to show amounts that are less than 0, such as low temperatures, debts, and elevations below sea level. Negative numbers can also be used as exponents.

In Chapter 3, you learned that positive exponents mean repeated multiplication. For example, 3^4 means $3 \cdot 3 \cdot 3 \cdot 3$. But what does 3^{-4} mean?

In this lesson, you will see that you can extend what you know about positive exponents to help you understand and work with expressions involving negative exponents.

Think & Discuss

In Chapter 3, you used stretching and shrinking machines as a way to think about exponents.

Take a look at this hookup.

If you put a 1-inch stick of bubble gum through this hookup, how long will it be when it exits? Describe at least two ways you could figure this out.

Investigation 1 ▶ Shrinking with Negative Exponents

You will soon use shrinking machines to help you think about negative exponents. First, though, you will review hookups of machines with positive exponents.

Problem Set A

If possible, find a single repeater machine that will do the same work as the given hookup.

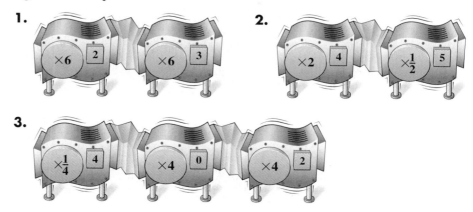

1. ×6 [2] ×6 [3]

2. ×2 [4] ×½ [5]

3. ×¼ [4] ×4 [0] ×4 [2]

Jack is a new employee at the resizing factory. He is assigned to all the machines with base 3. A clown troupe has asked Jack to stretch some ribbons for their balloons. To get familiar with the machines, Jack sends 1-centimeter pieces of ribbon through them and records the lengths of the outputs.

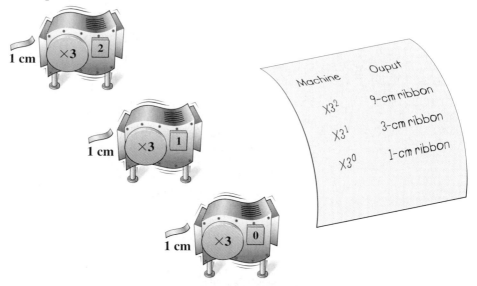

1 cm ×3 [2]

1 cm ×3 [1]

1 cm ×3 [0]

Machine	Ouput
X3²	9-cm ribbon
X3¹	3-cm ribbon
X3⁰	1-cm ribbon

Think & Discuss

Jack noticed a pattern in the outputs: Each time the number of repeats is reduced by 1, the resulting length is $\frac{1}{3}$ the previous length. Why?

Some of Jack's base 3 machines have negative exponents. When he puts 1-cm ribbons through a $\times 3^{-1}$ machine, a $\times 3^{-2}$ machine, and a $\times 3^{-3}$ machine, the pattern in the output length continues.

Problem Set B

1. Continue Jack's pattern in the chart below.

What Went In	Machine	What Came Out
1-cm ribbon	×3 [2]	9-cm ribbon
1-cm ribbon	×3 [1]	3-cm ribbon
1-cm ribbon	×3 [0]	1-cm ribbon
1-cm ribbon	×3 [−1]	
1-cm ribbon	×3 [−2]	
1-cm ribbon	×3 [−3]	

2. After conducting several experiments, Jack concluded that these three machines do the same thing:

a. Suppose you put an 18-inch length of rope into a $\times 3^{-1}$ machine. Describe two ways you could figure out how long the resulting piece would be.

b. Describe two other repeater machines—one that uses multiplication and one that uses division—that do the same thing as a $\times 3^{-2}$ machine.

c. Describe two other repeater machines—one that uses multiplication and one that uses division—that do the same thing as a $\times 3^{-3}$ machine.

3. Jack repeated the pattern shown in Problem 1 with two other bases.

a. With base 2, Jack found that each time the exponent was reduced by 1, the output length was half the previous output length. Complete the chart to show what happened to 1-cm ribbons.

b. A similar pattern appeared when Jack put 1-cm ribbons through base 4 machines. Complete the chart.

Machine	What Came Out
×2 [2]	
×2 [1]	
×2 [0]	
×2 [−1]	
×2 [−2]	
×2 [−3]	

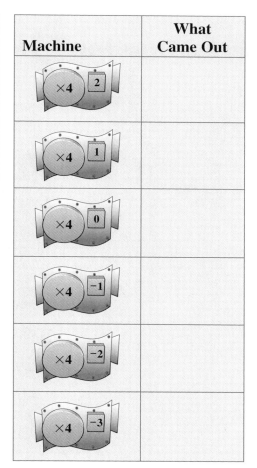

Machine	What Came Out
×4 [2]	
×4 [1]	
×4 [0]	
×4 [−1]	
×4 [−2]	
×4 [−3]	

Machines with other bases also follow the pattern. If you put a 72-inch pole through each machine, how long would each output be?

4.

5.

6.

7.

In general, when you put something in a multiplication machine with a negative exponent, like $\times 2^{-4}$, how can you find the output length?

Investigation 2 ▶ Evaluating Expressions with Negative Exponents

Thinking about shrinking machines can help you evaluate expressions with negative exponents.

EXAMPLE

Evaluate 4^{-3}.

Think about a $\times 4^{-3}$ machine. Putting something through a $\times 4^{-3}$ machine is equivalent to putting it through a $\times \left(\frac{1}{4}\right)^3$ machine, so $4^{-3} = \left(\frac{1}{4}\right)^3$, or $\frac{1}{64}$.

Evaluate $2^{-3} \times 2^2$.

Think about this hookup.

The $\times 2^{-3}$ machine halves the length of an input three times.

The $\times 2^2$ machine doubles the length twice.

The output is half the length of the original input. The hookup is the same as a $\times \frac{1}{2}$ or a $\times 2^{-1}$ machine. So $2^{-3} \times 2^2 = \frac{1}{2}$, or 2^{-1}.

Now you will practice evaluating expressions with negative exponents. Pay attention to how you solve the problems, and be ready to share your approach with others in your class.

Problem Set C

Evaluate each expression.

1. 3^{-4} **2.** 2^{-1} **3.** $3^{-3} \cdot 3^5$

4. $7^5 \cdot 7^{-7}$ **5.** $3^9 \cdot 3^{-6}$ **6.** $^-49 \cdot 7^{-2}$

7. $256 \cdot 4^{-3}$ **8.** $32 \cdot 1^{-5}$ **9.** $^-60 \cdot 6^{-1}$

10. $2^{-2} \cdot 3^{-1}$ **11.** $2^{-1} \cdot 3^{-1} \cdot 5^{-2}$ **12.** $2^{-2} \cdot 4^2 \cdot 2^{-1}$

13. Which is greater, 5^{-1} or 5^{-2}? Why?

14. Describe a hookup of two repeater machines, one with a negative exponent, that does the same work as a $\times 5^0$ machine.

15. Is it possible to shrink something to a length of 0 by using shrinking machines? Why or why not?

Problem Set D

You sometimes need to take extra care to be sure you understand a mathematical expression. For example, in expressions such as $^-3^{-2}$ and $(^-3)^{-2}$, the two negative symbols give two different kinds of information.

1. Why is $^-3^{-2} = ^-\frac{1}{9}$?

2. Why is $(^-3)^{-2} = \frac{1}{9}$?

3. What does $^-4^{-2}$ equal?

4. Write an expression that has a negative exponent and equals $^-\frac{1}{27}$.

Evaluate each expression.

5. $^-5^{-2}$ **6.** $(^-3)^3$ **7.** $^-1^{-5}$

8. $(^-1)^{-5}$ **9.** $(^-5)^{-3}$ **10.** $^-2^4$

Share & Summarize

1. Give two values of n for which $(^-2)^n$ is positive.

2. Give two values of n for which $(^-2)^n$ is negative.

3. Give two values of n for which $(^-2)^n$ is not an integer.

4. Write an expression with a negative exponent that equals $^-\frac{1}{64}$.

Investigation ▶3 Laws of Exponents and Scientific Notation

In Chapter 3, you learned about the laws of exponents, which describe how to simplify expressions with exponents.

Product Laws	Quotient Laws	Power of a Power Law
$a^b \times a^c = a^{b+c}$ $a^c \times b^c = (a \times b)^c$	$a^b \div a^c = a^{b-c}$ $a^c \div b^c = (a \div b)^c$	$(a^b)^c = a^{b \times c}$

Now you will apply these laws to expressions with negative exponents.

Problem Set E

The product laws work for negative exponents as well as for positive exponents. In Problems 1–6, use the product laws to rewrite each expression using a single base. Use a calculator to check your answers.

1. $50^{-6} \cdot 50^{12}$

2. $2^5 \cdot 2^{-6} \cdot 2^3$

3. $0.5^{-10} \cdot 0.5^{-10}$

4. $3^{-1} \cdot 9$

5. $^{-}3^{-3} \cdot 2^{-3} \cdot 4^{-3}$

6. $10^{-23} \cdot 10^{23}$

The quotient laws and the power of a power law also work for negative exponents. In Problems 7–12, use the appropriate law to rewrite each expression using a single base. Use a calculator to check your answers.

7. $7^3 \div 7^5$

8. $12^{-4} \div 4^{-4}$

9. $10^{-5} \div 10^3$

10. $(3^{-3})^2$

11. $2.3^2 \div 2.3^{-6}$

12. $(2^{-5})^{-5}$

Rewrite each expression using a single base or exponent.

13. $m^{23} \cdot m^{-17}$

14. $7^{-x} \cdot 7^x$

15. $a^{-3} \div b^{-3}$

Rewrite each expression using only positive exponents.

16. $\left(\frac{2}{3}\right)^{-1}$

17. $\left(\frac{1}{2}\right)^{-3}$

18. 10^{-7}

Simplify each expression as much as you can.

19. $^{-}3n \times {}^{-}3n$

20. $5n \div 2n^{-2}$

21. $4y^2 \times 3y^{-2}$

Remember

In *scientific notation,* a number is expressed as the product of a number greater than or equal to 1 but less than 10, and a power of 10. For example, 5,400 is written 5.4×10^3.

In Chapter 3, you used scientific notation to represent very large numbers. You can also use scientific notation to represent very small numbers. For small numbers, scientific notation involves negative powers of 10.

Problem Set F

A 5.3-cm drinking straw is sent through each machine. Give the length of the straw that emerges from each machine. Write your answers as decimals.

1.  ?

2. ?

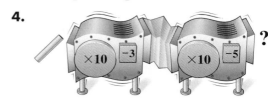

3. ?

4. ?

5. Write your answers for Problems 1–4 in scientific notation.

6. List your answers for Problem 5 from greatest to least.

Find each missing value.

7. $3 \times 10^? = 0.03$

8. $4.5 \times 10^? = 0.00045$

9. $? \times 10^{-3} = 0.0065$

10. $1 \times ? = 0.00000001$

11. There are 86,400 seconds in a day. How many days long is a second? Express your answer in scientific notation.

Share & Summarize

1. Without calculating the product, how can you tell whether $2^{-4} \times 2^3$ is greater than 0 or less than 0?

2. Which is smaller, 1×10^{-6} or 1×10^{-7}? How do you know?

On Your Own Exercises

Find the length of the output if an 144-inch input is sent through each machine.

1.

2.

3.

If possible, find a single repeater machine that will do the same work as the given hookup.

4.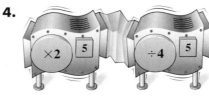

5.

6.

7.

Describe a machine with a negative exponent that does the same work as the given machine.

8.

9.

10.

11.

12.

13.

14. Without calculating, figure out whether 4^{-3} or 4^{-4} is greater. Explain how you decided.

impactmath.com/self_check_quiz

Evaluate each expression.

15. $^-343 \cdot 7^{-2}$

16. $^-1{,}375 \cdot 5^{-3}$

17. $128 \cdot 2^{-5}$

18. $^-72 \cdot 9^{-1}$

19. $48 \cdot 6^{-1}$

20. $243 \cdot 3^{-4}$

21. $^-9 \cdot 1^{-3}$

22. $4{,}096 \cdot 8^{-2}$

23. $^-11 \cdot 11^{-1}$

24. How could you rewrite Problem 22 as a division problem? As a multiplication problem using a positive exponent?

Decide whether each statement is true. If the statement is false, explain why.

25. $2^{-2} \cdot 2^{-1} = 2^{-3}$

26. $5^{-3} \cdot 5^{-1} \cdot 5^2 = 5^{-2}$

27. $3^{-1} \cdot 4^{-3} = 12^{-4}$

28. $3^{-2} \cdot 3^{-3} = 9^{-5}$

29. $x^{-5} \div x^7 = x^{-12}$

30. $k^{-2} \div k^{-5} = k^3$

31. $w^{-2} \div w^{-5} = w^{-7}$

32. $b^{-8} \div g^{-4} = bg^{-2}$

Without evaluating, tell whether each expression is greater or less than 1.

33. $3^{-4} \times 3^5$ **34.** $2^{-1} \times 2^{-3}$ **35.** $1^{-3} \div 2^{-3}$ **36.** $10^{-5} \div 10^3$

Evaluate each expression.

37. $^-4^{-2}$ **38.** $(^-5)^{-1}$ **39.** $^-2^3$ **40.** $(^-10)^4$

Rewrite each expression using a single base and a single exponent.

41. $\frac{1}{2^{-2}} \times 4^{-2}$

42. $100^{-20} \div ^-25^{-20}$

43. $^-3^{100} \times ^-\frac{5}{6^{100}} \times 10^{100}$

44. $\left(\frac{1}{3}\right)^{-a} \times \left(\frac{1}{5}\right)^{-a}$

45. $^-1{,}000^0 \div 3^0$

46. $15^{-5} \times \left(\frac{1}{5}\right)^{-5}$

Find the value of n in each equation.

47. $7.24 \cdot 10^n = 0.00724$

48. $5 \cdot 10^n = 0.5$

49. $n \cdot 10^{-4} = 0.000104$

50. $9.2 \cdot n = 0.0092$

51. A mile is 1.609×10^5 centimeters. How many miles is a centimeter? Express your answer in scientific notation.

52. There is 7.8125×10^{-3} gallon in an ounce. How many ounces are in a gallon? Express your answer in scientific notation.

Challenge Simplify each expression.

53. $^-6a^{-2} \div 4a^{-4}$

54. $^-\frac{1}{2}x^{-3} \times x^4$

55. $z^2 \times 2z^{-3} \div ^-6z^{-2}$

56. $9b^4 \times 9b^{-8}$

Connect & Extend

57. Challenge Describe a machine that has a negative exponent and that does the same work as the machine shown at right.

Find the length of the output if a meterstick is sent through each hookup.

58.

59.

60.

61.

62. A 1-inch stick of gum went into each machine. Complete the chart.

Machine	What Came Out	Machine	What Came Out
$\times\frac{1}{2}$ 2		$\times\frac{1}{2}$ −1	
$\times\frac{1}{2}$ 1		$\times\frac{1}{2}$ −2	
$\times\frac{1}{2}$ 0		$\times\frac{1}{2}$ −3	

63. Challenge A strand of wire is sent through this hookup. It emerges 2 inches long. What was the starting length?

Without calculating, figure out whether each expression is positive or negative and explain how you decided.

64. $(^-4)^{-1}$ **65.** $(^-4)^{-2}$ **66.** $(^-4)^{-3}$ **67.** $(^-4)^{-4}$

68. $(^-2)^{-3}$ **69.** $(^-2)^{-2}$ **70.** $(^-5)^{-1}$ **71.** $(^-10)^{-6}$

72. Prove It! It is true that $\left(\frac{1}{2}\right)^{-1} = (2^{-1})^{-1} = 2^1 = 2$. In Parts a–c, you will explain each step in this calculation.

     a. Why does $\left(\frac{1}{2}\right)^{-1} = (2^{-1})^{-1}$?

     b. Why does $(2^{-1})^{-1} = 2^1$?

     c. Why does $2^1 = 2$?

73. Physics Protons, neutrons, and electrons are *stable* particles—they can exist for a very long time. But many other particles exist for only very short periods of time. A *muon,* for example, has an average life of 0.000002197 second.

     a. Write this length of time in scientific notation.

     b. A particular muon existed for half the average life of this type of particle. How long did it exist?

     c. Another muon existed for three times the average life of this type of particle. How long did it exist?

Mixed Review

Evaluate.

74. $11 \cdot 33$ **75.** $11 \cdot 44$

76. $11 \cdot 55$ **77.** $3.5 + 0.251$

78. $3.54 + 0.754$ **79.** $3.5 - 1.6$

80. Draw a flowchart to represent the expression $\frac{4x + 8}{2}$. Then use backtracking to find the solution to the equation $\frac{4x + 8}{2} = 10$.

Complete each table, and write another expression that gives the same values.

81.

n	1	2	3	4	5	100
$\frac{n-1}{2}$		$\frac{1}{2}$				

82.

q	0	1	2	3	4	100
$\frac{q+8}{2}$						

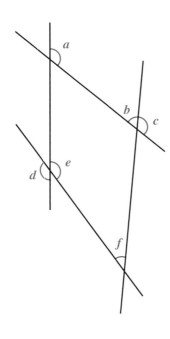

83. Use a protractor to measure the angles on the figure at left. Give your answers to the nearest 5°.

84. In June 2002, you could have received about 48.984 Indian rupees for 1 U.S. dollar.

    a. How many dollars could you have received in exchange for 100 Indian rupees?

    b. How many Indian rupees could you have received in exchange for 100 U.S. dollars?

85. The table shows how long a person born in the given year was expected to live, on average. For example, a boy born in 1950 was expected to live 65.6 years.

Years of Life Expected at Birth

Year	Life Expectancy (years)	
	Males	**Females**
1900	46.3	48.3
1910	48.4	51.8
1920	53.6	54.6
1930	58.1	61.6
1940	60.8	65.2
1950	65.6	71.1
1960	66.6	73.1
1970	67.1	74.7
1980	70.0	77.5
1990	71.8	78.8
2000	74.1	79.5

Source: Reprinted with permission from *The World Almanac and Book of Facts 2003.*

    a. By how many years did the life expectancy of a baby girl rise from 1900 to 2000? By how many years did the life expectancy of a baby boy rise in that time?

    b. By what percentage did the life expectancy of a baby girl rise from 1900 to 1990? By what percentage did the life expectancy of a baby boy rise in that time?

    c. What might have contributed to the dramatic rise in life expectancy between 1900 and 2000?

    d. Estimate the life expectancies of a baby boy and a baby girl in the year 2010.

VOCABULARY
**distance formula
quadrant**

Chapter Summary

You began this chapter learning about operations with signed numbers. Using the number-line model, you thought about addition and subtraction as facing a particular direction on a number line and moving forward or backward. You discovered some rules for these calculations, and then used them to develop rules for multiplication and division.

You extended your knowledge of positive and negative numbers by plotting points in all four quadrants. Next you looked at ways to calculate lengths of line segments on a graph, using two related formulas: the Pythagorean Theorem and the distance formula.

Finally, you learned what negative numbers mean as exponents. You learned that the laws of exponents apply when the exponents are negative. You also practiced writing very small numbers in scientific notation.

Strategies and Applications

The questions in this section will help you review and apply the important ideas and strategies developed in this chapter.

Adding and subtracting with signed numbers

Using the "walk the number line" model or any other strategy you know, describe how to compute each sum or difference.

    1. $2 - {}^-9$ **2.** ${}^-9 + 3$

    3. Write a word problem that can be solved in two ways, one involving subtraction with signed numbers and one involving addition with signed numbers. Then write an equation to represent each solution method.

Multiplying and dividing with signed numbers

Describe how to compute each product or quotient.

    4. ${}^-3.5 \times 4$ **5.** ${}^-9 \div {}^-4.5$

Without computing, determine whether each expression is greater than 0, less than 0, or equal to 0. Explain how you know.

    6. $(^-33)^5 \times (^-27)^2$ **7.** $\dfrac{^-5^{58} \times (^-12)^{23}}{^-5^3}$

Solve each equation.

8. $^-3x + 2 = ^-4$

9. $2a + 20 = 6$

10. The product of two integers is 10. What could the integers be? Name all the possibilities.

Working with points in all four quadrants

11. The graph shows how a hiker's elevation changed during a hike.

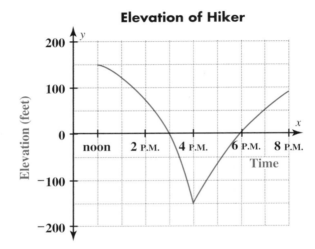

Elevation of Hiker

a. What was the lowest elevation the hiker reached, and when did she reach it?

b. What was the highest elevation the hiker reached, and when did she reach it?

c. At what time or times was she at sea level (elevation 0)?

d. At what elevation did she begin her hike?

e. Between what times was the hiker above sea level?

f. Between what times was she below sea level?

12. Think about all points (x, y) where $y = ^-2x$—that is, where the y-coordinate is equal to the product of $^-2$ and the x-coordinate. Examples are $(1, ^-2)$, $(^-2, 4)$, and $(^-3, 6)$.

a. Plot and label the three points listed above.

b. Plot and label four more points in which $y = ^-2x$.

c. Imagine plotting all points for which $y = ^-2x$. In what quadrants or on what axes would these points lie? Explain how you know.

Calculating lengths using the Pythagorean Theorem and the distance formula

13. Before moving into her new house, Susan is trying out furniture arrangements on graph paper. She wants to make sure there is enough room to walk around the furniture. She decides she needs at least 3 feet between pieces of furniture—except for the coffee table, which can be directly in front of the couch. Each square on her grid represents 1 square foot.

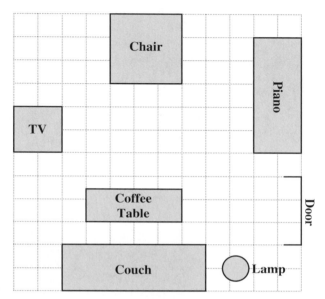

a. How many feet are there between the closest corners of the TV and the chair?

b. How many feet are there between the closest corners of the coffee table and the piano?

c. Are any pieces of furniture too close together?

14. Use the Pythagorean Theorem to find the missing side length.

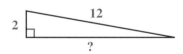

15. The distance formula is $\sqrt{(x_2 - x_1)^2 + (y_2 - y_1)^2}$.

a. When you calculate $(x_2 - x_1)$, what does the resulting number represent?

b. When you calculate $(y_2 - y_1)$, what does the resulting number represent?

c. How is using the distance formula like using the Pythagorean Theorem?

Evaluating expressions involving negative exponents

16. What are two ways to figure out $\left(\frac{1}{2}\right)^{-1} \times \left(\frac{1}{2}\right)^{-1}$?

17. Use the facts that $2^2 = 4$, $2^1 = 2$, and $2^0 = 1$ to explain why $2^{-1} = \frac{1}{2}$.

18. How can you use the laws of exponents to tell whether $3^4 \times 3^{-6}$ is greater than 1 or less than 1?

19. Imagine putting a 5-foot fishing rod into this hookup of shrinking machines.

a. Write this problem as a multiplication problem.

b. Write this problem using negative exponents.

c. What is the length of the exiting fishing rod?

Demonstrating Skills

Compute each sum, difference, product, or quotient.

20. $3 \times {}^-3$

21. ${}^-2.3 - 7.9$

22. $5.2 \div {}^-2.6$

23. $3.4 + {}^-5$

24. ${}^-3.2 - {}^-0.9$

25. ${}^-3 \div {}^-12$

26. ${}^-7 \times {}^-8.3$

27. $({}^-2)^4 \times {}^-5$

28. $2 \times {}^-3^2$

29. ${}^-0.2 + 9.1$

Without computing each sum, difference, product, or quotient, tell whether it will be less than 0, equal to 0, or greater than 0.

30. $9 - {}^-3$

31. ${}^-64 + 43$

32. $4 \times {}^-6$

33. ${}^-4 + {}^-3$

34. ${}^-6.2 - {}^-8.3$

35. ${}^-98 \div {}^-3$

36. Plot and label each point on a four-quadrant graph.

$T\,({}^-4, 3)$ $\qquad$ $U\,(2, 5)$ $\qquad$ $V\,({}^-2, {}^-6)$ $\qquad$ $W\,(6, {}^-5)$

$X\,({}^-6, 0)$ $\qquad$ $Y\,(0, {}^-2)$ $\qquad$ $Z\,(0, 0)$

Use the Pythagorean Theorem or the distance formula to find the length of each segment.

37. $(4, 3)$ to $({}^-1, 1)$

38. $(5, {}^-2)$ to $({}^-3, 0)$

Find each missing value.

39. $8.3 \times 10^? = 0.0083$

40. $3.7 \times ? = 0.00037$

Simplify each expression as much as possible.

41. $256 \cdot 2^{-5}$

42. $3y^2 \cdot y^{-3}$

43. $4^{-1} \cdot 4^{-3}$

44. ${}^-1^{-3}$

45. $({}^-5)^{-2}$

46. ${}^-4^3$

CHAPTER 5

Looking at Linear Relationships

Real-Life Math

At Any Rate Linear relationships always involve a constant rate. One of the most common types of rates is *speed*. The speed of an object—like a train—is a relationship between time and distance.

Imagine yourself behind the controls of the British Eurostar, the fastest train in Europe. What makes this train unique is that it runs through the Channel Tunnel, or Chunnel, an expansive tunnel drilled under the English Channel connecting Britain to France. The Eurostar, which first traveled through the Chunnel on June 30, 1993, can reach speeds of 186 mph on land and 80 mph in the Chunnel. That's more than twice the rate of an average train! People can now travel from Paris to London in just 3 hours—a journey that used to take days.

Think About It How long would it take to go 160 miles on the Eurostar through the Chunnel?

Family Letter

Dear Student and Family Members,

Chapter 5 introduces linear relationships—where a change in one variable results in a fixed change in another variable. In class we will describe situations, make tables of data about the situations, graph the data, and write linear equations that describe the relationships. Here is an example of one kind of problem we will be working with. See if you can solve any parts of the problem before we start the chapter.

Three telephone companies have long-distance rates:

Company	Rate
Easy Access Company	$1.00 for the first minute; 25 cents per minute thereafter
Call Home	20 cents per minute
Metro Communication	$3.00 for the first minute; 15 cents per minute thereafter

Here are some questions we will consider:

- For each company, calculate the amount due for calls lasting 5 minutes, lasting 15 minutes, and for two other lengths of time. Show your results in a table.

- Is there any company that is always the best buy? If so, which one? If not, tell when you would use each company to get the best buy.

We will also learn to predict which equations have graphs that are straight lines by looking at tables or algebraic rules. Then, from a graph or table, we will be able to determine the slope and *y*-intercept.

Vocabulary
Along the way, we'll be learning about these new vocabulary terms:

coefficient	**rate**	**variable**
constant term	**slope**	**velocity**
linear relationship	**speed**	***y*-intercept**
proportional		

What can you do at home?
During the next few weeks, your student may show interest in linear relationships or in different ways linear equations appear in the world outside of school. Together, you might enjoy using linear equations to calculate payments for jobs using different hourly rates.

5.1

Understanding and Describing Rates

V O C A B U L A R Y
rate

In the statement "Carlos types 30 words per minute," the rate *30 words per minute* describes the relationship between the number of words Carlos types and time in minutes. The speed of light, 186,000 miles per second, is a rate that describes the relationship between the distance light travels and time in seconds. In general, a **rate** describes how two unlike quantities are related or how they can be compared.

Here are some other statements involving rates:

Franklin's resting heart rate is 65 beats per minute.

On June 24, 2003, the exchange rate from U.S. to British currency was 0.602031 pound per dollar.

When she baby-sits, Yoshi earns $3.50 per hour.

At the Better Batter donut shop, donuts cost $3.79 per dozen.

Just the facts

One type of spider spins an entire web in just 20 minutes. It weaves at the rate of 1,000 operations per minute.

Think & Discuss

Each rate above involves the Latin word *per,* which means "for each." Try using the phrase "for each" to explain the meaning of each rate. For example, the exchange rate is the number of pounds you receive for each dollar you exchange.

Work with a partner to write three more statements involving rates. Explain what your rates mean using the phrase "for each."

You can use rates to write algebraic rules relating variables. For example, the equation $d = 186,000t$ uses the speed of light to describe the distance d in miles that light travels during a particular number of seconds t.

In this lesson, you will explore lots of situations involving rates, and you will look at tables, graphs, and algebraic rules for these situations.

Investigation ◤1◢ Understanding Rates

You have probably used rates often, right in your own kitchen. Working with cooking measures will help you understand different ways of thinking about and describing rates.

MATERIALS
graph paper

Problem Set A

Here is a rate expressed in words:

A teaspoon contains 5 milliliters.

This rate could also be stated this way:

There are 5 milliliters per teaspoon.

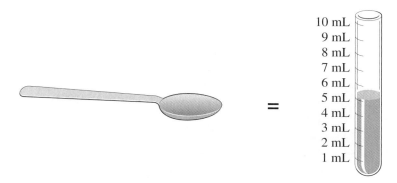

1. Suppose you double the number of teaspoons of some ingredient in a recipe. Does that double the number of milliliters? If you triple the number of teaspoons, do you triple the number of milliliters?

2. Using *m* for the number of milliliters, write a rule that tells how many milliliters are in *t* teaspoons.

3. A table of values can help you check that you have written a rule correctly. Copy this table. Without using your rule, complete the table to show the number of milliliters in *t* teaspoons. Use the table to check that your rule for *t* and *m* is correct.

Converting Teaspoons to Milliliters

Teaspoons, *t*	0	1	2	3	4	5	10
Milliliters, *m*							

4. On axes like those below, plot the data from your table.

Converting Teaspoons to Milliliters

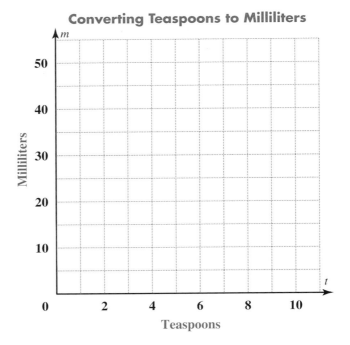

5. Does it make sense to connect the points on your graph with a line? Explain. If it does make sense, do it.

6. Does it make sense to extend your graph beyond the points? Explain. If it does make sense, do it.

Think & Discuss

Darnell and Maya wrote rules relating pints and quarts.

Whose rule is correct? Explain your answer.

The relationship between teaspoons and milliliters is called a **linear relationship** because the points on its graph lie on a straight line. In a linear relationship, as one variable changes by 1 unit, the other variable changes by a set amount. The amount of change per unit is the *rate*. In the rule $m = 5t$, the 5 shows that m changes 5 units per 1-unit change in t.

Because you can have 1.5 teaspoons, 2.25 teaspoons, and so on, it is sensible to connect the points you plotted in Problem 4 using a straight line. It is also sensible to extend the line beyond the points from the table since you can have 11, 16, 27, or even more teaspoons of an ingredient.

There are situations for which it is not sensible to connect the points on a graph. For example, this graph shows the relationship between the number of skateboards and the total number of wheels.

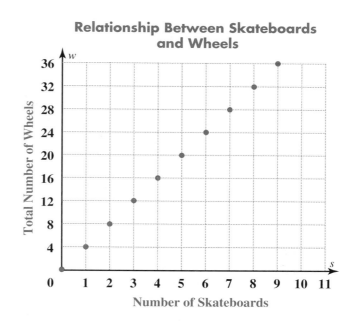

Relationship Between Skateboards and Wheels

Connecting the points doesn't make as much sense in this case, because the points in between would represent partial skateboards. For example, you wouldn't connect the points for 5 skateboards and 6 skateboards, because you couldn't have 5.5 skateboards or 5.976 skateboards.

Even when it doesn't make sense to connect the points on a graph, sometimes you may want to do so anyway to help you see a relationship. In this case, it's best to use a dashed line. You will learn more about this later.

Problem Set B

Some recipes give quantities in *weight.* However, in the United States, most recipes specify quantities by *volume,* using such measures as teaspoons, tablespoons, and cups. If you know how much a cup of a particular ingredient weighs, you can write a rate statement to calculate the number of cups for a given weight. For example, it is a fact that 1 cup of sugar contains about a half pound of sugar.

1. Rewrite the fact above using the word *per.*

2. Write an algebraic rule relating the number of cups of sugar c and weight in pounds p. Check your rule by completing the table.

Relationship Between Cups and Pounds

Cups, c	0	1	2	3	4	5	10
Pounds, p							

3. Use the data in your table to help you draw a graph to show the relationship between the number of cups of sugar and the number of pounds. Use a set of axes like the one shown.

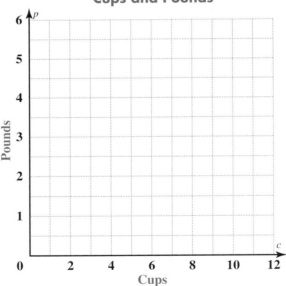

Relationship Between
Cups and Pounds

4. Does it make sense to connect the points or to extend the graph beyond the points? If so, do these things. Explain your reasoning.

5. From your graph, find the value of p when c is 7. Is that value what you would expect from your rule?

6. From the graph, find how many cups of sugar you would need if a recipe calls for 1.4 pounds. Is your answer what you would expect from your rule?

7. Do you find it easier to work from the graph or the rule? Explain.

Share & Summarize

1. You may have noticed that on many visits to the doctor, someone measures your pulse, or heart rate, to determine how fast your heart is beating. Measure your pulse yourself by putting your fingers to the side of your neck and counting the number of beats you feel. Record the number of beats in 1 minute. Explain what makes the number of beats per minute a *rate*.

2. Use your heart rate to write an equation relating the number of beats *b* to any number of minutes timed *t*. Explain what your equation means.

Investigation 2 Describing Rates

Rate relationships can be described in many ways.

Think & Discuss

Are these three students thinking of the same or different relationships? Why do you think so?

Problem Set C

Mario works in a market. His rate of pay is $10 per hour.

1. If Mario works twice as many hours one week than he does another week, will he earn twice as much? If he works three times as many hours, will he earn three times as much?

2. Copy and complete the table to show what pay Mario should receive for different numbers of hours worked.

Mario's Rate of Pay

Hours Worked, h	0	1	2	3	4	5	10	15
Pay (dollars), p								

3. Write a rule in words that relates Mario's pay in dollars to the number of hours worked. Begin your rule, "The number of dollars earned is equal to. . . ." Then rewrite the rule using the symbols p for pay and h for hours.

4. What part of your rule shows Mario's rate of pay?

5. Mario usually works a 35-hour week. How much does he earn in a typical week?

6. One week Mario earned $300. How many hours did he work that week?

Alec works in a local take-out restaurant. His rate of pay is $7 per hour.

7. If Alec works twice as many hours one week than he does another week, will he earn twice as much? If he works three times as many hours, will he earn three times as much?

8. Complete the table to show what Alec would be paid for different numbers of hours worked.

Alec's Rate of Pay

Hours Worked, h	0	1	2	3	4	5	10	15
Pay (dollars), p								

9. Write a rule for Alec's pay in words and in symbols.

10. What part of your rule shows Alec's rate of pay?

11. One week Alec worked 30 hours. How much did he earn that week?

12. Compare your symbolic rules for Mario and Alec. How are they the same? How are they different?

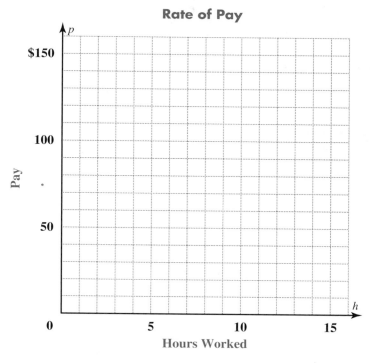

MATERIALS
graph paper

Problem Set D

In addition to symbols, tables, and words, graphs are useful for comparing rates of pay.

1. Draw a graph of the data for Mario's pay. Use a set of axes like the one below.

Rate of Pay

2. Does it make sense to connect the points on your graph? Does it make sense to extend the graph beyond the points? If so, do these things. Explain your reasoning.

3. On the same grid, draw a graph to show how much Alec earns. Label your graphs so you know which is for Mario and which is for Alec. Write each symbolic rule from Problem Set C next to the appropriate graph.

4. Compare the graphs for Mario and Alec. In what ways are they the same? In what ways are they different? Which is steeper?

5. Joelle earns $15 per hour. If you added a line to your graph to show Joelle's pay, how would it compare to Mario's and Alec's graphs?

Problem Set E

Tamsin lives in the country and works for an automobile association. Every second weekend, she is "on call." This means she must be available all weekend in case a car breaks down in her area. She is paid a fixed amount of $40 for the weekend, even if she doesn't have to work. If she is called, she earns an additional $10 per hour worked.

1. Complete the table to show what Tamsin would be paid for different numbers of hours worked during a single weekend.

Tamsin's Pay

Hours Worked, h	0	1	2	3	4	5	10	15
Pay (dollars), p	40	50						

2. If Tamsin works twice as many hours one weekend as she does another weekend, will she earn twice as much? Explain.

3. Write the rule for Tamsin's pay for a single weekend in words and in symbols.

4. What part of your rule shows the amount Tamsin earns for being on call? What part shows her hourly rate?

5. Use the data in your table to draw a graph showing how much Tamsin earns for various numbers of hours worked. Use the same grid you used for Problem Set D or a similar one.

Share & Summarize

1. Look again at the graphs that show how much Alec, Mario, and Tamsin earn. How do the graphs show differences in pay? How is a higher rate shown in a graph?

2. How do the symbolic rules show the differences in the rates at which the three people are paid? How is a higher rate shown in a symbolic rule?

Investigation ▶3▶ Proportional Relationships

Alec and Mario, from Investigation 2, are paid at a simple hourly rate. If they work double the number of hours, they receive double the pay. If they work triple the hours, they receive triple the pay, and so on.

VOCABULARY
proportional

The word **proportional** is sometimes used to describe this kind of relationship between two variables. We can say that for Alec and Mario, the number of dollars earned is proportional to the number of hours worked. The graphs for Alec's pay and Mario's pay are shown below. They are each a straight line that begins at the point (0, 0).

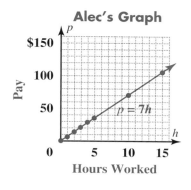

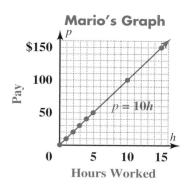

Tamsin is not paid at a simple hourly rate. She receives a fixed amount plus an hourly rate. For Tamsin, doubling the number of hours worked does not double the total pay received. The number of dollars earned is *not* proportional to the number of hours worked. The graph for Tamsin's pay is also a straight line, but it does not begin at the point (0, 0).

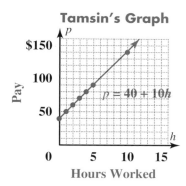

Although the relationships for Alec and Mario are proportional and the relationship for Tamsin is not, all three relationships are *linear relationships* because their graphs are straight lines.

Think & Discuss

Many companies issue cards for charging telephone calls. For which cards below are the charges proportional to the number of minutes? For which are they not proportional? Explain.

ACCESS Card No monthly fee!! Only 25¢ per minute

Easy Card Only 10¢ per minute plus $4 per month

Cheap Card 15¢ per minute — NO monthly fee —

FANTASTIC! Card ONLY 10¢ per minute month $5 per month $5 per month $5 per

DOLLAR CARD $3 per month 10¢ per minute

Budget Card NO monthly fee 20¢ per minute

Problem Set F

In Problems 1–4, the rule describes two variables that are proportional to each other. Rewrite each rule in symbols.

1. The circumference of a circle c is 3.14 times the diameter d.

2. The perimeter of a square p is 4 multiplied by the length of a side L.

3. At a certain time of day, the length of a shadow s cast by an object is twice the length of the object L.

4. Each length h in a copy is $\frac{1}{100}$ the length k in the original.

5. Compare the four rules you developed in Problems 1–4. In what ways are they similar? In what ways are they different?

MATERIALS

graph paper

Problem Set G

Ms. Cruz gave her class a test with 40 questions.

1. Latisha got 100% correct; Kate got only 50% correct. How many of the 40 questions did each answer correctly?

2. Lupe got fewer than 20 questions correct. If she could retake the test and double the number right, would that double her percentage?

3. Does halving the number of correct answers out of 40 halve the percentage?

4. Is the number of correct answers out of 40 proportional to the percentage?

5. One evening Ms. Cruz had to convert her class's test results into percentages. She had left her calculator at school and decided to use a graph to help her.

a. She first calculated a few values and put them in a table. Copy and complete her table.

Class Test Results

Number Correct (out of 40), n	0	10		30	
Percentage Correct, p	0		50		100

b. Ms. Cruz then drew a graph on graph paper by plotting the points from her table and drawing a line through them. She made the horizontal axis the number of correct problems out of 40, with one grid unit on the axis equal to 2 questions correct. She put percentage points on the vertical axis, with one grid unit on the axis equal to 5 percentage points. She could then quickly read percentages from the graph for the various scores out of 40.

Draw Ms. Cruz's graph. Does it make sense to extend the line beyond the point (40, 100)? Explain your thinking.

6. Use your graph to convert each score out of 40 into a percentage.

a. 38   **b.** 36

c. 33   **d.** 29

e. 25   **f.** 23

7. Lucas, Opa, and Raheem scored 65%, 75%, and 85%, respectively, on the test. How many problems out of 40 did each get correct?

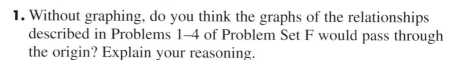

Share & Summarize

1. Without graphing, do you think the graphs of the relationships described in Problems 1–4 of Problem Set F would pass through the origin? Explain your reasoning.

2. A graph of a linear relationship goes through the point (0, 4). Is this a proportional relationship? Explain your reasoning.

3. Is $y = 5x + 7$ a proportional relationship? Explain your thinking.

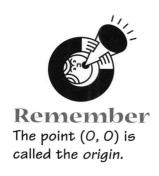

Remember
The point (0, 0) is called the origin.

Lab Investigation ▶ Rolling Along

MATERIALS

- 2 cylindrical pens or pencils (without edges) with different diameters
- lined paper
- hardcover book
- graph paper

Just the facts

Historians believe that this method of rollers was used to move the very heavy stones used to build the Egyptian pyramids.

In this investigation, you will conduct an experiment and collect some data. You will use your data to try to understand the relationship between two variables.

Work with Your Class

Watch how a heavy box rolls on three drink cans. The rollers will keep coming out from behind the box!

1. Do the rollers go backward?

2. Does the box go faster than the rollers?

3. Why do the rollers keep coming out the back?

Try It Out

With your group, conduct an experiment that involves rolling one object on top of another. Use a cylindrical pen or pencil (one that is perfectly round), a book, and a sheet of lined paper (wide lines work best). Put the pen on the paper so that it lies along one of the lines. Put the book on top of the pen so the edge of the book also aligns with one of the lines.

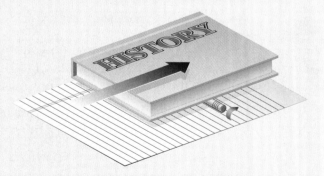

4. Roll the book on the pen until the pen has moved 2 line spaces. How far has the book moved?

5. Continue to roll the book so you can complete the table. When you collect your data, count and measure accurately! Otherwise, you might not see the patterns that show the relationship between the variables.

Number of Line Spaces Moved

Distance Moved by Pen, *p*	0	2	4	6	8	10
Distance Moved by Book, *b*	0					

Analyze Your Data

6. With your group, describe in words how the distance the book moves is related to the distance the pen moves. Then write the relationship using the symbols *p* and *b*.

7. Draw a graph with the distance moved by the pen on the horizontal axis and the distance moved by the book on the vertical axis. Plot the values you found in Problem 5. If you think it makes sense to do so, join the points with a line, and extend the line.

Apply Your Results

8. Using your rule, predict how far the book will move when the pen moves 16 line spaces. Check by doing it.

9. How many line spaces does the book move if the number of line spaces moved by the pen is 0.5? 1.5? 7.5? 1,350?

10. If the book moves 9 line spaces, how far does the pen move? Test it with your equipment.

11. Describe in words how the distance the pen moves is related to the distance the book moves. Write this rule in symbols.

Try It Again

Replace the pen with a different-sized pen or some other cylinder, and repeat the experiment.

12. Complete a new table.

Number of Line Spaces Moved

Distance Moved by Pen, *p*	0	2	4	6	8	10
Distance Moved by Book, *b*	0					

13. Are your results the same?

14. Is this a proportional relationship? How can you tell?

What Did You Learn?

15. Write a short report describing your experiments and the results.

On Your Own Exercises

Practice & Apply

1. **Measurement** A kilogram is equivalent to 1,000 grams.

   a. Describe this relationship in words using the word *per.*

   b. Write a rule in symbols that relates the two measures. Use k to represent the number of kilograms and g the number of grams.

   c. Copy and complete the table without using the rule. Then use your table to check that your rule is correct.

   Converting Kilograms to Grams

Kilograms, k	1	2	3	4	5
Grams, g					

   d. Use the data in the table to help you draw a graph to show the relationship between the measures. If it makes sense to connect the points, do so. If it makes sense to extend the line beyond the points, do so.

   e. Use your table, formula, or graph to find the number of grams in 1.5 kg, 3.2 kg, 0.7 kg, and 0.034 kg.

2. **Measurement** A pound is equivalent to 0.454 kilogram.

   a. Describe the relationship in words using the word *per.*

   b. Write a rule in symbols that relates the two measures. Use p for the number of pounds and k for the number of kilograms.

   c. Complete the table without using your rule. Then use your table to check that your rule is correct.

   Converting Pounds to Kilograms

Pounds, p	1	2	3	4	5
Kilograms, k					

   d. Use the data in the table to help you draw a graph to show the relationship between the measures.

   e. Use your table, formula, or graph to find the number of kilograms in 7 lb, 40 lb, 100 lb, and 0.5 lb.

Remember

When you plot points from a table of data, connect the points and extend the line beyond the points, if it makes sense to do so.

impactmath.com/self_check_quiz

3. These three phone cards have different charge plans for their customers.

DOLLAR CARD
$3 per month
10¢ per minute

Easy Card
Only 10¢ per minute
(plus $4 per month)

FANTASTIC!
Card
ONLY 10¢ per minute
month $5 per month $5 per month $5 per

a. Write the rule for each card's charge plan for 1 month, using *d* for charge in dollars and *m* for minutes of calls. Underline the part of each rule that shows the rate of charge.

b. Copy the table. For each card, fill in the table to show the monthly bill for different numbers of minutes of calls.

Phone Card Charges

Minutes, *m*	0	10	20	30	40	50	100	150	200
Dollar Bill (in dollars)									
Easy Bill (in dollars)									
Fantastic Bill (in dollars)									

c. On one set of axes, draw a graph for each card's charge plan. Put minutes on the horizontal axis and dollars on the vertical axis. Label the graphs to identify which card goes with which graph.

d. Does it make sense to connect the points? Does it make sense to extend the graphs beyond the points? If so, do it.

e. How are the graphs similar? How are they different?

f. How much would the monthly bill for 85 minutes be with the Dollar card?

g. For $12, how many minutes of calls could you make in one month with the Dollar card? With the Easy card? With the Fantastic card? If your answers are more than 59 minutes, convert them to hours and minutes.

4. In Problem Sets C and D, you solved problems about Alec's and Mario's rates of pay. Suppose that Alec receives a raise to $8 per hour.

a. Write a rule in symbols relating the number of dollars p that Alec earns to the number of hours h that he works.

b. What would the graph that shows his new pay rate look like? How would it compare to the two graphs you have already drawn (for Mario's pay and Alec's pay at the old rate)?

c. Make a table to show what Alec would now earn for 0, 1, 2, 3, 4, 5, and 10 hours of work.

d. On your grid from Problem Set D, draw a graph to show how much Alec is paid at the new rate. Does the graph look as you predicted in Part b? If not, try to figure out where your thinking went wrong.

5. Before compact discs became popular, people listened to music and other sound recordings on record players. A record player that runs at 45 rpm turns a record 45 times per minute. The unit rpm is an abbreviation for *revolutions per minute.*

a. Complete the table to show the number of times a record turns in a given number of minutes.

Playing Records

Minutes, t	1	2	3	4	5	6
Revolutions, r	45					

b. Write in words a rule that gives the number of times a record turns in a given number of minutes. Then write the same rule in symbols. Underline the part of your rule that shows the constant rate of turning.

c. Using your rule, predict the number of revolutions in 12 seconds (0.2 minute).

d. Draw a graph to show how the number of revolutions is related to time. Put time on the horizontal axis.

e. Use your graph to estimate how long it would take for a record to revolve 100 times and 300 times.

6. A painter needs about 1 fluid ounce of paint for each square foot of wall she paints.

   a. Write a rule in symbols for this situation, using p for fluid ounces of paint and f for square feet. Underline the part of your rule that shows the constant rate of paint use.

   b. Complete the table.

Painting Walls

Paint (fluid oz), p	0	10	20	30	40	50	80	100	140	170	220
Area (sq ft), f											

   c. How many fluid ounces of paint are needed to paint a wall with area 120 sq ft?

   d. How many gallons of paint are needed to cover 450 sq ft? There are 128 fluid ounces in 1 gallon.

   e. Draw a graph to show how the amount of paint depends on the area to be painted. Put area on the horizontal axis.

   f. Use your graph to estimate the area covered with 1 gallon of paint, 2 gallons of paint, and 3 gallons of paint.

   g. If a painter uses twice as much paint, can she cover twice as much area? If she uses three times as much paint, can she cover three times as much area? Is the relationship between area and amount of paint proportional?

Connect & Extend

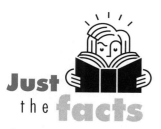

Just the facts

The metric system was first proposed in 1670 and is now used in most countries around the world.

7. Measurement 24 ounces is equivalent to 680 grams.

   a. Describe the relationship in words using the word *per*.

   b. Complete the table.

Converting Ounces to Grams

Ounces, z	0	1	2	3	4	5	6
Grams, g							

   c. Write a rule in symbols that relates the two measures. Use your table to check that your rule is correct.

   d. Use the data in the table to help you draw a graph to show the relationship between the measures. If you think it makes sense to do so, join the points with a line, and extend the line.

Measurement A kilogram (kg) is equivalent to 1,000 grams (g). A pound (lb) is equivalent to 0.454 kilogram (kg).

8. Find the number of grams in 1 lb, 2 lb, 0.6 lb, and 0.1 lb.

9. Find the number of pounds in 3 kg, 17 kg, 0.5 kg, 900 g, 300 g, and 56 g. Round to the nearest hundredth.

10. Challenge Darnell bought a gallon of spring water. The label indicated that a serving size is 1 cup, or 236 mL. Use the conversions at the left to answer the questions. Round to the nearest hundredth.

> 1 fl oz = 29.573 mL
> 1 oz = 28.350 g
> 1 gal = 3.785 L

   a. What is the volume of the serving size (1 cup) in fluid ounces?

   b. The mass of 1 mL of water is 1 g. What is the mass of 1 cup of water in grams?

   c. What is the weight of 1 cup of water in ounces?

   d. What is the volume of 1 cup of water in liters?

   e. What is the volume of 1 cup of water in gallons?

   f. There are 16 ounces in a pound. What is the weight of 1 cup of water in pounds?

   g. How many cups are in 1 gallon?

11. Challenge An architect is designing a building with two movie theaters. The Green Theater will be rectangular in shape, with each row holding 16 seats. The Blue Theater will be wider at the back: the first row will have 8 seats, the second row will have 16 seats, and all other rows will have 20 seats. The architect will decide how many rows to put in each theater on the basis of the number of people it is meant to hold.

   a. Make a table showing how the number of seats for each theater relates to the number of rows. Assume each theater will have at least 3 rows.

   b. On one set of axes, draw graphs to show how the number of seats in each theater depends on the number of rows.

   c. Does it make sense to connect the points on the graphs? Explain.

   d. For each theater, write a rule in symbols that tells how the number of seats depends on the number of rows.

   e. How many rows should be in each theater to give the theaters an equal number of seats?

   f. Which theater will have more seats if both have 11 rows?

   g. Is the number of seats in each theater proportional to the number of rows? Why do you think so?

Just the facts

The very first movie theaters were built in the late 1890s. By the 1920s and 1930s, grand "picture palaces" were springing up everywhere. And the first drive-in movie opened in 1933.

In your **own** **words**

How can you find the rate of a linear relationship using a graph? Using a symbolic rule? Using a table? Which representation do you prefer for this task? Why?

12. Carmen started to read a book at 10:20 in the morning. She usually reads a page in 2 minutes. Half an hour later, her friend Yolanda started to read. She reads 15 pages per hour.

a. How many pages did Carmen read before Yolanda started to read?

b. Create a table showing how many pages Carmen and Yolanda read by 10:50, 11:00, 11:10, 11:20, 11:30, 11:40, 11:50, and 12:00.

c. Based on your table, draw graphs on the same axes showing how many pages Carmen and Yolanda read between 10:50 A.M. and noon. Does it make sense to connect the points on the graphs? Does it make sense to continue the lines beyond the 12 o'clock mark? Explain your answers.

d. Write a rule in symbols for the number of pages Carmen read since 10:50. Write a rule in symbols for the number of pages Yolanda read since 10:50. Use m for number of minutes passed after 10:50. Underline the parts of the rules that show the constant rates of reading.

e. Is the relationship for Carmen proportional? Is the relationship for Yolanda proportional?

f. How many pages would each girl read in 5 min? In 16 min? In 48 min?

g. At what time will Carmen finish 40 pages of her book? At what time will Yolanda finish 40 pages of her book?

h. At what time will the number of pages Carmen has read be three times the number of pages Yolanda has read?

Mixed Review

Order each group of numbers from least to greatest.

13. $\frac{2}{5}$, 0.2, $\frac{3}{5}$, 0.4, $\frac{5}{3}$ **14.** 0.75, 0.18, 0.25, 0.025, 0.7

Fill in the blanks to make each a true statement.

15. $\frac{1}{8} + \frac{1}{3} +$ _____ = 1   **16.** $\frac{2}{7} + \frac{3}{10} +$ _____ = 1

17. Geometry Find the area and perimeter of the circle.

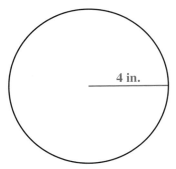

4 in.

18. Find 33% of 140.   **19.** Find 14% of 31.

20. What percent of 250 is 15?   **21.** What percent of 12 is 0.12?

Use the distributive property to rewrite each expression with parentheses.

22. $18 - 4f$ **23.** $13 - 26m$

Find each product. Express your answers in scientific notation.

24. $(5 \times 10^{14}) \times (11 \times 10^{13})$ **25.** $(8 \times 10^{9}) \times (6 \times 10^{6})$

Find the distance between the endpoints of the segment drawn on each number line.

26.

$$-3 \qquad 0 \qquad 3$$

27.

$$-3.5 \qquad 0 \quad 1$$

Rewrite each expression as addition, and calculate the sum.

28. $4 - {}^-3$ **29.** $10 - {}^-4$ **30.** $13 - {}^-10$ **31.** ${}^-7 - {}^-8$

Rewrite each expression in the form c^m.

32. $5^3 \times 5^2$ **33.** $2^7 \times 2^2$ **34.** $7^2 \times 7^7$ **35.** $a^2 \times a^3$

36. Number Sense Could the number 3,028,046,908 be some power of 4? Explain how you know.

5.2 Speed and the Slope Connection

Speed is a very common rate. It tells how the distance something travels depends on the time traveled. Speed can be measured in many ways. Here are a few common units of speed.

- miles per hour (mph)
- kilometers per hour (km/h or kph)
- feet per second (ft/s)
- meters per second (m/s)
- millimeters per second (mm/s)

MATERIALS
- stopwatch
- yardstick

Explore

Guess an average student's normal walking speed and record your guess. Then, one person in the class should walk at a "normal" speed from the back of the classroom to the front while another person accurately times him or her. Record that time and the distance the student walked, in feet.

Now here is a challenge: Several students should try to cross the classroom at a steady pace taking the amounts of time specified by the teacher—some greater and some less than it takes to cross at a "normal" speed.

You or a classmate should time how long it actually takes each student to cross the room. Then everyone in the class can compute the students' true speeds. To find the speeds, divide the distance walked by the time it took.

After you have an idea about how fast or slow the various speeds are, answer these questions:

- If a person travels 1.5 feet per second, is this a fast walk, a slow walk, or a run?

- If a second person moves with a speed of 9 feet per second, is this a fast walk, a slow walk, or a run?

- How much faster is the second person in comparison with the first?

Investigation Walking and Jogging

Some people walk quickly, taking several steps in a few seconds. Others walk more slowly, taking more than a second for each step. Some people naturally take longer strides than others. When you're walking with a friend, one or both of you probably changes something about the way you walk so you can stay side by side.

MATERIALS

graph paper

Problem Set A

Zach's stride is 0.5 meter long. Imagine that he's walking across a room, taking one step each second.

1. At what speed is Zach walking?

2. Copy and complete the table to show Zach's distance from the left wall at each time.

Time (seconds), *t*	0	1	2	3	4	5	6
Distance from Left Wall (meters), *d*							

3. At this constant speed, is the distance traveled proportional to the time? How do you know?

4. Write a rule that shows how to compute *d* if you know *t*.

5. What would Zach's speed be if he lengthened his stride to 1 meter but still took 1 second for each step? Make a table of values, and write a rule for this speed.

6. Imagine that Maya is also crossing the room, but she is taking two steps each second and her steps are 1 meter long. What is her speed? Make a table of values, and write a rule for her speed.

7. For all three tables, plot each set of points in the table and then draw a line through them. Put all three graphs on one grid, with time on the horizontal axis. Label each graph with its speed.

Think & Discuss

Compare the three graphs you drew in Problem 7.

• In what ways are they the same? In what ways are they different?

• What does the steepness in these distance-time graphs show?

MATERIALS

graph paper

Just the facts

The maximum speed a human being has ever run is about 27 miles per hour. The fastest animal on Earth, the cheetah, has been clocked at about 60 miles per hour.

Problem Set B

Many joggers try to jog at a steady pace throughout most of their runs. This is particularly important for long-distance running.

• Terry tries to jog at a steady pace of 4 meters per second.

• Maria tries to jog at a steady pace of 3 meters per second.

• Bronwyn doesn't know how fast she jogs, but she tries to keep a steady pace.

1. Make tables for Terry and Maria to show the distances they travel (*d* meters) in various times (*t* seconds).

2. Write rules that show how distance *d* changes with time *t* for Terry and for Maria.

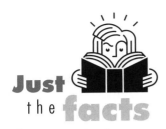
3. A timekeeper measured times and distances traveled for Bronwyn and put the results in a table.

Time (seconds), t	0	5	10	15	20
Distance (meters), d	0	17.5	35	52.5	70

How fast does Bronwyn jog? Write a rule that relates Bronwyn's distance to time.

4. On one grid, draw graphs for Terry, Maria, and Bronwyn. Put time on the horizontal axis. Label each graph with the name of the person and the speed.

5. Explain how you can tell from looking at the graph who jogs most quickly and who jogs most slowly.

All the points on each graph you drew are on a line through the point (0, 0). The steepest line is the one for which distance changes the most in a given amount of time—that is, when the speed is the fastest. The line that is the least steep is the one for which distance changes the least in a given amount of time—that is, when the speed is the slowest.

VOCABULARY
slope

Slope describes the steepness of a line. In this case, the slope tells how much the distance changes per unit of time. More generally, the **slope** of a line tells how much the y variable changes per unit change in the x variable.

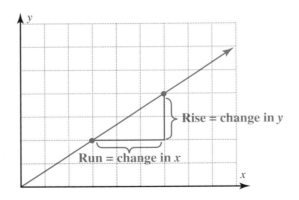

Sometimes slope is described as *rise* divided by *run*. This makes sense because y changes in the vertical direction, or "rises," and x changes in the horizontal direction, or "runs."

This graph shows how Terry's distance changed over time. To find the slope, choose two points, such as (10, 40) and (20, 80). From the left point to the right point, the *y* value changes from 40 to 80. The *rise* between these points is 80 − 40, or 40. The *x* value changes from 10 to 20, so the *run* between these points is 20 − 10, or 10. The slope—the rise divided by the run—is $\frac{40}{10}$, or 4.

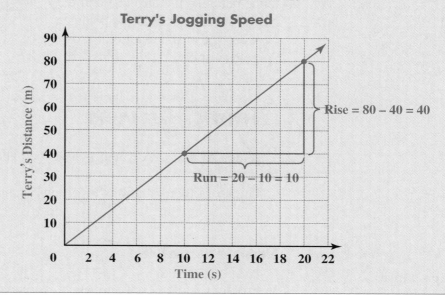

Terry's Jogging Speed

Think & Discuss

Look at your graphs for Maria and Bronwyn. What are the slopes of Maria's and Bronwyn's lines? What does the slope mean in Terry's, Maria's, and Bronwyn's graphs?

Share & Summarize

1. Javier walks at a speed of 5 feet per second. If you graphed the distance he walks over time, with time in seconds on the horizontal axis and distance in feet on the vertical axis, what would the slope of the line be?

2. Dulce walks at a speed of 7 feet per second. Suppose you graphed the distance she walks over time on the same grid as Javier's line. How would the steepness of her line compare to the steepness of Javier's line? Explain.

Investigation 2 Decreasing Distance with Time

An airplane flies from New York to Los Angeles. There are two distances that are changing: the distance between the airplane and the New York airport, and the distance between the airplane and the Los Angeles airport.

Think & Discuss

Which of the two distances described above is decreasing over time?

Think of other situations in which distance decreases over time.

MATERIALS

graph paper

Problem Set C

In Problem Set A, Zach and Maya were walking from the left wall of a room to the right wall. You figured out how far each person was from the left wall at different points in time. Suppose instead you want to know how far the person is from the *right* wall at each point in time.

1. When is the person closest to the right wall: at the beginning of the walk or at the end of the walk?

2. Suppose Maya walks at 1.5 meters per second across a room that is 10 meters wide. Copy and complete this table.

Maya's Walk

Time (seconds), t	0	1	2	3	4
Distance from Right Wall (meters), d	10				

3. Use the data in Problem 2 to draw a graph that shows the relationship between Maya's distance from the right wall and time.

4. What is the slope of the line you drew?

5. Use your graph to estimate when Maya would reach the right wall.

6. Explain how you can find the distance from the right wall if you know the time.

7. Write a symbolic rule that relates d to t.

Think & Discuss

Bianca and Lorenzo solved a problem on a quiz. Bianca wrote this rule: $d = {}^-2t + 20$. Lorenzo wrote this rule: $d = 20 - 2t$. Can they both be right? Explain your thinking.

Create a problem that can be described by one or both of these rules.

Problem Set D

Ruben and Kristen started walking away from a fence at the same time. Ruben walked at a brisk pace, and Kristen walked at a slow pace. They each measured the distance they had walked in 10 seconds. From this, they estimated how far from the fence they would have been at various times if they had continued walking. They drew distance-time graphs from their data.

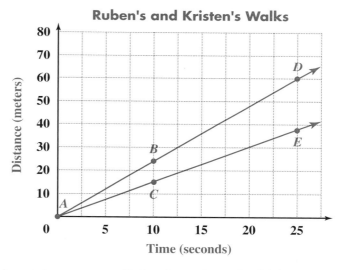

Ruben's and Kristen's Walks

1. Which graph represents Ruben's walk, and which represents Kristen's? Explain how you know.

2. What events in the story above match points *A*, *B*, and *C*?

3. What do points *D* and *E* tell you about the positions of Ruben and Kristen?

4. Use the graphs to estimate each person's walking speed in meters per second. Give your answers to the nearest tenth.

5. Which line has the greater slope, Ruben's or Kristen's? Explain why.

6. What are the slopes of the two lines? How are they related to Ruben's and Kristen's speeds?

Share & Summarize

1. How are the graphs in Problem Set D different from the graphs in Problem Set C?

2. How is the rule in Problem Set C different from the rules in Problem Set D? How are they the same?

3. Explain how the differences in the rules relate to the differences in the graphs.

Investigation 3 Describing Graphs

Some rates vary. For example, if you count your pulse for one minute, and then count it for another minute, you will probably get different results. It is normal for pulse rates to fluctuate, or change.

You would expect other rates to be fixed, or stay the same, at least for a while. For example, if your employer said your pay rate was $7 per hour, you would expect to earn that for each hour you work.

In this investigation, you will inspect the graphs below to find essential information about the movement of a group of cars along a particular highway: their directions, their speeds, and their relative locations.

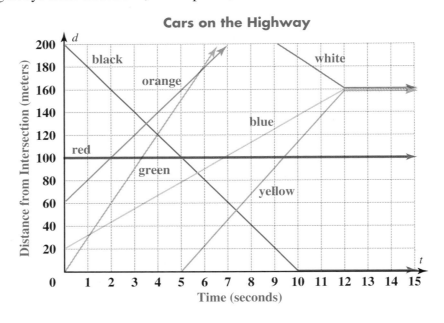

Cars on the Highway

Problem Set E

Suppose seven cars are all near an intersection. The graphs on page 328 show the distances between the cars and the intersection as time passes.

In your group, study the graphs carefully. Then discuss the following questions. Record your group's decisions, and be prepared to talk about them in class.

1. In what direction is each car moving in relation to the intersection?

2. Compare the cars' speeds.

3. Do any of the cars stop during their trips? If so, which cars?

4. Prepare a group report for one of the cars. Imagine you are in that car, and give the highlights of your trip for these 15 seconds. Include such observations as where and when you started the trip and what you saw going on around you—in front of the car, to the sides, and through the rearview mirror.

People often use one of two words when they describe how fast something is moving: *speed* or *velocity*. In fact, these two words mean mathematically different things.

VOCABULARY
speed
velocity

Speed is always positive. It shows how fast an object is moving, but it does not reveal anything about the object's direction.

Velocity can be either positive or negative (as slope can). The sign of the velocity shows whether an object is moving from or toward a designated point. The absolute value of the velocity is the same as the speed. While the black car is moving, for example, it has a positive speed of 20 meters per second. However, its distance to the intersection is decreasing, so its velocity is ⁻20 meters per second.

Share & Summarize

1. How can you determine from the graph whether a car is moving toward or away from the intersection?

2. How can you determine the speed of a car from the graph?

3. How can you determine from the graph whether a car is moving?

Investigation ▶ 4 ▶ Changing the Starting Point

You will now explore a situation in which runners start a race at different points. You'll see how rules and graphs show these differences.

MATERIALS

graph paper

Remember

When you compare two things, look at both similarities and differences.

Problem Set F

Alita ran a race with her younger sister, Olivia. Alita let Olivia start 4 meters ahead of the starting line. Alita ran at a steady rate of 3 meters per second while Olivia ran at a steady rate of 2 meters per second.

1. Make a table to show how many meters each sister was from the starting line at various times.

2. On the same grid, draw a graph for each sister showing the relationship between distance from the starting line and time. Compare the graphs.

3. Is Alita's distance from the starting line proportional to time? Is Olivia's distance proportional to time?

4. For each sister, write a rule in symbols to relate distance d and time t. How are the numbers in the rules reflected in the graphs?

Think & Discuss

In Chapter 4, you learned about elevations below sea level. For example, suppose Candace ended a hike at an elevation of ⁻150 feet. The number ⁻150 shows two things. First, the 150 tells Candace's distance from sea level. What does the negative sign show?

Describe some other situations in which you might use a negative sign along with a distance. Explain the meaning of the negative sign in each situation.

Problem Set G

Five brothers ran a race. The twins began at the starting line. Their older brother began behind the starting line, and their two younger brothers began at different distances ahead of the starting line. Each boy ran at a fairly uniform speed. Here are rules for the relationship between distance (*d* meters) from the starting line and time (*t* seconds) for each boy.

Adam: $d = 6t$

Brett: $d = 4t + 7$

Caleb: $d = 5t + 4$

David: $d = 5t$

Eric: $d = 7t - 5$

1. Which brothers are the twins? How do you know? Which brother is the oldest? How do you know?

2. For each brother, describe how far from the starting line he began the race and how fast he ran.

3. Which graph below represents which brother?

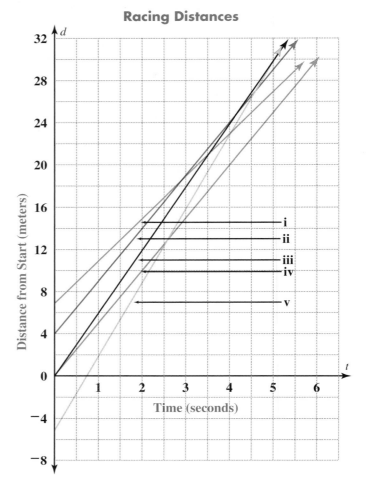

Racing Distances

4. What events match the intersection points of the graphs?

5. Use the graphs to help you find the order of the brothers at the given times.

   a. 2 seconds after the race began

   b. 3 seconds after the race began

6. Which two brothers stay the same distance apart throughout the race? How do you know?

7. If the finish line was 30 meters from the starting line, who won?

8. Which brothers' rules are proportional, and which are not?

The rules in Problem Set G have the form $d = rt + p$, with numbers in place of r and p. The value of r is the velocity—it gives the rate at which distance changes as time passes. The value of p tells the starting point. For example, Brett's rule is $d = 4t + 7$. His velocity is 4 meters per second, and he began 7 meters ahead of the starting line.

If you graph $d = rt + p$, with t on the horizontal axis, or x-axis, and d on the vertical axis, or y-axis, the coordinates of the starting point on the graph are $(0, p)$. When $t = 0$, $d = p$. Therefore, p is the value of d at which the graph intersects the y-axis. That is why p is also called the **y-intercept.** For Brett's rule, the y-intercept is 7.

VOCABULARY
y-intercept

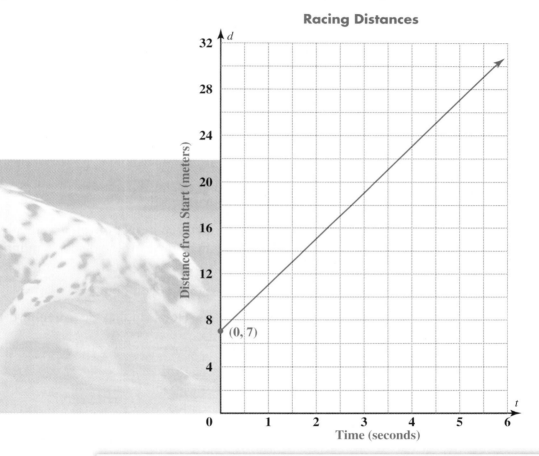

Racing Distances

Share & Summarize

1. In Problem Set G, who began behind the starting line? Which part of the graph shows this? Which part of the rule shows this?

2. The brothers' dog, King, ran the race with them. His rule was $d = {}^-15 - 2t$. What does the rule reveal about King's speed and starting position?

3. In what direction is King running?

On Your Own Exercises

Practice & Apply

1. **Sports** Suppose you are riding a bicycle at a steady rate of 16 feet per second.

   a. Imagine that you rode at this pace for various lengths of time. Copy and complete the table.

Time (s)	0	10	20	30	35	40	50	55	60	70
Distance (ft)	0									

   b. Write a rule relating the number of feet you travel *d* to the time in seconds *t*. Use your rule to find how long it would take you to travel 900 feet.

   c. From your table, find how many feet you ride in 1 minute. About how many miles is this?

   d. Use the distance in feet that you travel in 1 minute to find how far you could ride in 1 hour if you kept up your pace. What is your traveling speed in miles per hour?

   e. Write a second rule that relates the number of miles you travel *d* to the time in hours *t*.

   f. Draw a graph for the distance in feet traveled in relation to time in seconds. Draw another graph for the distance in miles traveled in relation to time in hours.

   g. What is the slope of each line you graphed?

   h. Compare the two slopes. How can they be different if the speeds are the same?

Remember

1 mile = 5,280 feet

impactmath.com/self_check_quiz

2. A 90-minute cassette tape plays for 45 minutes per side. The tape is about 26 meters long and moves at a steady rate.

a. Complete the table to show how many meters of tape play in given numbers of minutes.

Time (min), t	0	1	10	20	30	40	45
Length Played (m), L							26

b. Write a rule relating length of tape played to time.

c. What is the speed of the tape through the recorder in meters per minute? If you graphed the length of tape played in relation to time, what would the slope of the line be?

d. Find the length of a 120-minute cassette tape, which has 60 minutes per side. Assume that tapes of different lengths run at the same speed.

3. Sports A marathon is a race of 26.2 miles. Nadia and Mark are running a marathon. Nadia's speed is 5.2 miles per hour, and Mark's is 3.8 miles per hour. Assume their speeds are steady throughout the race.

a. Complete the table for Nadia and Mark. Distances are in miles.

Time (hours), t	0	1	2	3	4
Nadia's Distance from Start, N	0				
Mark's Distance from Start, M					
Distance between Nadia and Mark, d					

b. On a single grid, draw graphs that show these three relationships:

• time and Nadia's distance from start

• time and Mark's distance from start

• time and the distance between Nadia and Mark

c. What is the slope of each line? Which line is the steepest? Which line is the least steep? Why?

d. Write a rule for each of these three relationships.

e. Think about the distance between Nadia and Mark. How is the rate this distance changes related to Nadia's speed and Mark's speed?

Just the facts

The word *marathon* comes from the name of a famous plain in Greece. In 490 B.C., a runner was sent from this plain to Athens, about 25 miles away, to report the Athenians' victory in a battle against the Persians.

4. Sports Nadia, Helena, Bryson, and Mark start their marathon run, which is 26.2 miles long. Nadia's speed is 5.2 miles per hour, Helena's is 4.1 miles per hour, Bryson's is 4.85 miles per hour, and Mark's is 3.8 miles per hour. Assume their speeds are steady throughout the run.

a. Complete the table. Distances are in miles.

Marathon Results

Time (hours), t	0	1	2	3	4
Nadia's Distance from Finish, N	26.2				
Helena's Distance from Finish, H					
Bryson's Distance from Finish, B					
Mark's Distance from Finish, M					

b. On a single grid, draw graphs that show the relationships between time and distance from the finish for all four runners.

c. Compare the four graphs. In what ways are they similar, and in what ways are they different? What is the slope of each graph?

d. How much time will it take for each of the four runners to finish the marathon? You may want to extend your graph or table.

e. For each runner, write a rule in symbols that relates that runner's distance from finish (N, H, B, or M) to t. Compare your rules to those in Problem Set C. How are they different? How are they similar?

5. Geography Soon after Trina's trans-Australia flight left Adelaide for Darwin, the pilot announced they were 200 kilometers from Adelaide and cruising at a speed of 840 kilometers per hour. The plane was flying north from Adelaide, toward the South Australia border. To pass the time, Trina decided to calculate her distance from Adelaide at various times after the pilot's message.

a. Complete the table to show the distance (in kilometers) from Adelaide at various times (in minutes) after the pilot spoke.

Trans-Australia Flight

Time after Pilot's Message (min), t	0	1	15	30	60	90	120
Distance from Adelaide (km), d	200						

b. Draw a graph of the data.

c. The South Australia border is 1,100 km from Adelaide. Use your graph to determine about how long after the pilot's message the plane crossed the border.

d. Alice Springs is 1,400 km from Adelaide. How long after the message would Trina expect to pass over it?

e. Explain in words how to find the distance from Adelaide if you know the time after the pilot's message. Write the rule in symbols. Explain how each number in your rule is shown in the graph.

6. One Sunday, Benito, Julie, Sook Leng, Edan, and Tia visited a park at different times. They rented bikes of different colors and rode them along the park's bike route. The bike route goes from the rental shop to a cafe. It also goes in the opposite direction, from the rental shop to a lake.

The graphs show the friends' locations from noon until 2 P.M.

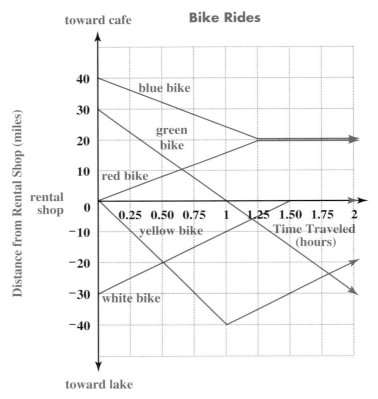

a. Describe the route taken by each bike. For example, the blue bike moves from the direction of the cafe toward the rental shop, and then stops.

b. Determine the speed of each bike. If a bike's speed changes, list all of its speeds.

c. Tia and Julie met about 20 miles from the rental shop and talked for an hour. Julie had come directly from the rental shop, and Tia had already visited the cafe and was returning her bike. What were the colors of their bikes?

d. From noon to 1 P.M. Benito had the highest speed, Sook Leng had the second highest, and Edan had the lowest. What color bikes were Benito, Sook Leng, and Edan riding?

e. Determine the slope of the line for each bike. If the slope changes, list all of them.

f. Are the bikes' speeds and the slopes of their lines always equal to each other? Why or why not?

For each equation, give the coordinates of the graph's intersection with the *y*-axis (in other words, give the *y*-intercept).

7. $y = 3x - 5$ **8.** $y = {}^-7x + 14$ **9.** $y = 1.5x + 2.4$

10. $y = 27 - 9x$ **11.** $y = {}^-35 + 3.1x$ **12.** $y = -\frac{3}{5} + \frac{3}{5}x$

Determine the slope of each line.

13. $y = 7x - 12$ **14.** $y = 31 - 4.6x$ **15.** $y = 8 + 3x$

16. $y = {}^-x - 1$ **17.** $y = {}^-0.21x + 98$ **18.** $y = \frac{1}{4}x - \frac{7}{2}$

19. Physical Science When you see lightning and then hear the clap of thunder, you are hearing what you have just seen. The thunder takes longer to reach you because sound travels much more slowly than light. Sound takes about 5 seconds to travel 1 mile, but light travels that distance almost instantly.

a. Complete the table to show the relationship between *d*, the distance in miles from a thunderstorm, and *t*, the time in seconds between seeing the lightning and hearing the thunder.

Seconds between Lightning and Thunder, *t*	0	5	10	15	20	25
Distance from Storm (mi), *d*						

b. If thunder and lightning happen at the same time, where is the storm?

c. How could you calculate how far away a storm is? Write a rule that relates distance *d* from a storm to the time *t* between seeing it and hearing it.

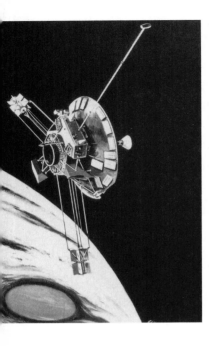

Graph a line with the indicated slope.

20. 1

21. 12

22. $^-3$

23. 0.37

24. $^-270$

25. **Astronomy** The spacecraft *Pioneer 10* was launched on March 3, 1972. Designed to last at least 21 months, it outlived and outperformed the fondest dreams of its creators. On April 25, 1983, *Pioneer 10* sent radio signals to Earth from Pluto, 4.58×10^9 km away.

  a. Radio signals travel at the speed of light (about 3.00×10^5 km/s). How long did it take signals to reach Earth from Pluto?

  b. **Challenge** What was *Pioneer 10*'s average speed, in kilometers per hour (kph), between March 3, 1972, and April 25, 1983?

In Exercises 26–30, do Parts a, b, and c for the given speed.

  a. Express the speed in feet per minute.

  b. Graph the distance moved over time, with time in minutes on the horizontal axis and distance in feet on the vertical axis.

  c. Determine the slope of the graph.

26. 3 meters per minute (1 foot is approximately 0.3 meter.)

27. 15 meters per second

28. 60 kilometers per hour

29. 45 feet per second

30. 75 miles per hour

31. Write the speeds given in Exercises 26–30 from slowest to fastest.

32. Redraw the two graphs for Exercises 28 and 30 on a single grid.

  a. What do you notice about the slopes of the two graphs? (Hint: Look at the angle each line makes with the horizontal axis.)

  b. Explain your result in Part a.

Remember
1 mile = 5,280 feet

33. A lion surveying his surroundings from a tall tree saw a horse half a kilometer to the south. He also spotted a giraffe 1 kilometer to the west. The lion jumped from the tree to chase one of these animals, and both the horse and the giraffe ran away from the lion with the maximum speed.

A lion can run 200 meters in 9 seconds, a horse can run 200 meters in 10 seconds, and a giraffe can run 200 meters in 14 seconds. The lion went after the animal that would take him less time to catch. After you solve this problem, you will know which animal he pursued.

a. Complete the table. For the distance between the lion and each of the other animals, assume the lion is running toward the animal. Distances are in meters.

Time (seconds)	0	50	100	150	200
Lion's Distance from Tree	0				
Horse's Distance from Tree	500				
Giraffe's Distance from Tree	1,000				
Distance between Lion and Horse	500				
Distance between Lion and Giraffe	1,000				

b. On a single set of axes, with time on the horizontal axis and distance on the vertical, draw these three graphs:
- the lion's distance from the tree
- the horse's distance from the tree
- the giraffe's distance from the tree

c. On another set of axes, draw these two graphs:
- the distance between the lion and the horse
- the distance between the lion and the giraffe

d. How are the five graphs similar, and how are they different?

e. What events match the points of intersection of the graphs for Part b?

f. Use the graphs to help you find how much time it would take the lion to catch the horse and how much time it would take the lion to catch the giraffe.

g. What are the slopes of the five lines? How are they different and why?

h. Write a rule in symbols for each line you graphed.

In your
own
words

Explain how the speed in a distance-time situation is related to the slope of the graph for the situation.

Fill in the blanks to make each a true statement.

34. $\frac{3}{12} + \frac{4}{12} +$ ____ $= 1$

35. $\frac{6}{8} + \frac{7}{8} +$ ____ $= 2$

36. Geometry Find the area and perimeter of the figure.

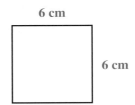

6 cm

6 cm

37. Geometry A rectangular prism has a base 5 inches long and 2 inches wide. The prism is 6 inches tall.

    a. What are the prism's volume and surface area?

    b. Another block prism is four times the height of the original prism. What are its volume and surface area?

    c. If a cylinder has the same volume as the original prism and the area of its base is 10 square inches, what is the cylinder's height?

38. Write this expression in two other ways: $x \div 7$.

39. Use the expression given to complete the table. Then write another expression that gives the same values. Check your expression with the values in the table.

m	1	2	3	4	5	100
$2(m - 1)$						

Find each product.

40. $4 \times \frac{1}{10^2}$ **41.** $61 \times \frac{1}{10^2}$ **42.** $842 \times \frac{1}{10^3}$

Rewrite each expression as addition, and calculate the sum.

43. $^-6 - {}^-12$ **44.** $7 - {}^-4$ **45.** $0 - {}^-4$

Write each number in standard notation.

46. 6×10^2 **47.** 8.6×10^3 **48.** 1.39×10^4

49. The product of two integers is 12. What could the integers be? List all possibilities.

50. Algebra Solve for y: $^-2y + 5 = 11$.

51. Suppose you put a 1-inch stick of gum through a $\times 6$ stretching machine, three times. Do you get the same result by putting a 1-inch stick of gum through a $\times 3$ machine, six times? Explain how you know.

Use exponents to write the prime factorization of each number. For example, $200 = 2^3 \times 5^2$.

52. 25   **53.** 72   **54.** 90

55. Find 13% of 2,375.   **56.** Find 178% of 312.

57. What percent of 200 is 35?

58. The table lists the average daily temperatures for one week in Algeville.

Average Daily Temperatures

Monday	Tuesday	Wednesday	Thursday	Friday	Saturday	Sunday
64°F	68°F	73°F	60°F	64°F	63°F	60°F

Use positive or negative numbers to express the change in temperature from

a. Monday to Tuesday   **b.** Tuesday to Wednesday

c. Wednesday to Thursday   **d.** Thursday to Friday

e. Friday to Saturday   **f.** Saturday to Sunday

Recognizing Linear Relationships

You have learned about linear relationships between variables such as pay earned and time worked. In this lesson, you will look at both linear and nonlinear relationships—but it won't always be obvious which is which!

Just the facts

Tiles are typically made from clay that has been hardened by being fired, or baked at very high temperatures, in a special oven called a *kiln*. A glaze might then be applied to the tiles.

Explore

Pat and Tillie are patio tilers who specialize in two-color rectangular patios. All their patios are constructed from tiles measuring 1 foot by 1 foot. The border is one color, and the center is another color. The patio shown here is 9 feet by 6 feet.

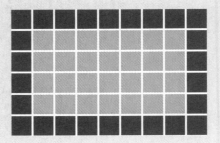

Their Totally Square line of patios contains a variety of square patio designs. Below is a Totally Square patio measuring 6 feet by 6 feet.

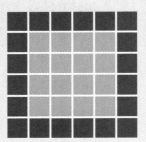

Find a rule that tells how many border tiles will be used on a Totally Square patio of a specified size. Then find a rule that tells how many nonborder tiles there will be. Tell what each variable in your rules represents.

Pat and Tillie call their patio designs that aren't square their ColorQuad line. Find a rule that expresses the number of border tiles in a ColorQuad patio with width w and length l. Then write a rule that expresses the number of nonborder tiles in a ColorQuad patio.

Investigation 1 ▶ Exploring and Describing Patterns

In the next set of problems, you will review and extend your ability to find patterns and to write rules in symbols.

Problem Set A

1. This hexagon pattern can continue in either direction.

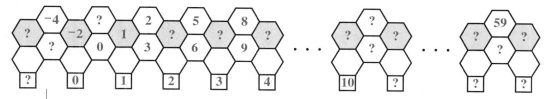

a. Find the missing numbers in the hexagon pattern. Then copy and complete the table. Discuss with your partner how you found the numbers.

Number in Square, s	0	1	2	3	4	10			
Number in Shaded Hexagon, h	-2	1							

b. Write a rule in symbols for computing h if you know s.

2. This sequence of squares is made from small tiles.

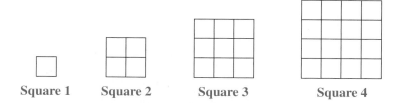

Square 1 Square 2 Square 3 Square 4

a. Find a relationship between the number of the square and its area. (The area is equal to the number of tiles in the square.)

b. Complete the table to show the area of each square.

Square Number	1	2	3	4
Area of Square	1			

Problem Set B

This pattern of Z-shapes is made from small squares.

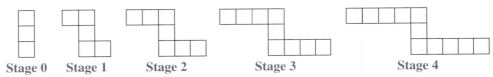

Stage 0   Stage 1   Stage 2   Stage 3   Stage 4

1. Complete the table.

Stage Number, n	0	1	2	3	4	5
Squares, s	3	5	7	9		

2. Zach found it took 21 squares to make Stage 9. How many will it take for Stage 10? Add two columns to your table for Stages 9 and 10.

3. Explain how the number of squares in a stage is related to the stage number. Be sure your explanation will work for all stages. Write a symbolic rule for finding s if you know n.

4. Use your rule to predict how many squares are needed for Stages 11 and 12. Draw the stages to test your rule.

5. Explain how the two numbers in your rule relate to the pattern of Z-shapes.

Problem Set C

Now you will investigate this pattern of T-shapes.

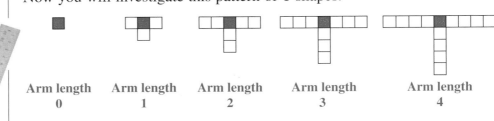

Arm length
0

Arm length
1

Arm length
2

Arm length
3

Arm length
4

1. Complete the table.

Arm Length, L	0	1	2	3	4	5	100	1,000
Squares, s	1							

2. How did you decide how many squares were needed to make a T-shape with arm length 100? With arm length 1,000?

3. Write a general formula (or rule) in words, and then in symbols, to relate the total number of squares to the arm length.

4. Use your formula to predict the number of squares for other arm lengths. Draw the shapes to check that your formula works.

5. Explain how the two numbers in your formula relate to the T-shapes.

6. Simon made a T-shape with 160 squares. Write and solve an equation to find its arm length.

7. Could Simon make a T-shape with 100 squares? Explain your reasoning.

8. Could Simon make a T-shape with 200 squares? Explain your reasoning.

Share & Summarize

Look at the tables you made for the Z-shapes and the T-shapes. The numbers in the second row show how each pattern grows.

1. How much does each pattern grow from stage to stage?

2. For each table, compare the amount of growth with the rule you wrote. How is the growth reflected in the rule?

3. How does this growth relate to the rate of change?

Investigation 2 ▶ Graphs and Rules from Patterns

In the last investigation, you wrote rules to describe visual patterns. As you know, graphs can help you find rules for patterns.

Think & Discuss

Look again at the T-shapes you studied in Problem Set C.

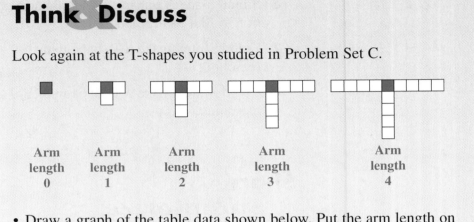

| Arm length 0 | Arm length 1 | Arm length 2 | Arm length 3 | Arm length 4 |

• Draw a graph of the table data shown below. Put the arm length on the horizontal axis. What is the shape of the graph?

Arm Length, L	0	1	2	3	4	5
Squares, s	1	4	7	10	13	16

• What is the rule for this pattern?

• Where do the two numbers in the rule appear in your graph?

• Does it make sense to connect the points in this graph? Why or why not?

As you've seen before, when you graph data for some situations, it can be helpful to sketch a line through the points even though you know not all of the points on the line make sense for the situation. The line can help you find other points that do make sense. It can also help you show the relationship between the variables.

If all the points on a line don't really make sense, draw a dashed line instead of a solid one. This will show that the line isn't really part of the graph.

Problem Set D

Look again at the Z-shape pattern from Problem Set B.

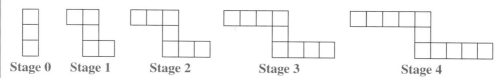

Stage 0   Stage 1   Stage 2   Stage 3   Stage 4

Stage Number, n	0	1	2	3	9
Squares, s	3	5	7	9	21

1. Would it make sense to connect points on a graph of these data? Explain.

2. Draw a graph of the table data. Try to draw a line through the points. Use a dashed line or a solid line, whichever makes more sense.

3. What kind of pattern do the points make?

4. Use your graph to predict the value of s if n is 8. Check with your table and your rule from Problem Set B to see if you are correct.

5. Use your graph to predict the value of n if s is 17. Check with your table and your rule to see if you are correct.

6. Does a point fall on the s-axis? What part of the pattern of Z-shapes does this show?

7. How does the slope of your line relate to the pattern of Z-shapes?

8. The rule for this pattern is $s = 2n + 3$. Explain where each number in the rule appears in your graph.

Problem Set E

You investigated this pattern in Problem Set A.

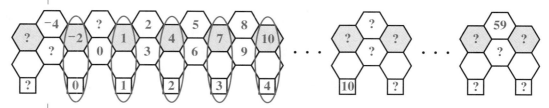

1. Five pairs of numbers have been circled: (0, ⁻2), (1, 1), (2, 4), (3, 7), and (4, 10). What is the next pair of numbers in this sequence? (Note: This part of the pattern is not visible.)

2. Plot the points from Problem 1 on a copy of the grid shown below. If the points lie on a straight line, draw the line using a dashed or a solid line, whichever makes more sense.

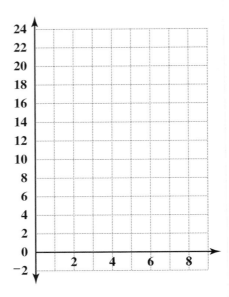

3. If you continue the sequence and graph the points, what would the coordinates of the ninth point on the graph be? Continue the pattern of hexagons, and check that your prediction is correct.

4. Use your graph to help you write a rule to link the two numbers in each pair. Choose letters to represent each variable. Check your rule with the pairs of values you already have.

5. Fill in the missing numbers in this part of the pattern. Use your rule to check your work.

Share & Summarize

1. Describe a situation in which it makes sense to connect the points on a graph. Then describe a situation in which it does not make sense to connect the points on a graph. Explain your thinking.

2. What is the slope of the line described by the rule $y = {}^-0.5x + 8$? What is the y-intercept of the line?

Investigation 3 ▶ From Rules to Graphs

In Investigation 2, you probably found that whenever a rule looked like

$$y = ax + b \qquad \text{(with numbers instead of } a \text{ and } b)$$

the graph of the points formed part of a straight line. In fact, whenever a rule looks like this, you can be sure that all the points in its graph will be on a straight line. This is why these relationships are called *linear*.

The table summarizes the relationships among the numbers in the rule $y = ax + b$, the features of the pattern described by this rule, and the features of the graph of this rule.

In Patterns	On Graphs
As the quantity x changes by 1 unit, the quantity y changes by a units.	The number a is the slope of the line. It shows the rate at which y changes as x changes.
b is the constant, or fixed, part of the pattern.	b is the constant. It shows where the line crosses the y-axis. b is called the y-intercept.

You can tell a lot about how a graph will look just by inspecting its rule.

Think & Discuss

Whose equation is correct? How could you explain to the other two students why they are incorrect?

Problem Set F

You and a partner are going to play a detective game. One of you will draw a graph for one of the rules below. The other will be the detective, looking for clues in the graph to decide which rule was used.

$$c = 2d + 3 \qquad c = 2d - 1 \qquad c = d + 4$$

$$c = d + 3 \qquad c = 10 - d \qquad c = 10 - 2d$$

$$c = {}^{-}2d + 3 \qquad c = 10 + d \qquad c = 10 + 2d$$

1. Take turns drawing and being the detective until each of you has drawn four graphs. Use d for the horizontal axis and c for the vertical axis.

2. Explain in a few sentences how you matched each graph with an equation.

Here is another set of equations that can be graphed. These equations aren't quite like the others, so you may need to look harder for clues.

$$x = 7 + y \qquad 2x = 2(6 + y) \qquad 3x = 3y - 9$$

$$x = 7 - y \qquad 2x = 2(6 - y) \qquad 3x = 3y + 9$$

3. Take turns drawing and being the detective until you and your partner have each drawn two graphs. Use x for the horizontal axis and y for the vertical axis.

4. Explain how you matched each graph with an equation.

Most of the equations you have encountered in this chapter that describe linear relationships were in the form $y = ax + b$ (for example, $y = 2x + 7$). Linear relationships can also be described by other forms of equations.

As you have just seen, a graph is a good way to figure out whether a relationship written in a form other than $y = ax + b$ is linear. A graph will also help you to find the answers to the next problem set.

Problem Set G

A fitness test has two parts: skipping, and running in place. For each part of the test, you can earn a whole-number score from 0 to 10. Your total score is given by the rule

$$\text{total score} = s + r$$

where s is the skipping score and r is the running score.

1. To be classified as "fit," you need a total score of at least 15. Use a grid like this one to represent all the ways a person doing this test could earn *just enough* points to be judged "fit."

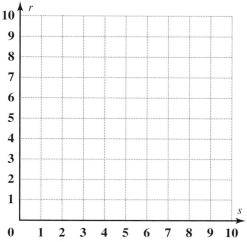

Remember

If some of the points on a line don't make sense, draw a dashed line instead of a solid one.

2. Use the same grid to represent *all* the ways a person could be judged fit. Use a different color from Problem 1.

3. Which of the following describe the possible scores for fit people? Explain your reasoning.

$$s + r = 15 \qquad s + r \le 15 \qquad s + r \ge 15$$

Share & Summarize

Consider these four equations.

$$10 - 2d = 2c \qquad 2c = 2(5 - d) \qquad 2(c + d) = 10 \qquad d = 5 - c$$

1. Draw graphs of the equations, and discuss what you notice. Put c on the horizontal axis.

2. Explain *why* what you noticed is happening.

3. Find some other equations that fit the pattern.

On Your Own Exercises

Practice & Apply

1. Shaunda made this pattern from toothpicks.

| Stage 0 | Stage 1 | Stage 2 | Stage 3 |

a. How many toothpicks are needed for Stage 4? For Stage 5?

b. Shaunda used 56 toothpicks to make Stage 11. How many would it take to make Stage 12?

c. Copy and complete the table.

Stage, *n*	0	1	2	3	4	5	11	12
Toothpicks, *t*	1	6					56	

d. Describe a general rule that shows how you can find the number of toothpicks from the stage number. Write the rule in symbols, and check that your rule works on all pairs of numbers in the table.

e. Use your rule to predict the number of toothpicks needed for some other stages. Test your prediction by building or drawing the stages.

f. Use the way the pattern grows from stage to stage to explain why your rule always works.

2. The patio tilers, Pat and Tillie, designed a tile pattern that can be adapted for floors of various sizes. The strip of tiles can be any length, as long as it begins and ends with a column of white tiles.

Write a rule for the pattern, giving the relation between the number of white tiles and the number of purple tiles for various lengths of the strip.

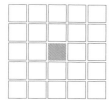

Side length 1

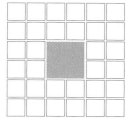

Side length 2

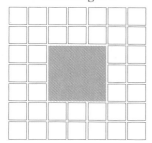

Side length 3

3. Sabrina and Rosalinda made a sequence of tile designs from square white tiles surrounding one square purple tile. The purple tiles come in many sizes. Three of the designs are shown at left.

a. Copy and complete the table.

Side Length of Purple Tile, s	1	2	3	4	5	10	100
White Tiles in Border, b							

b. Find a rule to link the number of white tiles in the border to the side length of the purple tile. Describe the rule in words and symbols.

c. Draw a graph using the first five pairs of numbers in your table. (When graphing an ordered pair, refer to the horizontal axis for the first number and to the vertical axis for the second number.)

d. Do the points lie on a line? Is this what you expected? Tell why or why not.

4. Rick and Cheri, who are also patio tilers, created a different type of patio.

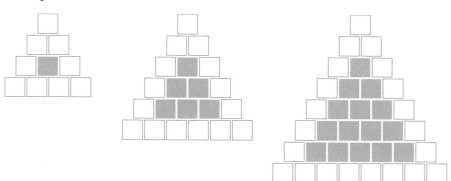

a. Complete the table.

Rows, r	4	6	8
White Tiles, w	9		
Purple Tiles, p	1		

b. Make a graph that shows how the number of rows relates to the number of white tiles. Make another graph that shows how the number of rows relates to the number of purple tiles. Put the number of rows on the horizontal axis.

c. Which graph is linear, the graph for the white tiles or the graph for the purple tiles?

d. Write a rule for the linear graph, and check it with the numbers in the table.

For each equation, draw a graph with x on the horizontal axis and y on the vertical axis. Draw four-quadrant graphs (include negative values for both x and y).

5. $0.4y = 3x - 7$

6. $y = 5(x + 5)$

7. $y - 3x = 6$

8. $2.5(y + x) = 9$

9. Match each equation with one of the graphs below.

a. $y = 2x$

b. $y = 0.5x$

c. $3y + 1.5x = 11$

d. $x = 0.3(y - 4)$

i.

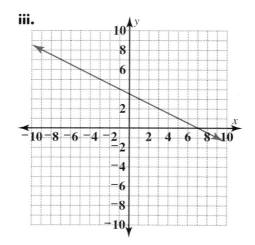

ii.

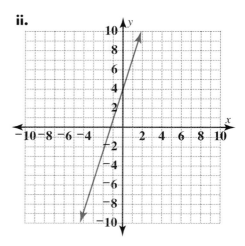

iii.

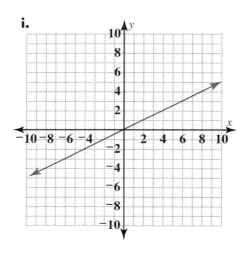

iv.

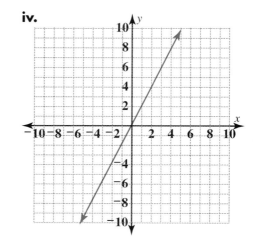

10. When Zoe's family checked into a campground, the manager gave them a 20-meter piece of rope. He said they could mark out any campsite they liked, as long as its border was no longer than the rope.

a. Zoe suggested they use all the rope to make a rectangular shape. She chose some values for the side of the rectangle they wanted to place facing the river. Then she calculated the other side of the rectangle. Copy and complete her table.

Length of River Side (meters), a	0	1	2.5	3	4.2	5.5	9.9	10
Length of Other Side (meters), b								

b. Draw a graph to show all possible side lengths for a rectangle with a 20-meter perimeter. Put the length of the river side on the horizontal axis.

c. Does it make sense to connect the points? Why or why not?

d. Use your graph to write a rule relating a and b.

e. The length of rope was the *maximum* border Zoe's family could use. They could also have used less rope. List the dimensions of five other rectangles that would be acceptable but that use less rope. Plot points on your graph to represent these rectangles.

f. What do you notice about the location of the points you chose, in relation to the line? On your grid, shade the region that includes these points.

g. Which equation or inequality below describes the shaded region of your graph? Explain in words what it means.

$$2(a + b) = 20 \qquad 2(a + b) \leq 20 \qquad 2(a + b) \geq 20$$

h. Choose some points in the shaded region, other than the points in Part e. Test these points in the equation or inequality you chose for Part g. What do you notice?

Connect & Extend

11. This tiling pattern can be any length as long as the pattern is complete (the first two columns and last two columns must look as shown here). Write a rule for the pattern, giving the relation between the number of white tiles and the number of purple tiles for various strip lengths.

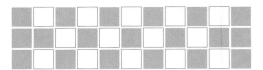

12. Invent a tiling pattern that shows a *linear* relationship between the number of purple tiles and the number of white tiles.

   a. Draw a few examples of your pattern.

   b. Create a table that shows the number of purple tiles and the number of white tiles for your examples.

   c. Write a symbolic rule for that relationship.

13. **Challenge** Invent a tiling pattern with green tiles and yellow tiles in which the relationship between the numbers of green tiles and yellow tiles is *nonlinear*.

   a. Illustrate your pattern with a few examples.

   b. Make a table for your pattern.

   c. Write a clear explanation of how you know the relationship is nonlinear.

   d. If possible, write a rule describing the relationship between the number of yellow tiles and the number of green tiles in your pattern.

14. Look back at your table for Problem 2 of Problem Set A. The pattern in that table is different from most of the patterns you worked with in this investigation.

   a. Draw a graph for the pairs of numbers in that table.

   b. How is this graph different from the graphs you drew in Problem Sets D through F?

15. Pat and Tillie often tile walkways by surrounding a single strip of purple tiles with white tiles. Here is one walkway they made.

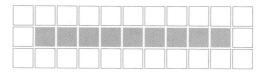

a. Draw this pattern for 1, 2, and 3 purple tiles.

b. Write a rule that relates the number of white tiles w to the number of purple tiles p.

c. Use your rule to make a table of values for the tiles needed for walkways of different lengths.

d. Draw a graph from your table. Put the number of purple tiles on the horizontal axis. Connect the points using a dashed line or a solid line, whichever makes more sense.

e. How are the two numbers in your rule shown in your graph?

f. How do these two numbers relate to the pattern of tiles in the walkway?

16. If you mark two points, only one straight line can be drawn through them. This means that when you have the rule for a linear relationship, you need to plot only two points to draw the line. Of course, even if you use this handy "two point" rule, it's not a bad idea to check a couple of more points to make sure you haven't made a mistake.

Here are five rules.

i. $y = 3x + 6$

ii. $y = x + 3$

iii. $y = x^2 + 3$

iv. $y = 12 - 2x$

v. $y = \frac{2}{x}$

a. Decide whether each of the rules is linear.

b. For each relationship you think is linear, find two points that will be on the graph, and use them to draw the graph. Then find a third point and check that it is on your line.

c. For each relationship you think is *not* linear, find several points and use them to draw the graph.

17. The three Pauli children all love licorice. Their mother often buys them a 30-centimeter strand of licorice. After the twins, Carmelo and Biagio, have had some, the rest goes to Maria, the youngest child. The twins have the same amount each.

   a. What is the least amount the twins could have? How much would Maria then get?

   b. What is the greatest amount each twin could have? How much would Maria then get?

   c. Choose some values for the length each twin has, and then calculate how long Maria's piece will be.

   d. Write a rule in symbols for finding the length of Maria's piece if you know the length of each twin's piece.

   e. Use your rule to find how long Maria's piece will be if Carmelo and Biagio have 5.5 cm each.

   f. Draw a graph of your data. Don't forget to label the axes. If it makes sense to draw a line through your points, do so.

   g. Use your graph to predict the length of the twins' pieces if Maria's piece is 9 cm long. Check your prediction with your rule.

Mixed Review

Simplify.

18. $\frac{1}{25} + \frac{4}{25}$ **19.** $\frac{1}{10} + \frac{6}{10}$ **20.** $\frac{2}{5} + \frac{1}{5} + 3$

21. Use the expression given to complete the table. Then write another expression that gives the same values. Check with the values in the table.

n	0	1	2	3	4	100
$\frac{n+1}{2}$						

Expand each expression.

22. $3(n + 7)$

23. $8(k + 11)$

24. Malik belongs to a regional drama club. There are a lot of students in the club, making it difficult for one person to contact everyone when necessary.

If the club director has to cancel a meeting, she calls Malik. He is in charge of starting a chain of phone calls to inform people of the cancellation. Malik is assigned to first call three people: Tom, Julia, and Corey. Then Malik, Tom, Julia, and Corey each call three people. This pattern continues until everyone has been informed.

a. Complete the table.

Round	0	1	2	3
Number of People Informed	1 (just Malik)	4		

b. After which round will at least 50 people know about the cancellation?

c. If 256 people need to be informed, how many rounds of calls must be made?

d. How many calls will Malik have made by the time 256 people have been informed?

Supply the missing exponent.

25. $24 \times 10^? = 24{,}000$

26. $2.8 \times 10^? = 28{,}000$

Rewrite each expression as addition, and calculate the sum.

27. $12 - {}^-4$ **28.** ${}^-8 - {}^-8$ **29.** ${}^-3 - {}^-2$

Find the distance between the endpoints of the segment.

30.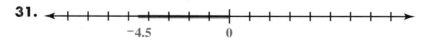

31.

Rewrite in the form c^m.

32. $6^9 \times 6$ **33.** $3^5 \times 3^5$ **34.** $10^{11} \times 10^5$

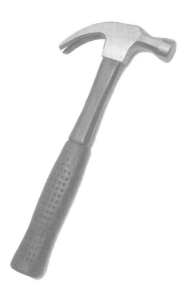

People who are skilled at something—like cooking or carpentry—learn or develop "tricks of the trade" that make their work more successful. For example, putting a little cooking oil in boiling water helps stop spaghetti from sticking. Many tricks of the trade seem like magic at first, but once you know them, you might wonder why you hadn't thought of them before!

Think & Discuss

Think of something you know how to do well, such as a sport, job, hobby, or household task. Can you think of a few tricks of the trade you might share with someone just learning?

In this lesson you will learn a trick of the mathematician's trade for helping to find rules for linear relationships.

Investigation 1 Finding Secret Rules

You can use some of the things you have learned about relationships to help you discover the secret rules in games of *What's My Rule?*

MATERIALS
graph paper

Problem Set A

Several students are playing *What's My Rule?*

1. In this game, Shaunda is the Rule Keeper and Luis is the Guesser.

Remember

One Rule Keeper and any number of Guessers can play *What's My Rule?* The Rule Keeper makes up a secret rule that takes inputs and produces outputs. Guessers give inputs to the Rule Keeper, who gives outputs until someone guesses the rule.

a. On a graph, plot the pairs of inputs and outputs. Decide whether the relationship could be linear.

b. Find a symbolic rule that fits the number pairs. Choose letters for the inputs and outputs, and be careful to make it clear which is which.

2. In this game, Darnell is the Rule Keeper and Kate is the Guesser.

a. On a graph, plot the pairs of inputs and outputs. Decide whether the relationship could be linear.

b. Find a symbolic rule that fits the numbers. Choose letters for the inputs and outputs, and make it clear which is which.

For each set of input/output pairs, tell whether the relationship could be linear. Then find a rule that fits the pairs.

3. (4, 16), (1, 1), $(\frac{1}{2}, \frac{1}{4})$, (3, 9)

4. (10, 0), (2, 8), (4.5, 5.5), (5, 5)

5. Jin Lee is the class *What's My Rule?* champion. Her secret is to plot points to see whether they lie along a line. If the rule describes a linear relationship, she can look at the graph and guess a rule that fits the points.

Explain how Jin Lee might derive a rule from the graph. Use illustrations, if you like.

Problem Set B

Jin Lee developed a quick strategy for finding rules in the game. She found that if she chooses consecutive integers as inputs, she can tell whether the relationship could be linear *just by looking at how the outputs change*. Look at the numbers she collected for four games she played.

Game 1

Input	1	2	3	4	5
Output	4	7	10	13	16

Game 2

Input	0	1	2	3	4
Output	1	6	11	16	21

Game 3

Input	0	1	2	3	4
Output	29	27	25	23	21

Game 4

Input	2	3	4	5	6
Output	10	15	21	28	36

1. Jin Lee concluded that Games 1, 2, and 3 could be linear, but that Game 4 couldn't be. Inspect her tables of input/output pairs. What pattern might have led Jin Lee to her decision?

2. For each game that Jin Lee thinks could be linear, write a symbolic rule that fits all the input/output pairs. Choose letters for the inputs and outputs, and be careful to say which is which.

3. Make a graph for each of the four games. Extend your graphs, and use them to estimate the output each game's rule would give for an input of $^-3$. Which estimates are you confident are most accurate? Explain.

Share & Summarize

1. Using just a graph, you can tell whether a rule could describe a linear relationship. How?

2. Using just a table with consecutive integers for inputs, you can tell whether a rule could describe a linear relationship. How?

3. Suppose a table shows some input/output pairs in a linear relationship. How can you use a graph of the table values to find other input/output pairs?

Investigation ▶2▶ The Method of Constant Differences

Jin Lee's method is a mathematical "trick of the trade." She makes her table in a special way, using consecutive integers for inputs. Then she finds the *differences* of neighboring pairs of outputs to see if they make a recognizable pattern.

If the difference between two neighbors is the same no matter which pair Jin Lee chooses, they are *constant differences*. In that case, a plot of the input/output pairs will lie along a line, and the relationship could be linear. Jin Lee can then use the constant difference to quickly find a rule that describes the linear relationship and fits the input/output pairs.

MATERIALS
graph paper

Problem Set C

Below are several tables you can use to try Jin Lee's trick. For each, if you think the relationship could be linear, find a rule and check it with the pairs in the table. If you think the relationship isn't linear, draw a graph to check. Pay attention to the order of the inputs!

1.

Input, p	⁻3	⁻2	⁻1	0	1	2	3
Output, q	⁻27	⁻17	⁻7	3	13	23	33

2.

Input, f	⁻1	0	1	2	3	4	5
Output, g	1	0	1	4	9	16	25

3.

Input, t	⁻2	⁻1	0	1	2	3	4	5
Output, u	1	0.5	0	⁻0.5	⁻1	⁻1.5	⁻2	⁻2.5

4.

Input, d	3	2	1	5	4
Output, s	15	25	35	45	55

5.

Input, x	0	2	1	5	4	3
Output, y	⁻2	16	7	43	34	25

6.

Input, m	4	1	3	2	5
Output, n	12	3	9	6	15

7. Imagine you are playing *What's My Rule?* and already know that the rule is linear. This means you can write the rule as $y = ax + b$, with numbers in place of a and b and with x as the input and y as the output. What input would you use so you could find the value of b? Why would you use that input? If you need ideas, look back at the tables you have seen and the rules you have written for them.

Problem Set D

You will now think some more about *why* Jin Lee's strategy for finding rules works.

1. The differences in the output values in this table are all 3.

x	1	2	3	4	5
y	4	7	10	13	16

Think of a situation—like those you have studied in this chapter—that might have this table. Describe what the difference of 3 means for your situation. Does it make sense that the relationship between these variables could be linear? Explain.

2. If you made a graph showing how y changes as x changes, what would its slope be? What does this slope mean for your situation in Problem 1?

3. Recall how we defined *slope*. Look back in your book if necessary. How does using consecutive integers for your inputs help you find the slope?

MATERIALS
graph paper

Problem Set E

Use the differences method to analyze the pattern in each table. Predict the missing outputs.

1.

Input, x	1	2	3	4	5	6	7	8
Output, y	1	2	4	8	16	32		

2.

Input, s	1	3	4	8	11	12	13	14
Output, t	8	18	23	43	58	63		

3.

Input, a	1	2	3	4	5	6	7	8
Output, b	1	2	4	7	11	16		

4.

Input, g	1	2	3	4	5	6	7	8
Output, h	7.5	6	4.5	3	1.5	0		

5.

Input, m	$^-3$	$^-1$	1	3	5	7	9	11
Output, n	2	$^-2$	$^-6$	$^-10$	$^-14$	$^-18$		

6.

Input, p	0	1	2	3	4	5	6	7
Output, r	$^-10$	12	34	56	78	100		

7. In which tables do the differences tell you the relationships could be linear? Write rules for those relationships.

8. Plot the data in each table. For the graphs that seem to be linear, draw a line through the points. Extend the linear graphs to show the value of the output when the input is 0, ⁻1, and ⁻2, and label those points with their coordinates.

9. Find the slope of each line you drew in Problem 8. Compare the slopes to the rules you wrote for Problem 7.

10. In Problem 7, did you choose all the relationships that could be linear? Did you write a correct rule for each of them? If not, figure out what went wrong, and write new rules for those relationships.

Nearly all the rules you have encountered in this chapter can be written in this form:

$$y = ax + b$$

The outputs y and the inputs x are called **variables** because their values can vary. The other two numbers also have special names.

A number that is multiplied by the input, like a, is called a **coefficient.** In the equation $y = ax + b$, we say that "a is the coefficient of x."

A number that stands by itself, like b, is often referred to as the **constant term.**

EXAMPLE

$y = 2x + 6$ The 2 is a coefficient and the 6 is the constant term.

$y = 4x - 3$ The 4 is a coefficient, but what is the constant term? It helps to rewrite the rule in the form $y = ax + b$. Since subtracting a number is the same as adding its opposite, this rule can be rewritten as $y = 4x + {}^-3$. So, the constant term is ⁻3.

$y = x + 12$ You can see that the constant term is 12, but there's no number being multiplied by x. Is there a coefficient? You can rewrite this rule, too: $y = 1x + 12$. Here, a is 1, so 1 is the coefficient.

Investigation ▶3 Understanding the Symbols

In this chapter, you have looked at relationships that describe real situations. Most of these relationships were *linear:* they could be graphed as a line and expressed in a rule that looks like $y = ax + b$.

Now you will extend what you know about the coefficient and the constant term, a and b, by investigating how they affect the quadrants the graph passes through.

MATERIALS
graph paper

Remember

The four quadrants of a graph are numbered like this:

II second quadrant	**I** first quadrant
III third quadrant	**IV** fourth quadrant

Problem Set F

1. What does the coefficient a tell you about the graph of $y = ax + b$?

2. What does the constant term b describe about a graph of $y = ax + b$?

3. For each equation in the table below, $a = 0$ and $b \neq 0$.

   a. Complete the table. Choose two more equations for which $a = 0$ and $b \neq 0$, and add them to your table.

x	-2	-1	0	1	2	3
$y = 0x - 2$						
$y = 3$						

   b. On one set of axes, graph all four equations.

4. How many quadrants does each graph pass through? Do you think this will always be true for $y = ax + b$ when $a = 0$ and $b \neq 0$? Test several more rules of your own to check.

5. For each equation in the table below, $a \neq 0$ and $b = 0$.

  a. Complete the table. Choose two more equations for which $a \neq 0$ and $b = 0$, and add them to your table.

x	⁻2	⁻1	0	1	2	3
$y = x$						
$y = {}^-2x$						

  b. On one set of axes, graph all four equations.

6. How many quadrants does each graph pass through? Do you think this will always be true for $y = ax + b$ when $b = 0$ and $a \neq 0$? Test some more rules to check.

7. How many quadrants do you think $y = ax + b$ will pass through if $a = 0$ and $b = 0$? Graph the equation $y = 0$. How many quadrants does the graph pass through?

8. Look back at the graphs in this chapter. How many quadrants can $y = ax + b$ pass through *at the most*? Explain your reasoning.

9. How many quadrants can $y = ax + b$ pass through *at the least*? Explain your reasoning.

10. Look at your graphs for this problem set for graphs that don't cross the horizontal axis. What do their rules have in common? Do you think all such rules will have graphs that don't cross the horizontal axis?

11. Are there any graphs in your examples, or elsewhere in this chapter, that do not cross the vertical axis? Can you think of a rule in the form $y = ax + b$ with such a graph?

The outputs in the two tables below start out the same way—both begin with 1, 2, 4, and 8—but they continue differently.

Input	1	2	3	4	5	6	7	8
Output	1	2	4	8	16			

Input	1	2	3	4	5	6	7	8
Output	1	2	4	8	15			

Even though these two tables match in four places, they do not match in the fifth place—and maybe will never match again! This shows that if someone gives you a sequence of numbers like 1, 2, 4, 8, . . ., you can't reliably say what number will come next without knowing more.

Problem Set G

In sequences like the one just mentioned, it helps to have information about what kind of relationship is being described. For example, if you know that a relationship is linear, you can quickly find the rule from just a few points.

1. This table describes a linear relationship.

Input	1	2	3	4	5
Output	3	1	$^-1$	$^-3$	$^-5$

a. Think about how many points you need to plot in order to draw the line for this relationship. What is the *fewest* number of points needed? Explain.

b. Draw the graph for the relationship using no more points than you need.

c. Use your graph to find a rule for this relationship.

Not all linear rules are in the form $y = ax + b$. If you can rearrange a rule into this form, you know that its graph will be a line.

2. Consider the equation $2y = 4x + 5$.

a. Graph the equation to show that it is a linear rule.

b. Write a new rule, using the form $y = ax + b$, that has the same graph as $2y = 4x + 5$.

Share & Summarize

1. Explain how the graph of a linear relationship can help you to guess the rule.

2. Explain how the method of constant differences can help you to guess the rule for a relationship.

On Your Own Exercises

Practice  **Apply**

In Exercises 1–5, do Parts a and b.

a. Plot the input/output pairs, and decide whether the rule could be linear.

b. If you think the rule could be linear, find a rule that fits the numbers and write it in symbols.

1.

Input	1	2	3	4	5
Output	3	20	37	54	71

2.

Input	1	2	3	4	5
Output	1,599	1,561	1,523	1,485	1,447

3.

Input	1	2	3	4	5
Output	12	8	4	$^-2$	$^-12$

4.

Input	0	1	2	3	4
Output	$^-26$	$^-21$	$^-16$	$^-11$	$^-6$

5.

Input	1	2	3	4	5
Output	17	1	$^-12$	$^-22$	$^-29$

Decide whether each relationship could be linear. If you think it could be, find a rule and write it in symbols, and then check your rule with each input/output pair.

6.

x	4	3	2	5	1
y	$^-7$	$^-5$	$^-3$	$^-9$	$^-1$

7.

z	$^-1$	1	3	5	7	9
w	$^-12$	$^-6$	0	6	12	18

8.

x	3	$^-1$	0	2	4	5	1
y	10	2	1	5	17	26	2

9.

m	4	7	5	3	6
n	31	19	26	35	22

10.

u	10	12	8	13	7
z	64	76	52	82	46

Remember
Watch whether the inputs are in order and consecutive!

11. The telephone service PhoneHome charges a fee of r dollars per minute. Customers also pay a service charge of s dollars every month, regardless of how long they use the phone. If m is the number of minutes you use the phone, then your monthly bill b in dollars can be computed with the rule $b = rm + s$.

a. Think about the four variables in this situation. What kinds of values can each variable have? For example, can r be 5? Can it be $^-3$? Can s be 3.24? Can it be 0? Can m be 5?

b. PhoneHome offers a special calling plan, the Talker's Delight plan. This plan has a monthly rate of s dollars and no additional charge per minute, no matter how long you talk. In this arrangement, what is the value of r? How many quadrants does the graph of $b = rm + s$ pass through for this charge plan?

c. PhoneHome also offers the TaciTurn plan. There is no monthly fee, but you *are* charged by the minute. What is the value of s in this pricing plan? How many quadrants does the graph of $b = rm + s$ pass through?

d. Assuming any values of r and s, how many quadrants can the graph of $b = rm + s$ pass through *at the most*, if m is time in minutes and b is your monthly phone bill? Explain your reasoning.

e. How many quadrants must $b = rm + s$ pass through *at the least*? Explain your reasoning.

12. Four graphs are drawn on the axes.

Graph A

Graph B   Graph C

Graph D

a. Point a has a negative first coordinate and a positive second coordinate. On which graphs could Point a be located?

b. Point b has two negative coordinates. On which graphs could Point b be located?

c. Which graphs contain points in both Quadrants II and IV?

d. Which graphs contain points on both the x-axis and the y-axis?

e. Which graph could contain points $(2, ^-6)$ and $(^-2, ^-6)$?

13. This set of squares is made from toothpicks.

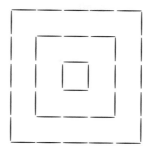

a. How many toothpicks are needed for each square shown?

b. Copy and complete the table.

Toothpicks along Side of Square	1	3	5	7	9	11		19
Toothpicks to Make Square							60	

c. The relationship shown in the table is linear. How could you prove this with a graph?

d. How could you prove the relationship is linear by inspecting the table?

e. Write a rule to calculate the number of toothpicks needed to make a square for a given number of toothpicks along the side of the square.

14. This pattern of triangular shapes is made from squares.

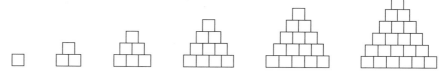

a. Complete the table. The *first* differences *d* are the differences between adjacent values of *A*. The differences between adjacent values of *d* are the *second* differences.

Base Length, *b*	1	2	3	4	5	6	7
Total Area, *A*	1	3	6				
First Differences, *d*		2	3				
Second Differences, *s*		1					

b. Is the relationship between *b* and *A* linear? How do you know?

c. Are the second differences constant?

d. Here's another geometric pattern. Make a table of base length, area, and first and second differences. Do you find any constant differences?

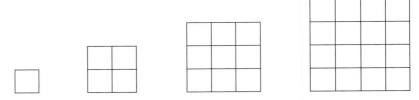

e. Write an equation for the area of the squares in Part d.

f. The equation for the triangle numbers in Part a is $A = \frac{b^2 + b}{2}$. What do you think constant second differences might tell you about a relationship's equation?

15. The last few lessons have been mostly about linear relationships, but there are several very important (and *common!*) sequences of output numbers that are not linear. Below are tables of some of these sequences. You'll see these patterns again and again as you continue to study mathematics.

a. Find the pattern and complete each table.

i.

Input	1	2	3	4	5	6	7	8	9	10	11
Output	1	4	9	16							

ii.

Input	1	2	3	4	5	6	7	8	9	10	11
Output	1	3	6			21					

iii.

Input	1	2	3	4	5	6	7	8	9	10	11
Output	1	2	3	5	8		21				

iv.

Input	0	1	2	3	4	5	6	7	8	9	10
Output	1	2	4		16						

b. These sequences are so important that each has its own special name. One sequence is related to triangular patterns, so it is called the *triangular numbers*. One is related to square patterns and is called the *square numbers*. The other two are called *powers of 2* and *Fibonacci numbers*. Try to match each name to its sequence.

c. The first differences of each sequence *also* have names. Try to name each set of first differences.

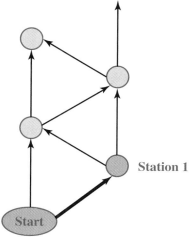

16. Firefighters, police, and other safety workers must have efficient routes from place to place within a city. They must also know about alternate routes. Finding the most efficient routes for snowplowing, street cleaning, meter reading, and patrolling requires an analysis of the city maps. Real city street plans are often messy, and counting the number of ways to travel efficiently from one place to another can be very complicated. An entire branch of mathematics is devoted to exactly such studies.

At left is an imaginary street plan with a pattern to explore. The arrows represent one-way streets, and the circles are stations where someone would stop (meters to read or hydrants to check, for example). You can see that there is only one path from Start to Station 1 along the one-way streets. It is indicated by a heavy arrow.

There are two paths from Start to Station 2, following the arrows.

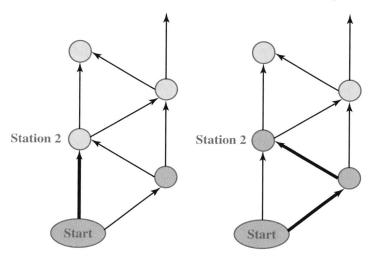

a. How many paths are there from Start to Station 3?

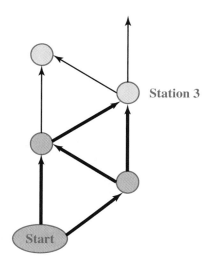

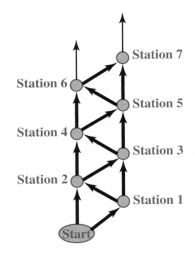

The drawing at left shows how the map continues.

b. Complete the table to show how many paths go from Start to each station.

Station Number	1	2	3	4	5	6	7
Ways to Get There	1	2					

c. Does the relationship between the station number and the number of ways to get there appear to be linear?

17. The equation $3y = 2x + 6$ is in the form $sy = tx + r$.

  a. Find a rule in the form $y = ax + b$ that has the same graph.

  b. Find a rule in the form $x = py + q$ that has the same graph.

  c. Find another way to write a rule that has the same graph.

Mixed Review

Insert parentheses to make each equation correct.

18. $3 \times 3 + 12 = 45$

19. $4 + 3 \times 12 = 84$

20. Use a factor tree to show all of the prime factors of 32.

21. Write the expression $\frac{m}{3}$ in two other ways.

Decide whether each equation is true or false. If false, rewrite the expression on the right side of the equation to make it true.

22. $2(x + 3) = 2x + 3$

23. $2(x + 9) = 2x + 18$

24. $3(2x + 1) = 2x + 3$

25. Which of the following expressions are equal to each other?

  a. $m + m + m + m + m$ **b.** $(5m)^5$ **c.** $5m$ **d.** m^5

  e. $3 \times 3 \times 3 \times 3$ **f.** 4×3 **g.** 4^3 **h.** 3^4

Write each number in standard notation.

26. 6×10^4 **27.** 4.8×10^3 **28.** 16.2×10^6

29. Social Studies In the 2000 U.S. presidential election, about 105,400,000 votes were cast.

  a. Write this number of votes using scientific notation.

  b. George W. Bush received about 48% of these votes. Approximately how many votes did he receive?

30. The product of two integers is ⁻12. List all possibilities for the two integers.

31. Algebra Solve for y: $⁻3y + 8 = 11$.

32. Number Sense Could the number 603,926,481 be a power of 6? Explain how you know.

33. Preview Geoff and Lina collect baseball cards. They visited a card dealer's shop together.

   a. Julián bought 5 packs of cards plus 3 single cards. Write an expression to show how many cards he bought. Let p stand for the number of cards in each pack.

   b. Lina bought 11 single cards plus 4 packs of cards. Write an expression to show how many cards she bought. Each pack has the same number of cards as the packs Julián bought.

   c. Julián and Lina bought the *same number* of baseball cards. How could you show this using the two expressions you wrote? Do it.

   d. Use your expressions to find how many cards are in a pack. Hint: The answer is greater than 5 but less than 10.

   e. Substitute your answer to Part d into your answer for Part c. What do you notice?

   f. What does your answer to Part e mean in terms of the situation?

Chapter Summary

In this chapter, you looked at linear relationships. The relationship between two variables is *linear* if all possible points, when graphed, are on a straight line. You began looking at linear relationships defined by a *rate,* which describes how two quantities change in relation to each other.

Most of the equations in this chapter are in the form $y = ax + b$. In this equation, y and x are the variables. The *coefficient a* represents the rate of change. When the line is graphed, a is the *slope* of the line. The *constant term b* doesn't change when x changes. The value of y when x is 0 is given by b, so b is also called the *y-intercept.* When b is 0, the relationship between y and x is *proportional.*

You may be able to guess whether a linear relationship will describe a situation just by looking at the equation. You can find an equation for a line from a graph or a table of values. From a graph, you can determine the slope a and the y-intercept b. From a table, you may be able to use the method of constant differences to find the slope. If you can't get the y-intercept directly from the table, you can use the slope and one data pair to calculate it.

Strategies and Applications

The questions in this section will help you review and apply the important ideas and strategies developed in this chapter.

Understanding and representing rates and proportional relationships

1. After her last haircut, Angelina's hair was 8 inches long. In 3 months it grew another inch. Assume that the hair growth was steady over the 3 months and that it continues at the same rate.

   a. Write the rate that shows Angelina's hair growth per month.

   b. If you drew a graph showing the relationship between Angelina's hair length and time, where would the graph start?

   c. Is Angelina's hair length proportional to the time it is allowed to grow since the haircut? Explain.

   d. Write the rule for this relationship. Be sure to define your variables.

impactmath.com/chapter_test

Recognizing linear relationships from tables and graphs

2. The thickness of a book depends on several things, including the number of sheets of paper it has and how thick the paper is. For one type of paper, a stack of sheets 1 inch thick contains 250 sheets.

a. Complete the table for this relationship.

Number of Sheets	0	250	500	750	1,000	1,250
Thickness of Stack (in.)		1				

b. From the table, do you think this is a linear relationship? Explain.

c. Plot the data points in your table. Put the number of sheets on the horizontal axis and the thickness of the stack on the vertical axis.

d. From your graph, do you think this is a linear relationship? Explain.

e. Is this relationship proportional? Explain.

Understanding slope and *y*-intercept in graphs

3. Dion left home in his car and drove 3 miles to the gas station. At a speed of 45 mph, he continued in the same direction to work. The drive from the gas station to work took 25 minutes.

a. Write a rule that represents Dion's distance from home at any time since he left the gas station until he arrived at work.

b. Graph this relationship.

c. What is the slope of the line? What does this number represent?

d. What are the coordinates of the point where the line intersects the vertical axis? What does this tell you about Dion's distance from home?

e. Is the relationship between Dion's distance from home and the time proportional?

Understanding the connection between a linear equation and its graph

4. Consider these rules for lines.

    i. $a = b + 1$ **ii.** $a = {}^-3b + 2$

    iii. $a = 2b - 1$ **iv.** $a = {}^-b - 2$

a. Predict what the graph of each rule will look like before you draw it. Will the line go up or down as you look from left to right? Where will it intersect the vertical axis?

b. On one grid, draw the graph for each rule. Test your predictions on a new rule and its graph.

5. The graph shows a linear relationship. Write an equation for the line, and explain how you found your equation.

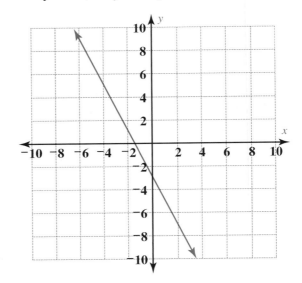

Using the method of constant differences to recognize and write equations for linear relationships

For each table in Questions 6–9, do Parts a and b.

a. Try to use the method of constant differences to decide whether the table could describe a linear relationship. If you can't use this method, explain why not, and find another way to decide whether the relationship could be linear.

b. For the relationships that could be linear, write an equation. For the relationships you think are not linear, plot the points to check.

6.

a	⁻2	⁻1	0	1	2	3	4
r	⁻3	⁻1	1	3	5	7	9

7.

x	⁻1	0	1	2	3	4	5
y	2	1	2	5	10	17	26

8.

m	0	1	2	3	4	5
n	⁻2.5	⁻0.5	1.5	3.5	5.5	7.5

9.

t	1	3	4	8	11	12
s	8	18	23	43	58	63

Demonstrating Skills

Graph each linear relationship.

10. $y = 3x + 19$ **11.** $y = {}^-1.7x + 7$ **12.** $y = {}^-5x - 1$

Decide whether you need two points or more than two points to graph each relationship.

13. $y = x^2 + 3$ **14.** $y = 8 + 15x - 7$

Decide whether each set of number pairs could describe a linear relationship.

15. $(0, 20); (2, 0); (1, 10); (3, {}^-10); ({}^-1, 30)$

16. $({}^-1, 21); (1, 25); (3, 29); ({}^-3, 17); (0, 23)$

Decide whether each linear relationship is proportional.

17. $f = 5g$ **18.** $p = {}^-15r + 1$ **19.** $u = {}^-7 + 3v$

Graph a line with each of the given slopes.

20. 7 **21.** $^-2.5$ **22.** 0.26 **23.** $^-348$

Determine the slope and give the coordinates of the y-intercept for each relationship.

24. $y = 3x + 9$ **25.** $y = {}^-15x + 39$ **26.** $y = x - 17$

Find a rule for each pattern.

27.

b	0	1	2	3	4	5	6
g	$^-5$	$^-1$	3	7	11	15	19

28.

b	0	1	2	3	4	5	6
g	13	7	1	$^-5$	$^-11$	$^-17$	$^-23$

Find a rule for each graph.

29.

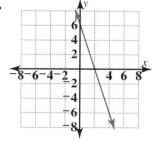

30.

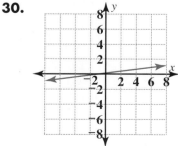

CHAPTER 6

Solving Equations

For Your Amusement Imagine you and your friends are going to an amusement park that was just built. This park has the most stomach-wrenching roller coasters and gravity-defying rides in the country. You are on your way to an exciting day, and you're leaving school and all your studies behind, right? But wait! You'll need to bring along your knowledge of equations so you get the most for your money. Think about all the fun you'll have, and just think: you'll be solving equations while you're at it.

Think About It A vendor offers to sell you one baseball cap for $2.75, or for $8 she'll sell you three. Which is the better deal?

Family Letter

Dear Student and Family Members,

Our next chapter is about solving equations. We will be reviewing the two methods of solving equations that we learned in previous chapters: guess-check-and-improve and backtracking (reasoning backward). Then we will learn to use a more efficient method: doing the same thing to both sides of the equation, which is probably the method you are most familiar with. Here is an example of the kind of problem we will explore.

Problem: Madeline and Neil collect autographs of basketball players. Madeline has 9 more autographs than Neil. They have 31 autographs in all. How many does each person have?

Let a represent how many autographs Neil has. Then write an equation that shows that Madeline and Neil have a combined total of 31 autographs: $a + a + 9 = 31$ or $2a + 9 = 31$.

Vocabulary Along the way, we'll be learning about these two new vocabulary terms:

conjecture   **model**

What can you do at home?

See if you and your student can find examples of equations in your work or in places you visit. Can you solve any of them?

Two Solution Methods Revisited

You already know at least two ways of solving equations: backtracking and guess-check-and-improve. *Backtracking,* which you used in Chapter 1, can help you solve many equations. For example, consider this equation.

$$\frac{5b + 3}{3} = 6$$

To find the solution, first set up a flowchart for the expression $\frac{5b + 3}{3}$. The flowchart shows the steps for calculating the value of the expression—the *output*—for any *input b.*

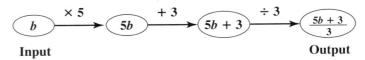

To solve $\frac{5b + 3}{3} = 6$, you need to find the input that gives an output of 6. Draw the flowchart again, this time entering 6 as the output and leaving the other ovals blank.

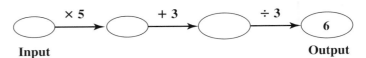

Now you can work backward, step by step, from the output to find the input. First ask yourself, "What number divided by 3 gives 6?" Write that number in the oval to the left of the 6.

Just the facts

Flowcharts are used in many professions. Builders use them to specify the steps in a house's construction, engineers to diagram how a part travels through the manufacturing process, and electricians to show how current flows in a circuit.

Think & Discuss

Continue backtracking from right to left to find the values for each oval in the flowchart. The number in leftmost oval is the solution of the equation.

What is the solution? Check your answer by substituting it into the equation.

You can use the *guess-check-and-improve* method to solve any equation, even equations that are difficult to solve by backtracking. To use this method, *guess* what the solution might be and then *check* your guess by substituting it into the equation. If your guess is incorrect, guess again. By comparing the results of each substitution, you can *improve* your guesses and get closer and closer to a solution.

Think & Discuss

Consider the equation $3b + 5 = b + 17$.

Suppose you begin with 1 as a guess for the solution. Substituting 1 for b gives 8 for the left side of the equation and 18 for the right side. You might enter these results in a table.

Guess	$3b + 5$	$b + 17$
1	8	18

Try 4 as your second guess. Substitute 4 for b in both sides of the equation, and record your answers in a copy of the table.

For a guess of 1, the difference between the two sides of the equation is 10. Are the two sides closer together or further apart when you substitute 4?

If 4 is not the solution, continue to adjust your guess until the two sides of the equation have the same value. What is the solution? Explain.

Investigation 1 Choosing a Solution Method

Backtracking is sometimes much easier or more efficient than guess-check-and-improve. You will now practice using these two methods to solve equations. As you work on the problems, think about how you could decide which method is easier for a particular equation.

Problem Set A

Use backtracking to solve each equation. Check your solutions by substituting them into the original equations.

1. $3f - 2 = 13$

2. $3(2g + 12) = 51$

3. $\frac{2a + 1}{3} = 5$

4. $2\left(\frac{3n + 4}{5} - 1\right) = 6$

Use guess-check-and-improve to solve each equation.

5. $u + 4 = 4u - 8$

6. $2q - 2 = 4q - 7$

7. $m(m - 6) = 91$ (There are two solutions; try to find both.)

For each equation, discuss with your partner which solution method—backtracking or guess-check-and-improve—would be easier or more efficient. Then solve the equations.

8. $\frac{5k - 4}{3} = 7$

9. $\frac{3n + 5}{2} = \frac{5n + 3}{3}$

10. $5B + 2(B + 1) = 16$

11. $4j + 74 = 478$

12. Look at the equations in Problems 8–11. Are any of them difficult to solve by backtracking? If so, explain why backtracking is difficult to use for these equations.

Solve each equation using any method you like. Check your solutions.

13. $4(5g - 1) = 56$

14. $\frac{3d + 1}{3} = 6d - 3$

15. $35 - 2s = 11$

16. $4x + 3 = {}^-3x - 4$

Algebra is a powerful tool for solving real and important problems. The situations that follow are generally much easier to solve than real problems, but they will help you develop the algebraic skills you need to solve them.

Problem Set B

1. Kim has $20 now and saves $6 per week. Noah has $150 now and spends $4 per week.

   a. How much money will Kim have *n* weeks from now?

   b. How much money will Noah have *n* weeks from now?

   c. Write an equation that represents Kim and Noah having the same amount of money after *n* weeks.

   d. Solve your equation for *n*. What does the solution tell you about this situation?

   e. How can you check your solution?

2. Inez's family is purchasing a new stereo system from Sharon's Sound Shop. The salesperson offers them a choice of payment plans.

Plan A: Make a $100 down payment and then pay $35 per month for 24 months.

Plan B: Make no down payment and pay $40 per month for 24 months.

   a. Which plan is more expensive? Explain.

   b. For each plan, write an expression for the amount Inez's family will have paid after *n* months.

   c. After how many months will they have paid the same amount no matter which payment plan they select?

In Problems 3–7, write an equation that fits the situation. Solve the equation using whatever method you like. Check your solution.

3. Finding 5 more than a given number gives the same result as tripling 1 more than that number. What is the number?

4. One angle of a triangle measures *f* degrees. One of the other angles of the triangle is twice this size, and the third angle is three times this size. Find the measures of all three angles.

5. A rectangular field is three times as long as it is wide. The fence around the field's perimeter is 800 meters long. How wide is the field?

6. An apartment building has a TV antenna on its roof. The top of the antenna is 66 meters above the ground. The building is 10 times as tall as the antenna. How tall is the antenna?

7. In a class of 25 students, there are 7 more girls than boys. How many boys are in the class?

Remember
The sum of the measures of the angles in any triangle is 180°.

Share & Summarize

1. Write an equation that *cannot* be solved easily by backtracking. Explain why backtracking would be difficult.

2. Together, Rachel and Zach have $14. If Rachel had $1 more, she would have twice as much money as Zach.

   a. Write and solve an equation to find how much money Rachel and Zach each have.

   b. Explain how you wrote your equation and how you decided which method to use to solve it.

Lab Investigation ▶ Using a Spreadsheet to Guess-Check-and-Improve

MATERIALS

computer with spreadsheet software (1 per group)

Guess-check-and-improve is a great method for solving simple equations, but it can be tedious when the equations are more complicated. The process is a lot faster and more fun when you have a computer do the calculations while you do the thinking.

Sonia decided to use a spreadsheet to help her solve this equation using guess-check-and-improve:

$$\frac{3n + 5}{2} = \frac{5n + 3}{3}$$

Setting Up the Spreadsheet

First, Sonia entered headings and formulas in the top two rows of the first four columns of her spreadsheet.

	A	B	C	D
1	Test Variable n	Left Side $(3n + 5)/2$	Right Side $(5n + 3)/3$	Difference
2		=(3*A2+5)/2	=(5*A2+3)/3	=B2-C2
3				

As Sonia entered guesses for n into her spreadsheet, she observed how the results changed so she could make a good choice for her next guess. When her guess is correct, the two sides of the equation will have the same value.

1. The formula in Cell D2 calculates the difference between the values in Cells B2 and C2. How do you think Sonia will use the difference to help make her next guess?

Sonia started guessing the solution by entering 0 in Cell A2. Here are her results.

Sonia suspected some results might not be whole numbers, so she formatted the numbers in Cells B2, C2, and D2 to show three decimal places.

	A	B	C	D
1	Test Variable n	Left Side $(3n + 5)/2$	Right Side $(5n + 3)/3$	Difference
2	0	2.500	1.000	1.500

She decided to try 1 for her next guess, so she entered 1 in Cell A2.

	A	**B**	**C**	**D**
1	Test Variable n	Left Side $(3n + 5)/2$	Right Side $(5n + 3)/3$	Difference
2	1	4.000	2.667	1.333

Try It Out

2. Set up a spreadsheet like Sonia's on your computer. Enter Sonia's first two guesses to make sure your spreadsheet works.

3. Use your spreadsheet to solve Sonia's equation. How do you know when you have found the solution?

4. Use your spreadsheet to solve each equation below. You will have to change the formulas in Cells B2 and C2 for each equation. Pay attention to the strategies you use to improve your guesses. You will be asked about them later!

    a. $4S - 70 = 18S$ **b.** $4n + 1 = \frac{3n + 8}{2}$

5. What problem-solving strategies did you use to help you improve your guesses?

Try It Again

Sonia's friend Teresa solved the equation by making a spreadsheet to try a series of 20 guesses at once. She started by setting up a spreadsheet like Sonia's. Teresa then used the spreadsheet's "fill down" command to copy the formulas in Cells B2, C2, and D2 into Rows 3–21 of Columns B, C, and D.

To make a series of guesses, Teresa modified Column A. She entered 0 in Cell A2 and the formula A2 + 1 in Cell A3. Then she copied this formula into Cells A4 to A21. The program created a column of numbers that increased by 1, from 0 in Cell A2, to 19 in Cell A21.

	A	**B**	**C**	**D**
1	Test Variable n	Left Side $(3n + 5)/2$	Right Side $(5n + 3)/3$	Difference
2	0	=(3*A2+5)/2	=(5*A2+3)/3	=B2−C2
3	=A2+1	=(3*A3+5)/2	=(5*A3+3)/3	=B3−C3
4	=A3+1	=(3*A4+5)/2	=(5*A4+3)/3	=B4−C4
5	=A4+1	=(3*A5+5)/2	=(5*A5+3)/3	=B5−C5

6. Try this on your spreadsheet. How will Teresa know which value of n gives a correct answer? Where is the solution on Teresa's spreadsheet?

Remember

Answers may be positive or negative, and they may be whole numbers or decimals.

Use Your Spreadsheet

Of course, not every equation has a solution that is an integer between 0 and 19.

7. Try to solve the equation $\frac{8x + 3}{3} = x - 3$ using Teresa's method. Make your first series of guesses the set of integers from 0 to 19.

   a. Notice that the differences in Column D increase as the values in Column A increase. These guesses don't seem to be getting any closer to a solution! How could you modify Teresa's strategy to get closer?

   b. Modify the formula in Column A to match your new strategy. Does 0 ever appear in Column D? If not, where does it look like the solution to this equation should be?

   c. Refine your guesses, if needed, and solve the equation.

Use a spreadsheet like Sonia's or Teresa's to solve each equation below. Some solutions may not be positive and may not be integers. If you use Teresa's method, you may need to try more than one series of test values.

Pay attention to the strategies you use to improve your guesses. You will be asked about them later.

   8. $15x + 32 = 3x - 16$

   9. $3n + 2 = n - 7$

  10. $\frac{7P + 34}{3} = \frac{13P - 4}{2}$

  11. Choose one of Problems 8–11 in Problem Set A. Use a spreadsheet to solve the equation.

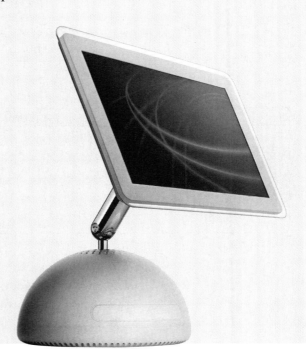

What Did You Learn?

12. When you solve equations with a spreadsheet, what strategies can you use to improve your guesses? Be sure to consider how to find negative and decimal solutions as well as positive integers.

13. Analyze this spreadsheet.

	A	B	C	D
1	Test Variable n	Left Side $4n + 2$	Right Side $7n + 12$	Difference
2		=4*A2+2	=7*A2+12	=B2−C2
3				

    a. What equation was this spreadsheet set up to help solve?

    b. What do the formulas in Cells B2, C2, and D2 mean?

    c. What results will you get if you enter 1 in Cell A2?

On Your Own Exercises

Practice & Apply

Use backtracking to solve each equation.

1. $4d + 7 = {}^-15$

2. $4(2w + 1) = 52$

3. $\frac{4x + 7}{5} = 3$

4. $3\left(\frac{2c + 1}{5} + 1\right) = 6$

Use guess-check-and-improve to solve each equation.

5. $b + 5 = 3b - 9$

6. $4t - 3 = 3t + 7$

7. ${}^-4r + 8 = 5r - 10$

Which would be a better way to solve each equation, backtracking or guess-check-and-improve? Use the method you think is best to find each solution.

8. $\frac{6y - 7}{3} = 5$

9. $4u + 7 = 5 + 3u$

10. $5t + 34 = 17$

11. Marcus has 20 comic books and gives 3 to his cousin Lelia every week. Mike has 10 comic books and buys 2 new ones each week.

   a. How many comic books will Marcus have after w weeks?

   b. How many comic books will Mike have after w weeks?

   c. Write an equation that represents Marcus and Mike having the same number of comic books after w weeks.

   d. Solve your equation to find after how many weeks the two boys will have the same number of comic books. Check your solution.

12. Kiran said, "Eleven more than my number is the same as 18 more than twice my number. What is my number?"

13. In a box of 64 red and blue whistles, there are 16 more blue whistles than red whistles. Write and solve an equation to determine how many are red.

14. Solve this equation by guess-check-and-improve. There are two solutions; find both.

$$m(m + 2) = 168$$

In your **own words**

Describe the types of equations that are better solved by backtracking, and the types that are better solved by guess-check-and-improve. Explain your reasoning.

impactmath.com/self_check_quiz

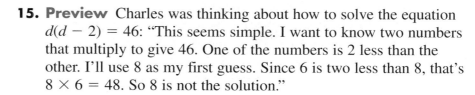

Connect & Extend

15. Preview Charles was thinking about how to solve the equation $d(d - 2) = 46$: "This seems simple. I want to know two numbers that multiply to give 46. One of the numbers is 2 less than the other. I'll use 8 as my first guess. Since 6 is two less than 8, that's $8 \times 6 = 48$. So 8 is not the solution."

   a. What should Charles choose as his next guess?

   b. Use guess-check-and-improve to find a solution for d to the nearest hundredth that gives a result within 0.1 of 46.

16. At a drive-in theater, admission is $4.00 per vehicle plus $1.50 per person in the car.

   a. Write an equation for this situation. Use a to represent the total admission, v the number of vehicles, and p the number of people.

   b. Use your equation to determine how much money the theater will collect if 43 cars carrying a total of 93 people see a movie.

   c. Use your equation to find the admission price for a car holding 3 people.

   d. The admission for one van was $16. Use your equation to find how many people were in the van.

17. Preview Try to solve this equation using any method you choose. What do you discover?

$$n^2 + 3 = 2$$

18. Jim and Lillian were playing *Think of a Number.* Jim said, "Think of a number. Triple it. Add 5. Multiply your result by 10." Lillian said she got 320.

   a. What equation could Jim solve to find Lillian's number?

   b. Use backtracking to solve the equation.

Just the facts

The first drive-in opened June 3, 1933, in Camden, New Jersey. The idea was a success, with close to 6,000 drive-ins operating by 1961. But as the car became a part of daily life, the thrill of sitting in one to watch a movie declined: by 2002, fewer than 500 drive-ins remained in the U.S.

Source: www.driveintheatre.com

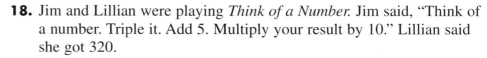

Find each sum.

19. $0.5 + 0.5$ **20.** $0.5 + 0.51$ **21.** $0.5 + 0.6$ **22.** $0.5 + 1.05$

23. Simplify each expression as much as possible.

x^4	M
x^2	O
$2x^2$	R
1	D
x^6	A
x	N

a. $x^2 + x^2$ **b.** $(x^2)^3$ **c.** $x^3 \div x^2$

d. $\frac{2x^2}{x^2} - 1$ **e.** $2x^2 - x^2$ **f.** $\frac{(x^2)^3}{x^2}$

g. Match each simplified expression to an answer in the grid at left, and write the corresponding letters on lines like those below. What word do you find?

$\underline{\hspace{2em}}$ $\underline{\hspace{2em}}$ $\underline{\hspace{2em}}$ $\underline{\hspace{2em}}$ $\underline{\hspace{2em}}$ $\underline{\hspace{2em}}$
  a   b   c   d   e   f

24. Consider this pattern of squares.

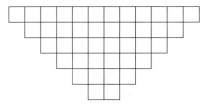

Stage 1

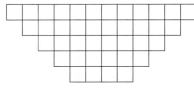

Stage 2

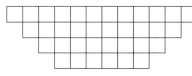

Stage 3

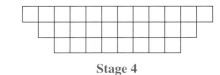

Stage 4

a. For the stages shown, tell how many squares are removed to get from one stage to the next.

b. How many squares do you remove to go from Stage s to Stage $s + 1$?

c. How many squares in all are in each stage shown?

d. How many squares are needed for Stage 5? For Stage 7?

e. Which stage has 100 squares?

f. Write an expression for the number of squares in Stage s.

6.2 A Model for Solving Equations

In this lesson and the next, you will learn an equation-solving method that is more efficient than guess-check-and-improve, and that can often be used when backtracking won't work.

First, you will learn to use a model to help you think about solving equations. The word *model* has several meanings. A model can be something that *looks* like something else. For example, you have probably built or seen model cars or airplanes that look just like the real vehicles, with all the parts of the model proportional to the actual parts.

V O C A B U L A R Y
model

In mathematics, a **model** is something that has some of the *key characteristics* of something else. In Chapter 1, you used bags and blocks as a model for algebraic expressions. Just as you can put any number of blocks into a bag, you can substitute any value for the variable in an expression. In Chapter 2, you used blocks as models for people of different sizes. Although the block models didn't look like people, the relationship of surface area to volume in the models was similar to that in human beings.

Mathematical and scientific models help people understand complex ideas by simplifying them or making them easier to visualize. In this lesson, bags and blocks are again useful as models for algebra expressions. By adding a balance, you can model algebraic *equations* and find more efficient ways to solve them.

Explore

On this balance, each bag contains the same number of blocks. The bags weigh almost nothing compared to the blocks.

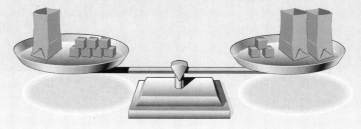

How many blocks are in each bag? Explain how you found your answer.

Check to make sure your answer is a solution to the puzzle.

Investigation Making and Solving Balance Puzzles

Now it's your turn to draw some balance puzzles to try with your classmates. Here are the rules for creating balance puzzles.

- In any puzzle, all bags must hold the same number of blocks.

- The two sides of the balance must have the same total number of blocks, with different combinations of bags and blocks on each side.

Problem Set A

1. Work by yourself to draw a puzzle with no more than 10 blocks on each side of the balance, including the blocks hidden in the bags. Exchange puzzles with your partner, and solve your partner's puzzle.

2. Now draw a more difficult puzzle, using a total of 12 or more blocks on each side. Exchange puzzles with your partner, and solve your partner's puzzle.

As you work with the balance puzzles in Problem Set B, try to think of more than one strategy for solving each puzzle.

Problem Set B

1. Zoe and Luis challenged their classmates to solve this balance puzzle.

a. How many blocks are in each bag? Check your solution.

b. What strategy did you use to solve the puzzle?

Zoe and Luis's classmates created similar puzzles. Solve each puzzle by finding how many blocks are in each bag.

2. Tanya and Craig's puzzle

3. Margaret and Ana's puzzle

4. Imran and Zach's puzzle

5. Lisa and Trent's puzzle

Share & Summarize

Select one of the balance puzzles from Problem Set B. Explain your solution method so that someone in another class could understand it. Be sure to justify your reasoning.

Investigation 2 Keeping Things Balanced

By now you have probably seen and used a variety of strategies to solve balance problems. In this investigation, you will focus on solving puzzles by "keeping things balanced." Soon you will see how the thinking involved in this strategy can help you solve equations.

EXAMPLE

Here's how Malik solved Problem 5 of Problem Set B.

I'm trying to keep the pans balanced. If I remove the same thing from each side, what is left will still balance. I could remove a bag from each side and have this.

I could then take 2 more bags and a block from each side, like this.

What is left must be equal, so each bag holds 5 blocks. Check it: 3 bags and 6 blocks is $3 \cdot 5 + 6 = 21$ blocks.

And 4 bags and 1 block is $4 \cdot 5 + 1 = 21$ blocks. It works!

Malik kept things balanced by taking the same number of bags or blocks from each side until he had only 1 bag on the right and 5 blocks on the left.

Problem Set C

Solve each puzzle by keeping things balanced. For each step, record what you do and what is left on each side of the balance. You may draw each step if you prefer. Be sure to check your solutions.

1.

2.

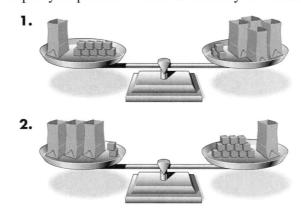

In Chapter 1, you wrote algebraic *expressions* to represent situations involving bags and blocks. In the same way, you can write algebraic *equations* to describe balance puzzles.

EXAMPLE

Here's how Kate made a balance puzzle.

On this side, Kate put 2 bags of blocks and 5 single blocks. If n stands for the number of blocks in each bag, then the number of blocks in "two bags and five blocks" is $2n + 5$.

On this side, Kate put 3 bags of blocks and 2 single blocks. The number of blocks in "three bags and two blocks" is $3n + 2$.

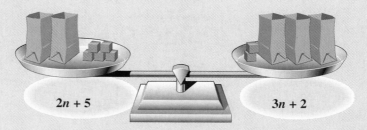

Since the sides balance, the number of blocks on the left, $2n + 5$, is equal to the number of blocks on the right, $3n + 2$. This can be expressed with the equation

$$2n + 5 = 3n + 2$$

When you find the number of blocks in each bag, you have found the value of n.

Problem Set D

1. Look at Kate's balance.

    a. Find the number of blocks in each bag.

    b. Check that your answer is a solution of the equation $2n + 5 = 3n + 2$.

2. Nguyen created this balance puzzle.

a. Let *b* stand for the number of blocks in each bag. Write an equation that fits Nguyen's balance puzzle.

b. Use the drawing to help you find the value of *b*.

c. Check that your answer to Part b is a solution of your equation.

Think & Discuss

Shamariah created this balance puzzle.

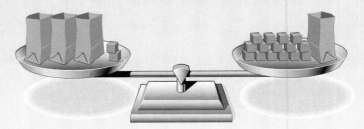

• Write an equation to represent Shamariah's puzzle.

• How many blocks are in each bag?

• Check that your answer is a solution of your equation.

Joel said that Shamariah wouldn't be able to create her puzzle with real blocks, because each bag would have to include a half of a block. Shamariah said she was thinking of a slightly different model—one with blocks made of clay so they could be easily divided. That way, her balance puzzles can include fractions of blocks.

Problem Set **E**

In these puzzles, the bags might hold fractions of blocks.

1. Helena made this puzzle.

a. Write an equation that fits Helena's puzzle.

b. How many blocks are in each bag? Explain how you found your solution.

c. Check that your answer to Part b is a solution of your equation.

2. Toby made this puzzle.

a. Write an equation that fits Toby's puzzle.

b. How many blocks are in each bag? Explain how you found your solution.

c. Check that your answer to Part b is a solution of your equation.

Share & Summarize

1. Make up a balance puzzle for which the solution is not a whole number of blocks. Exchange puzzles with your partner, and solve your partner's puzzle.

2. Write an equation that describes your puzzle. Explain how the equation fits the puzzle.

3. Check that the solution to your puzzle is also a solution of the equation.

Investigation ▶3 Solving Problems with Balance Puzzles

You have seen that the solution of a balance puzzle is the same as the solution of an equation that represents the puzzle. Now you will solve equations by creating and solving balance puzzles. In some cases, the solutions may not be whole numbers.

Problem Set F

1. Consider the equation $3n + 8 = 5n + 2$.

   a. Draw a balance puzzle to represent the equation. Explain how you know your puzzle matches the equation.

   b. Use your puzzle to solve the equation. Check your solution by substituting it into the equation.

2. Consider the equation $4t + 6 = 13 + 2t$.

   a. Draw a balance puzzle to represent the equation. Explain how you know your puzzle matches the equation.

   b. Use your puzzle to solve the equation. Check your solution by substituting it into the equation.

Try to solve these equations by imagining a balance puzzle, rather than actually drawing one. Check each solution by substituting it into the equation.

3. $2h + 3 = h + 10$
4. $4t + 6 = 13 + 3t$

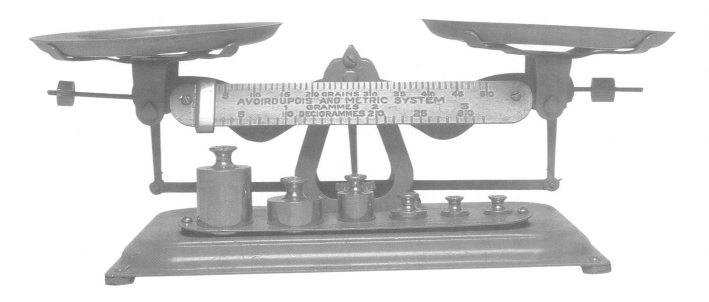

If you can represent a problem situation with an equation, you can sometimes draw or imagine a balance puzzle to find the solution.

Problem Set G

1. Jenny was filling bottles from a tank of water. When she had filled two bottles, 7 liters of water remained in the tank. When she had filled four bottles, 2 liters of water were left in the tank.

   a. Let v represent the amount of water each bottle holds. Write an equation you could solve to find the value of v.

   b. Solve your equation by imagining a balance puzzle. Check your solution.

2. Maya said, "I am thinking of a number. If I multiply my number by 6 and add 5, I get the same result as when I multiply it by 3 and add 17."

   a. Use M to represent Maya's number. Write an equation you could solve to find the value of M.

   b. Solve your equation by imagining a balance puzzle. Check your solution.

Share & Summarize

Explain how you can use a balance puzzle to solve an equation. Your explanation should be clear enough that someone in another class could understand it.

On Your Own Exercises

Practice **Apply**

The students in Ms. Avila's class made some balance puzzles. In Exercises 1–4, tell how many blocks are in each bag.

1. Chondra and Toshiro's puzzle
2. Alberto and Elena's puzzle

Remember
In a balance puzzle, every bag holds the same number of blocks.

3. Benita and Ayana's puzzle
4. Molly and Jonas's puzzle

5. Make up three balance puzzles with 11 blocks total on each side. Write an equation for each puzzle, and solve each equation.

6. Raphael and Gary made this balance puzzle. How many blocks are in each bag? Write down each step you take to solve this puzzle.

7. Consider this balance puzzle.

    a. Write an equation to fit this puzzle. Let *n* represent the number of blocks in each bag.

    b. Use the drawing to find the value of *n*.

    c. Check that your answer to Part b is a solution of your equation.

 impactmath.com/self_check_quiz

8. Consider the equation $3p + 10 = 5p + 3$.

    a. Draw a balance puzzle to represent the equation. Explain how you know your puzzle matches the equation.

    b. Use your drawing to solve the equation. Check your solution by substituting it into the equation.

9. Nicky said, "The sum of my number and 21 is the same as 4 times my number. What is my number?"

    a. Use N to represent Nicky's number. Write an equation you could solve to find the value of N.

    b. Solve your equation by imagining a balance puzzle. Check your solution.

10. The balances below each hold different sets of blocks, jacks, and marbles. Each marble has a mass of 10 grams.

    a. Find the mass of each jack and the mass of each block. Describe the steps you take.

    b. This balance has blocks, marbles, and jacks from Part a, plus some toothpicks. Find the mass of each toothpick.

11. Sareeta and her brother Khalid put candies in bags to give away at Sareeta's birthday party. They started with an equal number of candies and put the same number in each bag. Sareeta filled five bags and had one candy left. Khalid filled four bags and had seven candies left (which he decided to eat).

  a. Let n represent the number of candies in a bag. Write an expression that describes the way Sareeta distributed her candies.

  b. Write an expression that describes Khalid's four bags and seven extra candies.

  c. Because Sareeta and Khalid started with the same number of candies, it is possible to balance their distributions of candies. Draw a balance puzzle to show this.

  d. Write an equation that matches your balance puzzle.

  e. Use your balance puzzle to find how many candies are in each bag.

  f. Show that your answer to Part e is a solution of the equation.

12. Read the story about Sareeta and Khalid in Exercise 11.

  a. Make up a similar story that can be represented by the equation $3n + 2 = 2n + 9$.

  b. Draw a balance puzzle for this equation, and find the value of n.

  c. Show that your answer to Part b is a solution of the equation.

13. Consider the equation $2v + 50 = 7v + 30$.

  a. Describe a situation that matches this equation.

  b. Find the value of v. Check your solution.

  c. Explain what your solution means in terms of the situation you described in Part a.

In y o u r

own

words

Susan said, "When I think about solving equations using a balance puzzle, I think about doing the same thing on each side." What do you think she means?

14. Study the pattern of shapes.

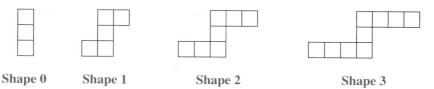

Shape 0   Shape 1   Shape 2   Shape 3

a. How many squares will be in Shape 4? In Shape 5?

b. Write an expression for the number of squares in Shape *n*.

c. How many squares will be in Shape 20?

d. Write an equation to find which shape will have 25 squares. Solve your equation.

15. Challenge Sergei and Hilary were cutting streamers of a particular length for a party. Sergei held up a strip of crepe paper and said he could cut one streamer from it and have 3 feet left over. Hilary said that if the strip of crepe paper were just 1 foot longer, she could cut exactly two streamers from it.

a. Find the length of the streamers Sergei and Hilary were cutting. Show how you found your answer.

b. Find the length of Sergei's strip of crepe paper.

Mixed Review

Find each product.

16. $^-4 \cdot 12$ **17.** $8 \cdot {}^-2.5$ **18.** $^-2 \cdot 4 \cdot {}^-8$

Find each quotient.

19. $^-50 \div 4$ **20.** $^-4 \div 48$ **21.** $^-8 \div {}^-64$

Rewrite each number in scientific notation.

22. 800   **23.** 24,000   **24.** 31,500,000

25. Draw a factor tree to find all the prime factors of 12.

26. Solve the equation $7a + 5 = 2a + 15$ for a.

Graph each linear relationship. Put g on the vertical axis.

27. $g = \frac{3}{2}a - 4$ **28.** $g = -\frac{1}{4}a + 2$ **29.** $g = {}^{-}a - 1.5$

30. Statistics The table shows the approximate distribution of blood types in the U.S. population. "Positive" and "negative" indicate whether a person's blood has a particular component, called the *Rh factor.*

Blood Types in the U.S.

Blood Type	Percent
O positive	38
O negative	7
A positive	34
A negative	6
B positive	9
B negative	2
AB positive	3
AB negative	1

a. What percentage of people in the United States have Rh-negative blood?

b. The projected U.S. population in 2050 is 391,000,000 people. If the distribution of blood types remains fairly constant, how many people in 2050 will have AB negative blood?

c. In 2050, about how many more people will have Rh-positive blood than Rh-negative?

d. Suppose the percentage of people with B negative blood doubles by the year 2050. How many people in 2050 will have this blood type?

Just the facts

The Rh factor is a blood protein that is present on the red blood cells of about 85% of people. It is called Rh because it was first identified in the blood of rhesus monkeys.

Thinking with Symbols

Kate solved the equation $2k + 11 = 4k + 3$ by using a balance puzzle.

> I'll show $2k + 11 = 4k + 3$ as bags and blocks on a balance. One side of the balance has 11 blocks and 2 bags. The other side has 4 bags and 3 blocks.

> I can remove 2 bags from each side, and the scale will still balance. Taking 2 bags of k from each side, I have $11 = 2k + 3$ left.

> If I take 3 blocks from each side—what will be left? Let's see...I'll have $8 = 2k$ left. I can see that 2 bags must hold 8 blocks...so 1 bag must hold 4 blocks: $4 = k!$

Kate summarized her solution in symbols.

$$2k + 11 = 4k + 3$$
$$11 = 2k + 3 \quad \text{after taking 2 bags from each side}$$
$$8 = 2k \quad \text{after taking 3 blocks from each side}$$
$$4 = k \quad \text{after taking half of what's left on each side}$$

Explore

Think about how you would solve the equation $14 + 3a = 7a + 2$ by using a balance puzzle. For each step in your solution, use an equation to express what is left on the balance. Then summarize your solution using symbols like Kate did. If it's helpful, draw what's left on the balance after each step.

Investigation ▶ 1 Writing Symbolic Solutions

You have solved equations by drawing and imagining balance puzzles. Now you will practice using symbols to record the steps in your solutions.

Problem Set A

Solve each equation by drawing or imagining a balance puzzle. Summarize your solution in symbols as Kate did.

1. $2x + 7 = 5x + 4$

2. $4r + 20 = 10r + 5$

For the balance at the beginning of this lesson, Kate labeled her solution to describe how the sides of the *balance* changed at each step.

Here is Kate's solution again, but this time the labels describe the *mathematical operations*. They show how the *equation* changed at each step.

> **EXAMPLE**
>
> $2k + 11 = 4k + 3$
> $11 = 2k + 3$ after subtracting $2k$ from each side
> $8 = 2k$ after subtracting 3 from each side
> $4 = k$ after dividing each side by 2

Problem Set B

1. Copy your symbolic solutions for both problems of Problem Set A. Next to each step, explain how the *equation* changed from the previous step.

Try to solve each equation below by working with just the symbols and doing the same operations on both sides. Try to make the equation simpler each time. Next to each step in your solution, explain how the equation changed from the previous step. If you have trouble, draw or imagine a balance puzzle.

2. $3y + 17 = 4y + 6$

3. $3x + 4 = x + 14$

Share & Summarize

Share Summarize

Explain how doing the same mathematical operation to both sides of an equation is like doing the same thing to both sides of a balance puzzle.

Investigation ▶2 Doing the Same Thing to Both Sides

Balance puzzles and equations can be solved by doing the same thing to both sides. Each step simplifies the problem, until the solution is clear.

Problem Set C

Try to solve each equation by doing the same mathematical operation to both sides *without* thinking about a balance puzzle. Next to each step in your solutions, explain how the equation changed from the previous step.

1. $5x = x + 4$

2. $r + 12 = 3r + 4$

3. $2P + 20 = 5P + 6$

4. $5y + 3 = 3y + 15$

5. Kenneth made up a number puzzle.

I'm thinking of a number. If I multiply the number by 5 and subtract 9, I get twice my original number.

Write an equation you can solve to find Kenneth's number. Solve your equation by doing the same thing to both sides. Next to each step in your solution, explain how the equation changed from the previous step. Be sure to check your solution.

Kenneth's number puzzle is an example of when thinking about a balance puzzle would not be helpful. This is because it's difficult to show subtracting 9 blocks from 5 bags using a balance puzzle—try it and see!

Doing the same thing to both sides works for a great many equations. From now on you can use this method to solve any equation—whether or not the equation fits the model of a balance puzzle.

Problem Set D

Solve each equation by doing the same thing to both sides. If it helps, explain how the equation changed at each step. Check your solutions.

1. $2P + 7 = 4P + 10$

2. $3r - 2 = r$

3. $2a - 6 = 0$

4. $y - 3 = 2y - 6$

5. $1.5x + 3.5 = 8 + x$

6. $\frac{4}{5}x = 12$

7. Zach, Luis, and Shaunda found the correct solution to this equation:

$$5r - 16 = r$$

Zach started by subtracting $5r$ from both sides. Luis started by adding 16 to both sides. Shaunda started by dividing both sides by 5.

a. Solve the equation by starting with Zach's first step.

b. Solve the equation by starting with Luis's first step.

c. Solve the equation by starting with Shaunda's first step.

d. Whose method seems easiest? Explain why.

Share & Summarize

1. Consider the equation $2x - 1 = \frac{1}{2}x + \frac{7}{2}$.

a. Explain why the equation is difficult to solve by thinking about a balance puzzle.

b. Explain why the equation can be solved by doing the same thing to both sides.

2. Solve $2x - 1 = \frac{1}{2}x + \frac{7}{2}$ by doing the same thing to both sides. Explain how the equation changed at each step.

Investigation Solving More Equations

You have seen that by doing the same thing to both sides, you can solve equations that are difficult to solve with balance puzzles. Now you will practice solving some more equations.

Problem Set E

1. Solve the equation $5 - 2x = 3x$ by first adding the same amount to both sides. Check your solution.

2. Solve the equation $10 + 4z = 2z - 2$. Check your solution.

3. Solve the equation $\frac{1}{2}B = B - 4$ by first multiplying both sides by the same amount. Check your solution.

4. Solve the equation $\frac{k}{3} = 8 - k$.

Solve each equation by doing the same thing to both sides.

5. $4b - 2 = 6b - 4$

6. $2c + 1 = 9 + 3c$

7. $2d + 1 = 14 - d$

8. $5e + 2.5 = 2.5 - 7.5e$

9. Maya's hamster had a litter of babies. After Maya gave away three of the babies, $\frac{3}{4}$ of the litter remained. How many baby hamsters did Maya have *left*?

Just the facts

Hamsters average 6 to 8 babies per litter, but occasionally have 15 or even more! Babies are born furless, and their eyes don't open for one to two weeks.

When you solve an equation by doing the same thing to both sides, you usually try to make the equation simpler at each step. What if instead you made an equation *more complicated* at each step?

EXAMPLE

Start by writing the solution.	$x = 7$
Multiply both sides by 3.	$3x = 21$
Add $2x$ to both sides.	$5x = 21 + 2x$
Subtract 2 from both sides.	$5x - 2 = 19 + 2x$
Divide both sides by 2.	$2.5x - 1 = 9.5 + x$

In Problem Set F, you will create complicated algebra problems by starting with a simple equation and doing the same thing to both sides to make it more complex. When you are satisfied that your equation is sufficiently difficult, challenge a partner to solve it. See if you can stump your classmates!

Problem Set F

1. Check to make sure the solution of the final equation in the Example is 7.

2. Explain why 7 is the solution of *each* equation in the Example.

3. Make up your own "complicated" equation. Start by choosing a number. Write the equation $x = $ *your number*. Then make the equation more complicated by doing the same thing to both sides three or four times. Verify that the number you first chose is the solution of your final equation. (If it isn't, retrace your steps and fix the problem.)

4. Exchange equations with your partner, and solve your partner's equation.

Share & Summarize

Explain why the solution of an equation does not change when you alter the equation by doing the same thing to both sides.

On Your Own Exercises

Practice
Apply

Solve each equation by drawing or imagining a balance puzzle. Use symbols to record your solution steps. Check your solutions.

1. $2x + 9 = 6x + 1$ **2.** $6z + 28 = 12z + 10$

3. $29 + 6w = 11w + 4$ **4.** $9 + 5v = 2v + 12$

Solve each equation by working with the symbols. Next to each step in your solution, explain how the equation changed from the previous step. Check your solutions.

5. $2a + 10 = 6a + 2$ **6.** $9b + 28 = 12b + 7$

Solve each equation by doing the same thing to both sides. Check your solutions.

7. $3 + 9c = 12c$ **8.** $18 + d = 7d + 6$

9. $2k - 6 = 4k + 5$ **10.** $\frac{1}{4}m - 3 = 0$

11. $9.25n + 3.3 = 1.3 - 0.75n$ **12.** $0.1p + 6 = 7$

13. $17r + 4 = 34r + 2$ **14.** $2s + 7 = 10s - 7$

15. Darnell said, "If you multiply the number of dogs I have by 8 and then subtract 10, you get the same number as when you triple the number of dogs I have and add 5."

   a. Write an equation you can solve to find how many dogs Darnell has.

   b. Solve your equation by doing the same thing to both sides. Check your solution.

16. Karl said, "If I divide my dog's age by 4 and subtract 2, I get the same result as when I divide my dog's age by 8."

   a. Write an equation you can solve to find the dog's age.

   b. Solve your equation by doing the same thing to both sides. Check your solution.

17. Consider the equation $27q - 100 = 2q$.

    a. Solve the equation by first adding 100 to both sides.

    b. Solve the equation by first subtracting $2q$ from both sides.

    c. Solve the equation by first subtracting $27q$ from both sides.

    d. Which of your solutions was easiest?

Solve each equation by doing the same thing to both sides.

18. $\frac{1}{2}q + 3 = 10.5 + q$ **19.** $\frac{1}{3}c = 3c - 9$

20. $\frac{d}{7} = 12 + d$ **21.** $\frac{2}{7}d - 2 = 3d - 4\frac{5}{7}$

22. Kate is making lunch for her three best friends, but the only juice she has is in the little bottles her sister likes. She has 16 small bottles of apple juice. She fills a 10-ounce glass for each friend and one for herself, and she still has 6 bottles of juice left over. Write an equation you could solve to find how many ounces of juice a bottle holds. Solve your equation.

23. Write three equations with a solution of $^-5$.

Connect & Extend

24. Yago tried to solve this equation using a balance puzzle.

$$5r + 7 = 13 + 2r + 3r$$

What solution do you think Yago found?

25. Odell tried to solve the equation $6(x + 7) = 4x - 3$, and he recorded his steps as follows:

$6(x + 7) = 4x - 3$
       $6x = 4x - 10$ after subtracting 7 from both sides
       $2x = {}^-10$ after subtracting $4x$ from both sides
         $x = {}^-5$ after dividing both sides by 2

When Odell checked his solution, it didn't work. What was his mistake?

Challenge Solve each equation by doing the same thing to both sides.

26. $\sqrt{x} = 3$ **27.** $\sqrt{v} - 2 = 1$ **28.** $\sqrt{5x - 9} = 4$

In your
own
words

Describe two or
more types of
equations that are
difficult to model
with balance puzzles. Do you think
those equations
can be solved easily
by doing the same
thing to both
sides? Explain.

29. You can solve many equations by drawing two graphs and finding the point where they intersect. Consider the equation $2x + 3 = 1 + 3x$.

   a. On the same set of axes, graph $y = 2x + 3$ and $y = 1 + 3x$.

   b. Give the coordinates of the point where the graphs intersect.

   c. Now solve $2x + 3 = 1 + 3x$ by doing the same thing to both sides.

   d. How does the x-coordinate of the intersection point of the equations $y = 2x + 3$ and $y = 1 + 3x$ compare to the solution of $2x + 3 = 1 + 3x$? Explain why this makes sense.

   e. Solve $3x - 4 = 6 - 2x$ by drawing two graphs.

Solve each equation.

30. $3(x + 4) = {}^-5$ **31.** $7(x - 4) = 2(x + 1)$ **32.** $-\frac{x}{8} = \frac{3x + 4}{4}$

33. Kate has four packages of balloons and 8 single balloons. Her brother gave her 20 more balloons and three more packages. She now has twice the number of balloons she started with.

   a. Write and solve an equation to find how many balloons are in a package.

   b. How many balloons did Kate start with?

34. **Geometry** The volume of a cylinder is given by the formula $V = \pi r^2 h$, where r is the radius and h is the height.

   a. Cylinders A and B have the same volume. Write an equation you could solve to find the height of Cylinder B. Explain how you wrote your equation.

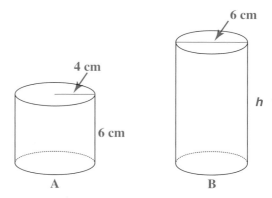

   b. Solve your equation to find the height of Cylinder B to the nearest 0.1 cm.

Evaluate each expression.

35. $\frac{2}{13} \cdot {}^-\frac{26}{7}$

36. $\frac{3}{8} \div {}^-\frac{1}{5}$

37. $0.05 \div {}^-40$

38. ${}^-7.4 - {}^-12.1$

39. $\frac{9}{4} - \frac{2}{7} + \frac{1}{6}$

40. ${}^-0.7 \div 2$

Write the prime factorization of each number using exponents. For example, $200 = 2^3 \cdot 5^2$.

41. 98

42. 54

43. 36

44. Copy and complete the chart.

Fraction	$\frac{1}{2}$	$\frac{3}{5}$		
Decimal	0.5		0.65	
Percent	50%			21%

Order each group of numbers from least to greatest.

45. $0.05, 0.001, 5, 1, \frac{1}{5}$

46. $\frac{1}{2}, \frac{3}{4}, \frac{1}{4}, \frac{3}{2}, \frac{1}{3}$

47. Calculate the area and circumference of a circle with radius 2 cm.

48. **Challenge** Calculate the area of a square with a side length of s cm and a diagonal of 1.5 cm. Show how you found your answer.

49. The United States covers about $2.38 \cdot 10^9$ acres, including water, and has a population of about 281,000,000 people.

a. Express the population of the United States using scientific notation.

b. Express the total area of the United States using standard notation.

c. Compute the number of people per acre in the United States.

50. **Probability** Gregory put a blue chip, a green chip, and two red chips into a box. He asked Mai to pull out a chip without looking into the box.

a. What is the probability Mai will draw a red chip?

b. Suppose Gregory adds three more blue chips to the original four chips in the box. Now what is the probability Mai will draw a blue chip? A red chip?

Remember

The Pythagorean Theorem says that if a and b are the legs of a right triangle and c is the hypotenuse, then $a^2 + b^2 = c^2$.

6.4

Simplifying Equations

Many problems that people encounter every day can be tackled by using mathematical techniques such as writing and solving algebraic equations. Sometimes the equation that is written to solve a problem looks quite complicated. However, equations can often be simplified to make them easier to solve.

Think & Discuss

Rewrite this equation to make it easier to solve. Explain each step of your simplification.

$$2n + 3(n + 1) = 43$$

Solve your simplified equation, and check that your solution is a solution of the original equation.

Investigation Building and Solving Equations

You have solved many problems by writing and solving equations. Sometimes, after you write an equation for a situation, you can simplify it before you solve. You already know some techniques for rewriting expressions to make them simpler.

EXAMPLE

Jonna and Don collect autographs of baseball players. Jonna has 5 more autographs than Don. If they have 19 autographs altogether, how many does each friend have?

Here's how Maya solved this problem.

Let n be the number of autographs Don has. Then Jonna has $n + 5$.

$n + (n + 5) = 19$	Combine Don's and Jonna's autographs.
$n + n + 5 = 19$	Remove the parentheses.
$2n + 5 = 19$	Add n and n.
$2n = 14$	Subtract 5 from both sides.
$n = 7$	Divide both sides by 2.

So Don has 7 autographs and Jonna has 12. This adds to 19, so it checks.

Problem Set A

1. Zoe thought about the problem in the Example differently.

   "I'll let J stand for the number of autographs Jonna has. Don has 5 fewer than Jonna, so he must have $J - 5$ autographs."

   a. Write the equation Zoe had to solve.

   b. Do you think Zoe will get the same numbers of autographs for Jonna and Don as Maya did? Why or why not?

   c. Solve Zoe's equation. Check your solution.

2. Ivan and Shanee have 48 model planes altogether. Ivan has three times as many model planes as Shanee. Write and solve an equation to find how many model planes each student has.

3. Melissa has 1 more than twice the number of beanbag animals Rachel has. Latanya has a third as many beanbag animals as Melissa has. Together, the three friends have 38 beanbag animals. Write and solve an equation to find how many beanbag animals each friend has.

In Problem Sets B and C, you will practice rewriting expressions to make them simpler and easier to handle.

Problem Set B

Veronica is sorting groups of baseball cards by brand: Players, Best, and Atlantic. In Problems 1–4, write and solve an equation to find the number of baseball cards in each pile. Check that your solutions make sense.

1. Veronica begins with 103 baseball cards, all Players and Best. There are 11 more Best cards than Players cards.

2. The next group Veronica sorts has a total of 98 Players and Atlantic cards. It has 36 fewer Atlantic cards than Players cards.

3. The third group has 112 cards, which Veronica divides into two piles. The Atlantic pile has three times as many cards as the Best pile.

4. The last pile, a mix of all three brands, contains 136 cards. Once Veronica sorts it into three piles, the Players pile has 7 more than twice the number in the Best pile, and the Atlantic pile has three times as many as the Players pile.

Just the facts

The first baseball cards were printed in the late 1800s. In 1933 they began to be packaged with bubble gum—a stick of gum and one card for a penny!

Source: www.baseball-almanac.com

A bookstore clerk is sorting boxes of books to be shelved. From each box, he makes three stacks of books: fiction, nonfiction, and children's. In Problems 5–7, write and solve an equation to find the number of books in each stack.

5. The first box has 31 books. The nonfiction stack has 5 fewer books than the fiction, and the children's has 15 more than the fiction.

6. From the next box, the fiction stack has three times as many as the children's, and the nonfiction has two fewer than the children's. There are 78 books altogether.

7. From a third box, the nonfiction stack has one more than double the number in the fiction stack, and the children's stack has double the number in the nonfiction stack. The difference between the numbers of books in the children's stack and the fiction stack is 47.

Problem Set C

Solve each equation.

1. $3n + 2 = n$

2. $5(s - 2) + {}^-22 = {}^-32$

3. $7(t - 3) + 1 = 3(t + 4)$

4. $3(2x + 1) + 2(x - 3) = \frac{1}{2}(x - 10)$

Share & Summarize

1. Simplify and solve the equation $3(n + 4) + 2(n - 12) = 2n$. Explain each step in your solution.

2. Solve the equation $3x + x + (x + 4) = 24$. Check your solution.

Investigation Subtracting with Parentheses

At the beginning of Investigation 1, Maya wrote an equation to show how two collections of baseball players' autographs were added. She used parentheses around the sum $n + 5$, because that sum represents the number of autographs one person had.

$$n + (n + 5) = 19$$

To add such an expression, you can simply remove the parentheses. For example, the equation above can be rewritten as

$$n + n + 5 = 19$$

In this investigation, you will explore what happens when you *subtract* expressions in parentheses.

Just the facts

Some of the rarest and most valuable baseball cards can actually be worth *less* if they are signed by the player than if they aren't!

Think & Discuss

How would you find the value of the expression $63 - (10 + 3)$?

Simon and Jin Lee evaluated $63 - (10 + 3)$ in different ways.

• Simon added 10 and 3, and then subtracted the sum from 63.

• Jin Lee subtracted one number at a time from 63: first she subtracted 10, and then she subtracted 3.

Which method do you prefer? Are both approaches correct? Explain.

Simon wondered whether *any* sum could be subtracted by subtracting the numbers one at a time. After testing several cases, Simon said he thinks $a - (b + c)$ and $a - b - c$ are *equivalent expressions*. (Remember: Two expressions are *equivalent* if they have the same value for any values of the variables.) Is he correct? Explain.

VOCABULARY
conjecture

Simon's statement is a **conjecture**—a statement he believes to be true based on some cases he has considered. If Simon's conjecture is true, the two expressions, $a - (b + c)$ and $a - b - c$, will give identical results no matter what numbers are substituted for the variables. If you can find even *one* case for which the two expressions give different results, you can conclude they are *not* equivalent and that Simon's conjecture is false.

Problem Set D

Test Simon's conjecture by trying at least four sets of values for a, b, and c. Record your results in a table like the one shown, which displays the results for $a = 50$, $b = 3$, and $c = 12$. Test a variety of values for a, b, and c. Consider cases in which

- a is greater than $(b + c)$
- a is less than $(b + c)$
- c is a negative number
- b is a negative number

a	b	c	$b + c$	$a - (b + c)$	$a - b - c$
50	3	12	15	35	35

Based on your research, do you think Simon's conjecture is correct?

You may already be convinced that Simon's conjecture is correct. To *prove* it is correct, though, you must show that it makes no difference what the values of the variables are. You can't test every possible combination, but you can use algebra and what you know about operations with negative numbers to show that the conjecture is true for all sets of values.

You know that subtracting a number is the same as adding its opposite, and that adding a number is the same as subtracting its opposite.

$$5 - {}^-2 = 5 + 2 \qquad\qquad 15 + {}^-8 = 15 - 8$$

You also know that the opposite of a number equals $^-1$ times the number.

$$^-3 = {}^-1 \cdot 3 \qquad\qquad {}^-({}^-11) = {}^-1 \cdot {}^-11$$

Think & Discuss

The series of steps below *proves* that $a - (b + c) = a - b - c$. Explain each step.

Step 1: $a - (b + c) = a + {}^-(b + c)$

Step 2: $\qquad\qquad\quad = a + {}^-1(b + c)$

Step 3: $\qquad\qquad\quad = a + {}^-1 \cdot b + {}^-1 \cdot c$

Step 4: $\qquad\qquad\quad = a - b - c$

So, $a - (b + c) = a - b - c$.

Problem Set E

Evaluate each expression.

1. $40 - (10 + 6)$

2. $35 - (5 + 8)$

3. $17.5 - (6.4 + 0.7)$

Rewrite each expression without parentheses. Then rewrite the expression again, if possible, to make it as simple as you can.

4. $7 - (7 + s)$ **5.** $12 - (r + 8)$ **6.** $(t - 7) + 3$

Shaunda has a strategy for quickly solving difficult subtraction problems in her head.

EXAMPLE

To compute $63 - 19$, Shaunda estimates the answer by first subtracting 20 from 63 to get 43. Then she corrects her estimate: she knows she subtracted 1 too many (20 is 1 greater than 19), so she adds back 1 to get 44 as the result.

Here is Shaunda's strategy in symbols.

$$63 - 19 = 63 - 20 + 1 = 43 + 1 = 44$$

Problem Set F

1. Solve each problem using Shaunda's method. First approximate by subtracting 20, and then add back the amount needed to correct the result. Write the steps in symbols as done in the Example.

     a. $63 - 18$ **b.** $63 - 17$ **c.** $63 - 15$

2. In these problems, the number being subtracted is close to 200. Modify Shaunda's method and apply it to find each difference. Write the steps in symbols.

     a. $351 - 197$ **b.** $351 - 195$ **c.** $351 - 190$

Zach said that after looking at his answers for Problems 1 and 2, he thinks $a - (b - c)$ is equivalent to $a - b + c$.

3. Look at the symbolic version of Shaunda's example: $63 - 19 = 63 - 20 + 1 = 43 + 1 = 44$. What do you think Zach would say are the values for a, b, and c?

4. Use the table to test the two expressions in Zach's conjecture with at least four sets of values for *a*, *b*, and *c*. One example is shown. Try cases in which

- *a* is greater than ($b - c$)
- *c* is greater than *b*
- *a* is less than ($b - c$)
- *b* is a negative number

a	*b*	*c*	$b - c$	$a - (b - c)$	$a - b + c$
50	12	3	9	41	41

Based on your results, do you think Zach's conjecture is correct?

5. Zach said the proof of his conjecture is very similar to the proof of Simon's conjecture in the Think & Discuss on page 423. Here's the first step of Zach's proof.

$$a - (b - c) = a + {}^{-}(b - c)$$

a. Explain how you know this first step is a true statement.

b. Finish the proof. If you need help, look back at the proof of Simon's conjecture. Be sure to give an explanation for each step in your proof.

Problem Set G

Evaluate each expression.

1. $50 - (10 + 6)$ **2.** $50 - (10 - 6)$ **3.** $50 - (6 - 10)$

Rewrite each expression without parentheses. Then rewrite the expressions again, if possible, to make them as simple as you can.

4. $12 - (r - 8)$ **5.** $5 - (5 + n)$

6. $3 - (7 - t)$ **7.** $15r - (3r - s)$

Share & Summarize

Select one of the expressions from Problems 4–7 of Problem Set G and explain how you thought about it. Write an explanation to convince someone that your simplified version is correct.

Investigation ▶3 More Practice with Parentheses

As you saw in Investigation 2, the two expressions below will always be equal, no matter what the values of x and y are. But maybe you're still not sure it really makes sense that they are equal.

$$10 - (x + y) \qquad 10 - x - y$$

Think about it this way: Imagine that you have $10. You owe your parents x dollars, and you owe your best friend y dollars. How much money will you have left if you can pay both your debts?

You can add your debts together, $x + y$, and then subtract the total from $10. The amount you have left is $10 - (x + y)$. Or you can subtract your debts one at a time, and so the amount you have left is $10 - x - y$. In either case, you will have the same amount of money left, so these two expressions must be equal.

Problem Set H

1. Luis owed his little sister x dollars. She wanted to buy a new book, so she asked Luis to pay back some of the money. He said he would give her y dollars.

   a. After paying back y dollars, how much did Luis still owe his sister?

   b. Luis had $15 after paying his sister the y dollars he promised. If he were to pay the rest of the money he owed, how much would he have left? Find two expressions to represent this amount.

2. Zoe wanted to borrow money from two of her siblings. Neither wanted to make her a loan, so she struck a deal with them. If she took more than a week to pay them back, she would pay them double the amount borrowed. Her brother loaned her x dollars and her sister loaned her y dollars. Poor Zoe  wasn't able to get any money for two weeks, when a friend paid her $25 that he had borrowed.

   Which expressions represent the amount of money Zoe had left after paying her siblings double the amount she had borrowed from them?

   a. $25 - 2(x + y)$ **b.** $25 - (2x + 2y)$ **c.** $25 - 2x - 2y$

Jena says that when she has to multiply an expression in parentheses by a quantity, she first uses the distributive property to multiply each item inside the parentheses. Then she uses the rules for adding and subtracting expressions in parentheses.

EXAMPLE

Here's how Jena simplified $5 - 2(3 - 8)$.

First she distributed the 2:	$5 - (6 - 16)$
Then she applied the rule for subtracting parentheses:	$5 - 6 + 16$
Then she subtracted and added:	15

Here's how Jena used the same approach to simplify $12k - 3(7 - 2k)$:

$$12k - 3(7 - 2k) = 12k - (21 - 6k)$$
$$= 12k - 21 + 6k$$
$$= 18k - 21$$

Now you will practice combining adding and subtracting with parentheses with some of the earlier skills you developed in order to become more skilled at simplifying expressions with parentheses.

Problem Set ▌

Simplify each expression as much as possible.

1. $5r - 2(r - 8)$ **2.** $32L - 3(L - 7)$

3. $12n - 2(n - 4)$ **4.** $4(3x - 2) - 3x$

5. $^-5(g - 3)$ **6.** $8t - \frac{1}{3}(12t - 30)$

For each expression, write an equivalent expression without parentheses.

7. $a + d(b + c)$ **8.** $a - d(b + c)$

9. $a + d(b - c)$ **10.** $a - d(b - c)$

11. In the Example, you saw how Jena simplified $5 - 2(3 - 8)$. Another way to simplify $5 - 2(3 - 8)$ is shown below. Explain the reasoning for each step.

   a. $5 - 2(3 - 8) = 5 + {}^-2(3 - 8)$

   b. $\qquad\qquad = 5 + ({}^-6 - {}^-16)$

   c. $\qquad\qquad = 5 + ({}^-6 + 16)$

   d. $\qquad\qquad = 5 + {}^-6 + 16$

   e. $\qquad\qquad = 15$

Use any method you like to solve each equation. Check your answers.

12. $12 - \frac{1}{2}(2n - 6) = 7$

13. $2(2q + 4) - 2q = 24$

14. $2x - (4 - x) = 5$

15. $4r - 3(r + 1) = 0$

16. Sunny Sports Shop sells beginner's tennis rackets for $40 each. Most customers buying one of these rackets also buy several cans of tennis balls, which cost $2.50 each. The customer's bill has three parts:

- the price of the racket

- the total price of the balls

- sales tax, which is 7% of the combined price of the racket and the balls

a. A customer buys a racket and x cans of balls. Write expressions for each of the three parts of the customer's bill.

b. Write an expression for the total bill, which includes the racket, x cans of balls, and the sales tax.

c. Aurelia bought a racket and some cans of balls. Her total bill came to $53.50. Write and solve an equation to find how many cans of balls she bought.

Just the facts

Until the 1980s, tennis rackets were made from wood. Wood tennis rackets—first made in the U.S. in the 1870s—are now a collector's item.

Share & Summarize

1. Simplify and solve this equation. Write an explanation of your work that would convince someone that your simplified equation is correct.

$$5n - \frac{1}{3}(3n - 6) = \frac{1}{4}(8n + 12)$$

2. Check your answer by substituting it into the original equation.

On Your Own Exercises

Practice & Apply

1. Together, Jon and Joanna have 19 computer games. Joanna has 3 more games than Jon. Write and solve an equation to find the number of games each friend has. Check your answer.

2. Malik and five of his friends donated change to a school fund-raiser to buy food for homeless children. Together they donated $4.40. Three of the friends gave 20¢ less than Malik, and two gave 10¢ more. Write and solve an equation to find how much Malik donated. Check your answer.

3. Peter had 3 piles of pickled peppers. He has a total of 52 pickled peppers. His second pile has 7 more peppers than the first pile. His third pile has 3 fewer than the first pile. Write and solve an equation to find how many pickled peppers are in each pile. Check your answer.

4. Mr. Álverez baked 60 fruit tarts for the school bake sale. He put them on four trays. The second tray had twice as many tarts as the first. The third tray had 1 more than the second. The fourth tray had 5 more than the first. Write and solve an equation to find how many tarts are on each tray. Check your answer.

Just the facts

The Winged Liberty Head dime was minted from 1916 to 1945. The coin's designer put wings on Liberty's cap to represent freedom of thought, but the image is often mistaken for the Roman god Mercury—and thus the coin is commonly known as the Mercury dime.

5. Gloria sorted her collection of Winged Liberty Head dimes into three stacks based on their quality. The second stack has twice as many coins as the first, the third stack has 36 more than the second, and the first stack has 43 fewer than the third. How many dimes are in each stack? Show how you found your answer.

Solve each equation.

6. $d + (2d - 3) = 13(d - 1)$ 7. $5(2x - 1) + 3x = 2(x + 5)$

Evaluate each expression.

8. $72 - (12 + 9)$ 9. $72 - (12 - 9)$ 10. $12.5 - (3.3 - 9.5)$

11. $38 - (7 + 28)$ 12. $291 - (84 - 45)$ 13. $5 - (2.8 + 5.9)$

Rewrite each expression without parentheses. Then rewrite the expressions again, if possible, to make them as simple as you can.

14. $8 - (x - 8)$ 15. $15 - (8 + y)$ 16. $x - (37 + x)$

17. $6 - (12 - n)$ 18. $5s - (6r + s)$ 19. $8x - (^-2x - 3y)$

For each expression, write an equivalent expression without parentheses.

20. $A + (B - C)$ **21.** $A - (B - C)$ **22.** $A - (B + C)$

Rewrite each expression without parentheses. Then rewrite the expressions again, if possible, to make them as simple as you can.

23. $2(a - 8) + 30$ **24.** $30 + 4(t - 7)$ **25.** $5 + 3(n + 5)$

26. $15r + 2(r + s)$ **27.** $3t + 3(10 - t)$ **28.** $7x + \frac{1}{2}(2x - 4y)$

29. $m + 4(m + 5)$ **30.** $4p + 7(2p + 2)$ **31.** $4(x + 2) + 5(3x + 2)$

Solve each equation.

32. $8 - (4 - n) = 18$ **33.** $8 - (n - 4) = 18$

34. $6x - 2(x + 1) = 10$ **35.** $2x - 3(x - 1) = 17$

36. $3y + 4(y + 2) = 78$ **37.** $2(x + 3) + x = 63$

38. $5(x + 2) = 6x$ **39.** $6 - 2(8 - 2r) = 20$

40. $3 - 2.5(m + 2) = \frac{m}{2} + 4$ **41.** $r + 8 = 3(1 - r) + 2$

42. GameWorld is selling SuperSpeed game systems for \$100. Game cartridges are \$40 each. Sales tax is 6% of the total price for all items.

   a. Write an expression for the total, with tax, of one SuperSpeed system with g game cartridges.

   b. One customer bought several games and had a final bill of \$360.40. Write and solve an equation to find how many games the customer bought.

43. Make up a problem, like those in Problem Set A, about two friends who collect postage stamps. Make sure your problem can be solved so that each friend has a different whole number of stamps. Include an explanation of how you made up your problem and how it can be solved.

Daspletosaurus   USA 32

Connect & Extend

44. In a town election, the three candidates receiving the most votes were Malone, Lawton, and Spiros. Malone received 60 more than twice as many votes as Lawton. Spiros received 362 votes more than Lawton. In all, 2,430 votes were cast. How many votes did each candidate get? Who won the election? Show how you found your answer.

45. This conversation occurred the day after the class election.

Desiree: I heard that Evita won our class election for student council.

Reynoldo: Yes, but it was really close! She got only 1 more vote than Mai.

Tom: It sure wasn't very close for third place. Mai got twice as many votes as Lu-Chan.

Lehie: Did everybody vote?

Ken: Yes, except James, who's still home with chicken pox.

Of the 27 students in the class, only Evita, Mai, and Lu-Chan received votes. How many votes did each of the three students receive? Show how you found your answer.

46. Nuna is y years old. In 30 years, she will be three times as old as she is now. Write and solve an equation for this situation.

47. Two caterpillars are crawling in and out of a hole in a fence. The caterpillars take turns being in the lead. The length of the longer caterpillar is L, and the length of the shorter caterpillar is S. The hole is 9 cm above the ground.

Look at Picture i. The expression $9 + L - S$ represents the height reached by the head of the lead caterpillar above the ground.

Remember that the hole is 9 cm above the ground.

Match each of the 12 expressions with the drawings. Some positions match more than one expression, and some expressions match more than one position.

a. $9 + (S - L)$ **b.** $9 + (S + L)$ **c.** $9 - (S + L)$

d. $9 - S - L$ **e.** $9 - L - S$ **f.** $9 - L + S$

g. $9 + (L + S)$ **h.** $9 - (L + S)$ **i.** $9 - S + L$

j. $9 + S - L$ **k.** $9 - (S - L)$ **l.** $9 - (L - S)$

Just the facts

Monarch caterpillars eat milkweed leaves. These leaves contain a poison that makes the caterpillars taste bad to birds and other predators.

Remember

Equivalent expressions give identical results when the same values are used for identical variables.

48. Find all the sets of equivalent expressions in Parts a–l of Exercise 47.

Replace each ☐ with + or − to make the statement correct.

49. $7 - (A + B) = 7 - A \,\square\, B$

50. $14 + (A - B) = 14 + A \,\square\, B$

51. $12 - (A - B) = 12 \,\square\, A \,\square\, B$

52. Graphing can often help you identify equivalent expressions. For example, $x - 5$ and $5 - x$ are not equivalent expressions because $y = x - 5$ and $y = 5 - x$ do not produce the same graph.

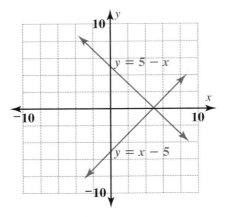

However, $8 - (x + 5)$ and $8 - x - 5$ *are* equivalent because $y = 8 - (x + 5)$ and $y = 8 - x - 5$ produce the same graph.

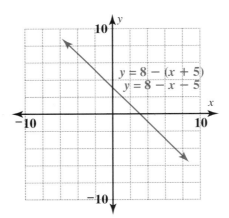

**In** your **own** words

Describe a situation that can be represented by both $a - (b - c)$ and $a - b + c$. Explain how both expressions apply to the situation.

For each pair of equations, plot several points on a coordinate grid to draw the lines, and determine whether the related expressions are equivalent.

a. $y = 4 - (x + 5)$ and $y = 4 + (5 - x)$

b. $y = 7 - (3 - x)$ and $y = 7 + x - 3$

53. Albert and Mary were trying to simplify this expression, which looks a bit complicated.

$$4 - 2(^-5x - 10)$$

Albert said the expression is equal to $2(^-5x - 10)$. Mary said it's equal to $4 + 10x - 20$. Is Albert correct? Is Mary correct? If they are not correct, what mistakes did they make?

Mixed Review

Evaluate each expression.

54. $\frac{7}{8} + \frac{1}{2}$

55. $\frac{3}{5} + \frac{1}{3}$

56. $\frac{2}{9} + \frac{1}{3}$

57. $\frac{5}{3} - \frac{1}{2}$

58. $\frac{7}{2} - 1$

59. $\frac{4}{3} - \frac{2}{6}$

60. $\frac{1}{4} \cdot \frac{1}{10}$

61. $\frac{1}{10} \cdot 2$

62. $\frac{7}{5} \cdot \frac{1}{10}$

63. Order these fractions from greatest to least.

$$\frac{1}{3} \qquad \frac{2}{7} \qquad \frac{3}{8} \qquad \frac{2}{9} \qquad \frac{3}{11} \qquad \frac{4}{19}$$

Without using a calculator, predict whether each product is less than 0, equal to 0, or greater than 0.

64. $^-45 \cdot {}^-2^6$

65. $^-n^2$

66. $(^-3)^7 \cdot 3^7$

Geometry Find the area and perimeter of each figure.

67. This is a rectangle.

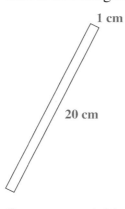

1 cm

20 cm

68. This is a trapezoid.

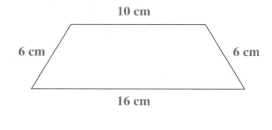

10 cm

6 cm   6 cm

16 cm

69. Geometry A block prism has length 8 in., width 1 in., and height 6 in.

a. What is the prism's volume? What is its surface area?

b. Another block prism is three times as tall as the first prism. What is the volume of this prism? What is its surface area?

c. A cylinder has the same volume as the original prism and a height of 12 in. What is the area of the cylinder's base?

Several students are playing the game *Think of a Number*.

It almost seems like magic that everyone gets the same answer even though they all started with different numbers. But it isn't magic. People who think of these games use algebra to make sure their tricks will work.

Explore

Play the *Think of a Number* game in the cartoon using different numbers. Do you always get 10?

Use the variable *n* instead of a number. Write expressions for the results after each step.

Use algebra to simplify the final expression. Then explain to your partner why you always have 10 left.

Investigation 1 ▶ Number Games

In this investigation, you will use algebra to understand and create your own *Think of a Number* games.

Problem Set A

1. Here's another *Think of a Number* game.

   • *Think of a number.*

   • *Add 5 to your number.*

   • *Subtract 5 from your original number.*

   • *Subtract the second result from the first result.*

   a. Try the game. What do you get?

   b. Try the game with different types of numbers: a number less than 5, a number greater than 100, a negative number, and a number that is not a whole number. What do you notice about your final results?

   c. Now try the variable n instead of a number. What do you get for the final expression? Use what you learned in the previous lesson to simplify the expression.

2. Here's another game.

   • *Think of a number.*

   • *Add 6 to your number.*

   • *Multiply your result by 2.*

   • *Subtract 4 from your result.*

   • *Halve your result.*

   • *Take away the number you first thought of.*

   Play the game with your partner a few times with different numbers. Then use algebra to explain what happens.

3. With your partner, create your own *Think of a Number* game so that you always get the same number as a result. Try it out on another pair of students. Make sure you can explain why it works.

Here is a different kind of *Think of a Number* game.

Problem Set B

1. Discuss this new game with your partner. How does it work? It may help to let a variable stand for the number and then simplify the expressions you get from following the game instructions.

2. Simon made up this *Think of a Number* game.

- *Think of a number.*

- *Subtract it from 20, and double your answer.*

- *Subtract the result from 40.*

When Monique said she had 14, Simon said she had started with 7. Peter finished with 6, and Simon told him he'd been thinking of 3. Marta's result was 8, and Simon smiled and said, "Four!"

a. How is Simon able to figure out the number each person started with?

b. Use algebra to show what the result of Simon's game will always be. Play the game with several numbers to check that your prediction works. Make sure some of the numbers are not whole numbers.

3. With your partner, make up a new *Think of a Number* game that allows you to guess someone else's number. Use algebra to see what the result must be, and then try your game on another pair of students.

Share & Summarize

Tara used this rule for her *Think of a Number* game.

$$\frac{3n + 15}{3} - 5$$

Tara says that when the person she is playing with tells her the result, she says, "That's the number you started with." She claims she is always right.

1. Write the words Tara might use to tell someone what to do.

2. Does Tara's trick always work? Write an explanation of why you think it will or won't.

Investigation 2 Simplifying Equations for Graphing

In Chapter 5, you graphed linear equations—equations whose graphs are straight lines. You know that an equation in the form $y = ax + b$ is linear. When equations are written in other forms, however, it may be difficult to know if they are linear. Sometimes it is easier to graph an equation, or to determine whether it is linear, if you simplify it first.

Problem Set C

1. Consider this equation:

$$y = x + \frac{1}{2}(2x + 5)$$

   a. Do you think the relationship between x and y is linear? Why or why not?

   b. Simplify the expression on the right side of the equation. Is the simplified equation linear? Why or why not?

   c. Make a table of values, and draw a graph of the relationship. Include some negative values of x in your table.

   d. How can you tell from your table and your graph whether the equation is linear?

   e. Use your simplified equation to find the slope of the graph. Check it with the graph.

2. Consider the equation $y = 3(x - 1) - 2(1 - x)$

    a. Do you think this represents a linear relationship? Why or why not?

    b. Simplify the equation. Is it linear?

    c. Make a table of values, and draw a graph of the equation. Include some negative values of x in your table.

    d. How can you tell from your table and your graph whether the equation is linear?

    e. Use your simplified equation to find the slope of the graph. Check it with the graph.

The next problem set will give you more opportunities to explore the relationship between linear equations and their graphs.

Problem Set D

1. Consider the equation $2y + 6x = 3$.

    a. Do the same thing to both sides of $2y + 6x = 3$ to get a new version of the equation with only y on the left side. How can you tell whether the equation represents a linear relationship?

    b. If you were to draw the graph, what would the slope be?

    c. Draw the graph to check your prediction.

2. Consider this graph.

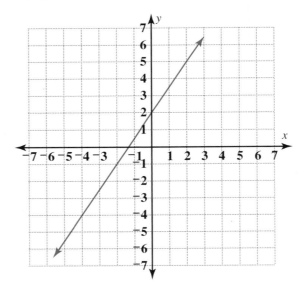

    a. Write an equation that would produce this graph.

    b. Do the same thing to both sides of your equation to obtain a different form of the same equation.

3. Consider these equations.

    i. $2y - 3x = 4$

    ii. $2y - 3x = 7$

    iii. $3x - 2y = 1$

  a. On one set of axes, draw graphs of these equations.

  b. What is the same about the graphs? What is different about them?

  c. Rewrite the equations with only y on the left side to show that the three relationships are linear. What is the slope of each graph?

Share & Summarize

Study the two lines on the graph.

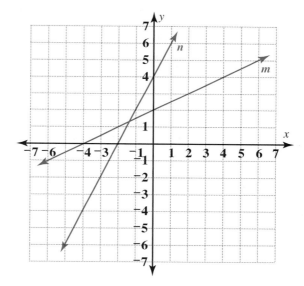

1. Match Lines m and n with these two equations.

  a. $4y - 2x = 8$

  b. $2y - 4x = 8$

2. Rewrite both equations so that y is alone on the left side.

3. Which version of each equation is easier to graph?

4. Which version can you use to tell more easily whether the equation is linear?

On Your Own Exercises

Practice & Apply

1. Consider this *Think of a Number* game.

- *Think of a number.*

- *Add 4 to your number, and record the result.*

- *Subtract 4 from your first number, and record that result.*

- *Add 8 to the first result, and then subtract the second result.*

a. Try this game using positive, negative, and fractional numbers. What happens?

b. Use algebra to explain your observation in Part a.

c. Suppose the last instruction was "Add 8 to the second result, and then subtract the first result." What would happen in this case? Use algebra to show why this happens.

2. Chondra thought of a number trick of her own.

- *Think of a number.*

- *Multiply it by 4 and then subtract 6.*

- *Divide that result by 2.*

- *Add 3.*

- *Tell me your result.*

a. Chondra always knew what the original number was, once she knew the final result. Try this with a few numbers, and find the relationship between the starting number and the final result.

b. Using *n* for your number, try Chondra's trick again. Simplify the final result to explain the relationship you found in Part a.

3. Choose any number. Subtract your number from 9, and double the result. Subtract that answer from 19. Subtract 1, and then subtract your original number.

a. Try this with several numbers. What do you notice?

b. Using *n* for the starting number, write an expression for the final result. Simplify your expression.

impactmath.com/self_check_quiz

4. Consider the equation $y = 2(x + 1) - 3(2 - x)$.

   a. Make a table of values, and draw a graph of the relationship.

   b. Is the equation linear? Explain how you know.

   c. Rewrite the equation in the form $y = ax + b$. Use the simplified equation to find the slope of the graph.

5. Consider these three equations.

     i. $4y - 5x = 8$

    ii. $4y - 5x = 12$

   iii. $5x - 4y = 2$

   a. On one set of axes, draw graphs of these relationships.

   b. What is the same about these graphs? What is different about them?

   c. Rewrite the equations to show that the three relationships are linear.

6. Crystal played this *Think of a Number* game with her classmates. She claimed she would always be able to give the starting number when a classmate gave an answer.

   • *Think of a number.*

   • *Subtract your number from 9 and then multiply by 10.*

   • *Add the result to your starting number.*

   • *Divide the result by 9.*

   a. Try the game with several numbers. Try one-digit, two-digit, negative, and fractional numbers. What do you discover?

   b. Now use n as the starting number. What do you get for the final expression? Simplify it.

   c. **Prove It!** Crystal said, "If you add another step to my game— Subtract your result from 10—players will always get their original number." Use your simplified expression from Part b to prove that this is true.

7. Find a page of a monthly calendar in your house, or copy this calendar.

a. Draw a rectangle around six dates on the calendar, not all in the same week. Add the pairs of dates in opposite corners of your rectangle. What do you notice about the two sums?

b. Draw more rectangles around different sets of six dates. Add the pairs of dates in opposite corners of each rectangle, and record your results. What do you notice?

c. Now you will use algebraic expressions to explore what you've discovered. Let the number in the upper left corner of each rectangle be n. Find expressions for the numbers in the other three corners in terms of n. Notice that there are two types of rectangles you need to consider, A and B.

d. For both types of rectangles, add the expressions for opposite corners and compare the sums. What can you conclude?

8. Consider these equations.

$$2y = x + 7 \qquad 4y = 2x + 14$$

a. Suppose x represents the number of blocks in a bag, and y represents the number of blocks in a box. Draw a balance puzzle for each equation.

b. Suppose there are 3 blocks in each bag. How many blocks must be in each box in the first puzzle? In the second puzzle?

c. Suppose there are 4 blocks in each box. How many blocks must be in each bag in the first puzzle? In the second puzzle?

d. Choose another number of blocks to be in each bag. For that number, how many blocks must be in each box in the first puzzle? In the second puzzle?

e. What do your observations in Parts b–d tell you about the relationship between x (representing the number of blocks in each bag) and y (representing the number of blocks in each box) for the two equations? In other words, for any given value of x, what do you know about the value of y? Explain.

f. The graph of $2y = x + 7$ is shown below. What do you think the graph of $4y = 2x + 14$ will look like? Plot points to check your answer.

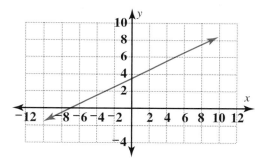

**Mixed
Review**

Find each difference.

9. $\frac{8}{5} - \frac{2}{5}$ **10.** $\frac{8}{3} - 1$ **11.** $\frac{3}{10} - \frac{1}{10} - \frac{5}{10}$

Find each product.

12. $2 \cdot \frac{1}{3}$ **13.** $\frac{7}{5} \cdot \frac{5}{7}$ **14.** $\frac{4}{7} \cdot \frac{1}{2}$

15. Lehie drew this flowchart. Tell what equation Lehie was trying to solve. Then copy and complete the flowchart.

16. Geometry Consider this cylinder.

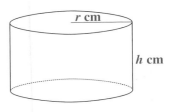

a. Find the volume of the cylinder.

b. Find the volume of a cylinder with a radius twice this cylinder's radius.

c. Find the volume of a cylinder that is twice as high as this cylinder.

17. Metal Manufacturing makes copper wire. Bars of copper are sent through presses that stretch the metal to different lengths.

a. A 1-meter bar of copper is sent through Machine X. A 4-meter strand of copper emerges from the other end. When the 4-meter strand is sent through the machine again, how long will the exiting piece of wire be?

b. Machine Y stretches copper to 8 times its original length. How many times would a 2-meter bar of copper need to travel through Machine Y to end up the same length as the final piece of wire in Part a?

c. A 1-centimeter bar of copper is sent through Machine X and then through Machine Y. How long is the wire that emerges?

d. A bar of copper was sent through Machine X and then through Machine Y, producing a 12.8-meter piece of wire. How many centimeters long was the original bar?

18. These triangles have the same shape but are different sizes.

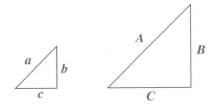

a. Measure the sides of both triangles. What do you notice?

b. Measure the angles of both triangles. What do you notice?

Chapter Summary

VOCABULARY
conjecture
model

In this chapter, you reviewed two methods of solving equations, back-tracking and guess-check-and-improve. You also learned new ways to solve equations, beginning with a *model*—balance puzzles. By using the model, you learned to do the same thing to both sides of an equation.

You learned some ways to simplify complicated equations, including rules for adding and subtracting expressions in parentheses.

Finally, you wrote expressions and equations for various situations and used your new skills to simplify and solve them. By rewriting expressions using your knowledge of algebra, you proved *conjectures* and found new ways to recognize linear relationships.

Strategies and Applications

The questions in this section will help you review and apply the important ideas and strategies developed in this chapter.

Solving equations using different methods

1. Solve each equation using guess-check-and-improve or backtracking. Explain why you chose the method you used.

   a. $(3x + 32) \div 4 = 7$

   b. $^-2x + 3(x - 5) = 2$

2. Describe each method for solving equations. Give two examples of equations for which each method works well.

   a. guess-check-and-improve

   b. backtracking

   c. using a balance puzzle

   d. doing the same thing to both sides

Writing equations to solve problems

3. Alisha's three cats are on special diets. Scout is the least active. Squeaky, the most active, is allowed twice as much cat food as Scout. The third cat, Picard, gets 90 g of food each day. Alisha gives the cats a total of 240 g of food each day. Write and solve an equation to find the amount of food each cat receives each day.

4. Francisco was copying some computer files. He had four color-coded floppy disks, but none of them were completely blank. The red disk had the least free space on it. The blue disk had 100 kilobytes (Kb) more free space than the red one. The green disk had 10 Kb more than three times as much free space as the blue one. The yellow disk had 78 Kb less than six times as much free space as the red one.

  a. Choose a variable, and use it to write expressions for the amount of free space on each disk.

  b. The blue and green disks together had the same amount of free space as the yellow disk. Write and solve an equation to find the amount of free space on each disk.

5. Kimiko divided her cassette tapes into three categories: rock, hip hop, and movie soundtracks. She had 10 more hip hop tapes as soundtracks, and 3 fewer than four times as many rock tapes as hip hop tapes. She also had 33 more rock tapes than hip hop tapes. Write and solve an equation to find how many tapes she had in each category.

Writing and simplifying expressions

6. Try this *Think of a Number* game.

- *Think of a number.*
- *Add 8, and multiply your result by 6.*
- *Subtract 6, and divide your result by 3.*
- *Subtract your original number.*
- *Subtract your original number again.*

a. Using n for the starting number, write an expression that shows all the steps in the game. Do not simplify your expression.

b. Prove that the final result is always 14. Explain each step of your proof.

7. Consider the relationship described by this equation:

$$2(x + y) - 3(5 - x) = 3$$

a. Rearrange this equation into the familiar form $y = ax + b$.

b. Graph the relationship. Choose at least two points on the graph, and substitute the coordinates into the original equation to check your work.

Demonstrating Skills

Simplify each expression as much as possible.

8. $a + 2(a - 3)$

9. $2b - (3 + b)$

10. $12 - (3r - 7)$

11. $2d(1 + c) - (3 + d)$

12. $^-4(3 - b) + 2(3b + 2) - 7(b + 2)$

Solve each equation.

13. $7s - 12 = 3s - 4$

14. $2(1 + c) - (2 + c) = 10$

15. $\frac{2}{3}M = M + 4$

16. $3t + 1.3 = 2.6 - 3.5t$

17. $\frac{3x + 5}{3} = \frac{8x + 12}{6}$

18. $4(5 - 2b) + 3(2b - 10) = ^-7(b + 2)$

CHAPTER 7

Similarity

Real-Life Math

It Really *Is* a Small World In California, there is a theme park based on a popular brand of plastic building blocks. One of the great attractions at the park is a detailed scale model of five areas of the United States—Washington, D.C., New Orleans, New York, New England, and the California coastline—built from 20 million plastic blocks. This attraction was created to celebrate the diversity of the United States and the people who live there.

Think About It The model of the U.S. Capitol building has a height of 7.2 feet—measured from its base to the top of the statue of freedom. The actual Capitol building is 40 times as tall. How tall is the actual Capitol building?

Family Letter

Dear Student and Family Members,

Our next chapter is about similarity between figures or shapes. Similar figures have the same shape, but not necessarily the same size. Congruent figures are figures that have the same shape *and* the same size. We will explore similar geometric figures in two-dimensional and three-dimensional models, and learn about scale and scale factors between similar figures.

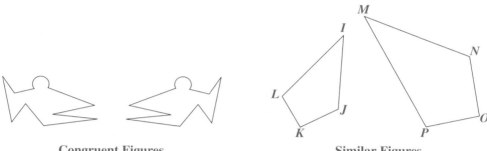

Congruent Figures　　　　**Similar Figures**

We will also investigate the relationships between the scale factor, area, and perimeter of similar figures. We will learn how to dilate two-dimensional drawings as well as three-dimensional objects. For instance, model trains are scale model of real trains and a globe is a scale model of Earth.

At the end of the chapter, we will apply what we've learned to solve an interesting problem: Could the bones of giants 12 times as large as we are really support their weight? Make a prediction and compare it to the answer at the end of the chapter.

Vocabulary Along the way, we'll be learning about these new terms:

congruent	counterexample	ratio
corresponding angles	dilation	scale factor
corresponding sides	equivalent ratios	similar

What can you do at home?

As you look around, you are likely to see many examples of similar shapes: scale drawings, reduced or enlarged photocopies, maps and the actual areas they show, and different-sized boxes of the same kind of cereal or other items. It might be fun for you and your student to point out what you think are similar shapes, and then measure each to verify that they actually are similar.

7.1 Are They the Same?

What does it mean to say that two figures are the same?

These figures are "the same" because they are members of the same *class of objects*. They are both rectangles.

Some figures have more in common than just being the same *type* of figure. One of the figures below, for example, is an enlargement of the other. They have the same *shape* but are different *sizes*. Two figures that have the same *shape* are **similar.**

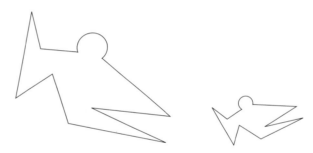

VOCABULARY
congruent

Of course, the most obvious way in which two figures can be "the same" is for them to be identical. Figures that are the same size *and* the same shape are **congruent.** The figures below are congruent.

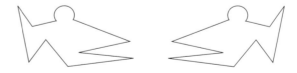

These rectangles are also congruent.

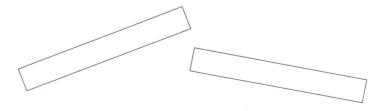

Notice that similarity and congruence don't depend on how the objects are positioned. They can be flipped and rotated from each other.

Explore

Your teacher will give you a sheet of paper with drawings of three figures. One or two other students in your class have figures that are congruent to yours. Find these students.

How did you determine which of the other students' figures were congruent to yours?

Investigation 1 Identifying Congruent Figures and Angles

To find who had figures congruent to yours, you needed to invent a way to tell whether two figures are congruent. Now you will use your test for congruence on more figures and on angles.

MATERIALS
• ruler
• protractor

Problem Set A

In each problem, decide whether Figures A and B are congruent. If they are not congruent, explain why not.

1.

2.

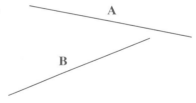

3.

4.

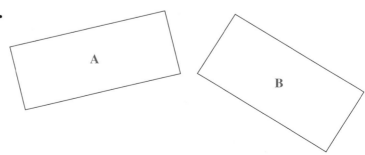

5.

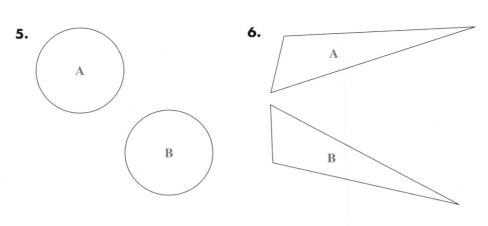

6.

In Problem Set A, you compared several pairs of figures. One important geometrical object is a part of many figures: the angle. What do you think congruent *angles* look like?

Think & Discuss

The angles in each pair below are congruent.

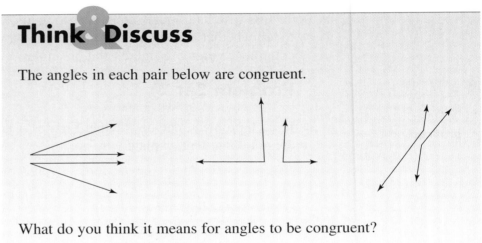

What do you think it means for angles to be congruent?

MATERIALS
• ruler
• protractor

Problem Set B

Decide whether the angles in each pair are congruent. If they are not congruent, explain why not.

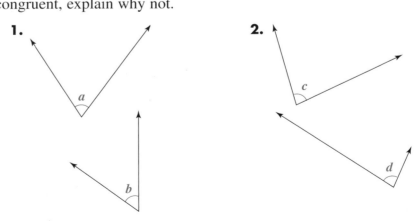

1.

2.

One way to test whether two figures are congruent is to try fitting one exactly on top of the other. Sometimes, though, it's not easy to cut out or trace figures, so it's helpful to have other tests for congruency.

MATERIALS
• ruler
• protractor

VOCABULARY
counterexample

Problem Set C

Each problem below suggests a way to test for the congruence of two figures. Decide whether each test is good enough to be *sure* the figures are congruent. Assume you can make *exact* measurements.

If a test isn't good enough, give a **counterexample**—that is, an example for which the test wouldn't work.

1. For two line segments, measure their lengths. If the lengths are equal, the line segments are congruent.

2. For two squares, measure the length of one side of each square. If the side lengths are equal, the squares are congruent.

3. For two angles, measure each angle with a protractor. If the angles have equal measures, they are congruent.

4. For two rectangles, find their areas. If the areas are equal, the rectangles are congruent.

MATERIALS
• ruler
• protractor

Share & Summarize

Decide which figures in each set are congruent. Explain how you know.

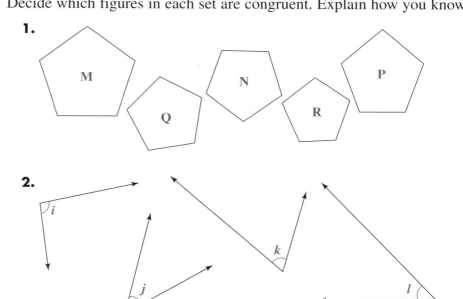

1.

2.

Investigation 2 ▶ Are They Similar?

You now have several techniques for identifying *congruent* figures. How can you tell whether two figures are *similar*?

MATERIALS

metric ruler

Remember

Figures are *similar* if they have the exact same shape. They may be different sizes.

Problem Set D

Work with a partner. To begin, draw a rectangle with sides 1 cm and 3 cm long. You need only one rectangle for the two of you.

1 cm ⬜ 3 cm

1. Now, one partner should draw a new rectangle whose sides are 7 times the length of the original rectangle's sides. The other partner should draw a new rectangle in which each side is 7 cm longer than those of the original rectangle. Label the side lengths of both new rectangles.

2. With your partner, decide which of the new rectangles looks similar to the original rectangle.

In Problem Set D, you modified a figure in two ways to create *larger* figures. Now you will compare two ways for modifying a figure to create *smaller* figures.

MATERIALS

metric ruler

Problem Set E

Work with a partner. Begin by drawing a rectangle with sides 11 cm and 12 cm long.

1. Now, one partner should draw a new rectangle whose sides are one-tenth the length of the original rectangle's sides. The other partner should draw a new rectangle in which each side is 10 cm shorter than those of the original rectangle. Label the side lengths of both new rectangles.

2. With your partner, decide which of the new rectangles looks similar to the original rectangle.

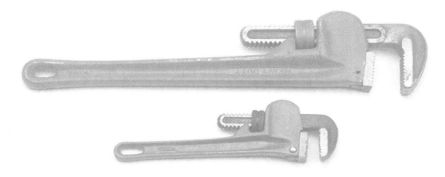

You have used two types of modifications to create rectangles larger and smaller than a given rectangle. You will now try these modifications on a triangle.

Problem Set F

Work with a partner. You each need a set of linkage strips and three fasteners. To find lengths on the linkage strips, count the *gaps* between holes. Each gap is 1 unit.

Separately, you and your partner should use your three linkage strips to construct a right triangle with legs 6 units and 8 units and hypotenuse 10 units. Trace the inside of your triangle on a sheet of paper.

1. One partner should follow the instructions in Part a, and the other should follow the instructions in Part b.

a. Construct a triangle whose side lengths are half those of the first triangle. That is, the lengths should be 3, 4, and 5 units. Trace the inside of the triangle on your paper.

b. Construct a triangle whose side lengths are each 2 units less than those of the first triangle. That is, the lengths should be 4, 6, and 8 units. Trace the inside of the triangle on your paper.

2. With your partner, decide which modification produces a triangle that looks similar to the original triangle.

Share & Summarize

You have modified rectangles and triangles in two ways to create larger and smaller figures.

- In one method, you multiplied or divided each side length by some number.

- In the other method, you added a number to or subtracted a number from each side length.

Which method produced figures that looked similar to the original?

Investigation ▶3 Ratios of Corresponding Sides

VOCABULARY
ratio

A *ratio* is a way to compare two numbers. When one segment is twice as long as another, the **ratio** of the length of the longer segment to the length of the shorter segment is "two to one."

One way to write "two to one" is 2:1. That means that for every 2 units of length on the longer segment, there is 1 unit of length on the shorter segment. For example, the ratio of the length of Segment *m* to the length of Segment *k* is 2:1.

Two other ways to write "two to one" are "2 to 1" and $\frac{2}{1}$.

It's possible to use different ratios to describe the same relationship.

EXAMPLE

Maya and Simon think about the ratios of the side lengths in triangles differently.

Two ratios are **equivalent ratios** if they represent the same relationship. Maya pointed out that the ratio 1:3 means that for every 1 cm of length on one segment, there are 3 cm of length on the other. Simon said the ratio 4:12 means that for every 4 cm of length on one segment, there are 12 cm of length on the other. These two ratios represent the same relationship: the length of the first segment is multiplied by 3 to get the length of the second segment. Therefore, 1:3 and 4:12 are equivalent ratios.

Problem Set G

1. Name at least two ratios equivalent to the ratio of the length of Segment *MN* to the length of Segment *OP*.

Decide whether the ratios in each pair are equivalent. Explain how you know.

2. 1:4 and 2:8

3. $\frac{2}{5}$ and $\frac{3}{9}$

4. 3:5 and 5:3

5. $\frac{1}{3}$:1 and 1:3

6. Darnell and Zoe were analyzing a pair of line segments. "The lengths are in the ratio 2:3," Darnell said. "No," Zoe replied, "the ratio is 3:2." Their teacher smiled. "You're both right—but to be clear, you need to give more information about your ratios."

   What did their teacher mean? Are 2:3 and 3:2 equivalent? How could Darnell and Zoe both be correct?

In Investigation 2, you created rectangles and triangles that were similar to other rectangles and triangles. For each shape, you used a part of the original figure to create the *corresponding part* of the new figure.

Corresponding parts of two similar figures are located in the same place in each figure. For example, Triangles *ABC* and *DEF* are similar. Sides *AB* and *DE* are **corresponding sides,** and ∠*B* and ∠*E* are **corresponding angles.**

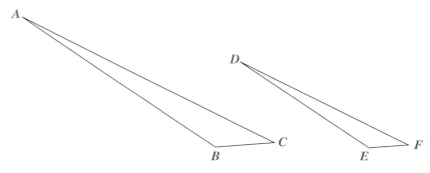

When you created similar rectangles and triangles, the ratios of the lengths of each pair of *corresponding sides* were equivalent. In fact, this is true for all similar figures: the ratios of the lengths of each pair of corresponding sides must be equivalent.

Problem Set H

The figures in each pair are similar. In each problem, identify all pairs of corresponding sides and all pairs of corresponding angles.

1.

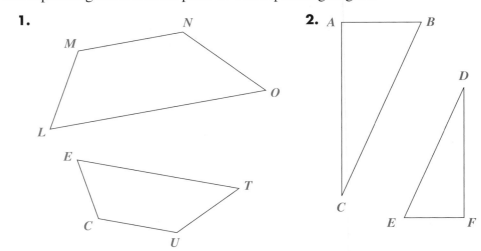

2.

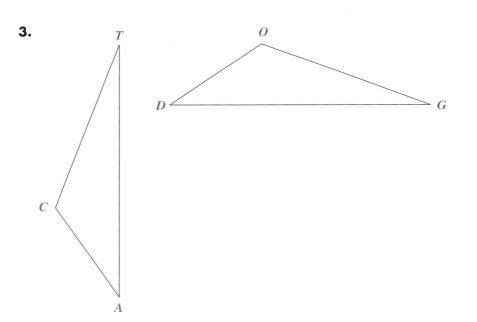

3.

Just the **facts**

The concept of similar triangles can be used to estimate dimensions of lakes, heights of pyramids, and distances between planets.

Problem Set I

If figures are similar, each pair of corresponding sides must have the same ratio. But if ratios of corresponding sides are the same, does that mean the figures *must* be similar? You will explore this question now.

1. Here are two quadrilaterals.

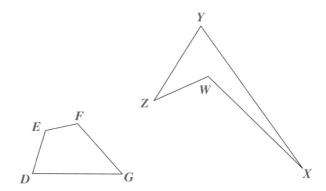

a. Copy and complete the table for Quadrilateral *DEFG*.

Description	Side	Length (cm)
longest side	DG	
second-longest side	FG	
third-longest side	DE	
shortest side	EF	

b. Now complete the table for Quadrilateral *WXYZ*.

Description	Side	Length (cm)
longest side	XY	
second-longest side	WX	
third-longest side	YZ	
shortest side	ZW	

c. Find the ratio of the longest side in Quadrilateral *WXYZ* to the longest side in Quadrilateral *DEFG*. Find the ratios of the remaining three pairs of sides in the same way:

- second longest to second longest

- third longest to third longest

- shortest to shortest

d. What do you notice about the ratios in Part c? Are Quadrilaterals *WXYZ* and *DEFG* similar? Explain your answer.

2. Here is a third quadrilateral.

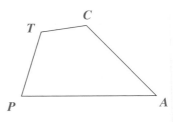

a. Complete the table for Quadrilateral *CAPT*.

Description	Side	Length (cm)
longest side	*AP*	
second-longest side	*CA*	
third-longest side	*TP*	
shortest side	*TC*	

b. Find the ratio of the longest side in Quadrilateral *CAPT* to the longest side in Quadrilateral *DEFG*. Find the ratios of the remaining three sides in the same way.

c. What do you notice about the ratios you found in Part b? Could Quadrilateral *CAPT* be similar to Quadrilateral *DEFG*? Explain.

3. The ratio of the corresponding side lengths of Quadrilaterals Y and Z to Rectangle A is 1:2.

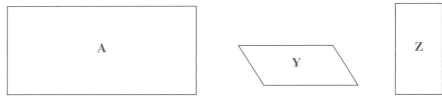

a. Are Quadrilaterals Y and Z both similar to Rectangle A? Explain.

b. How is Quadrilateral Y different from Quadrilateral Z?

c. What information—other than corresponding side lengths having the same ratio—might help you decide whether two polygons are similar?

Share & Summarize

1. Describe *equivalent ratios* in your own words.

2. Is the fact that corresponding sides are in the same ratio enough to guarantee that two polygons are similar? Explain your answer.

Investigation Identifying Similar Polygons

Remember

Two angles are congru-ent if they have the same measure.

▶ M A T E R I A L S

• ruler

• protractor

In Investigation 3, you found that when two figures are similar, the ratio of their corresponding side lengths is always the same. Another way of saying this is that the lengths of corresponding sides share a *common ratio*.

You also discovered that *angles* are important in deciding whether two figures are similar. However, you might not have found the relationship between corresponding angles. In fact, *for two polygons to be similar, corresponding angles must be congruent.* You won't prove this fact here, but you will use it throughout the rest of this chapter.

To test whether two polygons are similar, you need to check only that corresponding side lengths share a common ratio and that corresponding angles are congruent.

Problem Set J

Determine whether the figures in each pair are similar. If they are not sim-ilar, explain how you know.

1.

2.

3.

4.

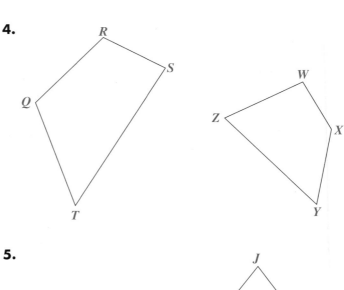

5.

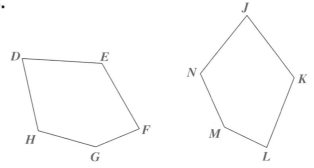

If two figures do not have line segments and angles to measure, how can you decide whether they are similar? One method is to check important corresponding segments and angles, even if they are not drawn in. For example, on these two spirals, you might measure the widest and tallest spans of the figures (shown by the dashed segments) and check whether they share a common ratio.

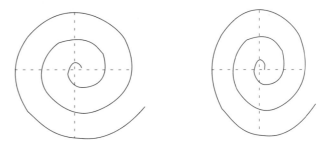

Just checking these two segments won't tell you for *sure* that the figures are similar, but it will give you an idea whether they *could* be. If the ratios aren't equivalent, you will know for certain the figures are not similar.

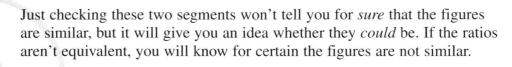

Problem Set K

Work with a partner. Try to figure out whether each pair of figures is, or could be, similar. Explain your decisions.

1.

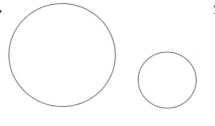

2.

3.

Share & Summarize

Try to stump your partner by drawing two pentagons, one that's similar to this pentagon and another that isn't. Exchange drawings with your partner. Try to figure out which of your partner's pentagons is similar to the original, and explain how you decided. Verify with your partner that you each have correctly identified the similar pentagon.

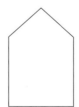

On Your Own Exercises

Practice & Apply

1. Look at the triangles below, but make no measurements.

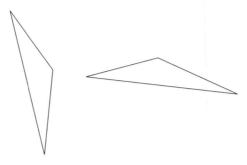

    a. Just by looking, guess whether the triangles are congruent.

    b. Check your guess by finding a way to determine whether the triangles are congruent. Are they congruent? How do you know?

2. Look at the triangles below, but make no measurements.

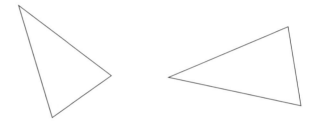

    a. Just by looking, guess whether the triangles are congruent.

    b. Check your guess by finding a way to determine whether the triangles are congruent. Are they congruent? How do you know?

3. Examine these rectangles.

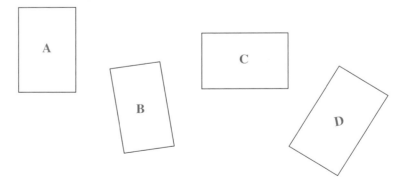

    a. Just by looking, guess which rectangle is congruent to Rectangle A.

    b. Find a way to determine whether your selection is correct. Which rectangle *is* congruent to Rectangle A? How do you know?

impactmath.com/self_check_quiz

4. Examine the figures below.

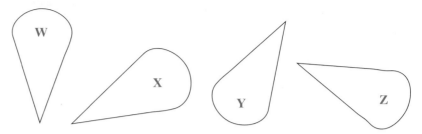

    a. Just by looking, guess which figure is congruent to Figure W.

    b. Find a way to determine whether your selection is correct. Which figure *is* congruent to Figure W? How do you know?

5. Rectangle R is 4.5 cm by 15 cm.

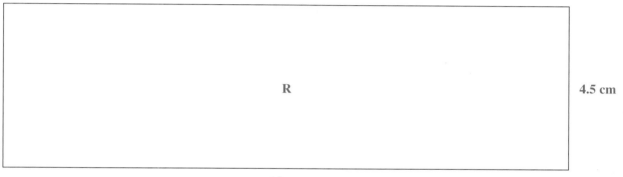

    a. Draw and label a rectangle with sides one-third as long as those of Rectangle R.

    b. Draw and label a rectangle with sides 3 cm shorter than those of Rectangle R.

    c. Which of your rectangles is similar to Rectangle R?

6. In Investigation 2, you explored two ways to modify rectangles and triangles. One method produces similar figures; the other does not. In this exercise, you will examine whether either of the methods will produce a similar figure when the original is a square.

    a. Draw a square that is 6 cm on a side. This is your *original* square.

    b. Make a new square with sides one-third as long as the sides of your original square.

    c. Make a new square with sides 3 cm shorter than those of your original square.

    d. Which of the methods in Parts b and c creates a square that is similar to your original? Explain.

Decide whether the ratios in each pair are equivalent. Explain how you decided.

7. 1:3 and 9:11

8. $\frac{1}{2}$ and $\frac{2}{3}$

9. 3:4 and 6:8

10. $a:b$ and $2a:2b$

Name two ratios that are equivalent to each given ratio.

11. 2:3

12. $\frac{6}{10}$

13. 50:50

Exercises 14 and 15 show a pair of similar figures. Identify all pairs of corresponding sides and angles.

14.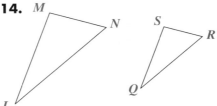

15.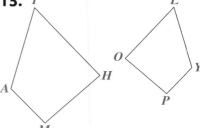

16. Examine these triangles.

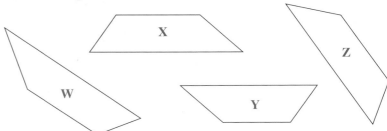

    a. Just by looking, guess which triangle is similar to Triangle A.

    b. Make some measurements to help determine whether your selection is correct. Which triangle *is* similar to Triangle A?

17. Examine these quadrilaterals.

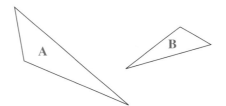

    a. Just by looking, guess which is similar to Quadrilateral Z.

    b. Make some measurements to help determine whether your selection is correct. Which quadrilateral *is* similar to Quadrilateral Z?

18. In Problem Set C, you found that calculating area is not a good way to test whether rectangles are congruent.

    a. Name types of figures for which calculating area *is* a good test for congruence. That is, if you have two figures of that type and you know they have the same area, you also know they must be congruent.

    b. Devise your own test of congruence for rectangles that *will* work.

For each pair of figures, explain what you would measure to test for congruence and what you would look for in your measurements.

19. two circles   **20.** two equilateral triangles

21. One way to determine whether two-dimensional figures are congruent is to lay them on top of each other. This test will not work, however, with three-dimensional figures.

    a. How could you determine whether two cereal boxes are congruent?

    b. How could you determine whether two cylindrical soup cans are congruent?

22. Challenge The word *bisect* means to divide into two equal parts. The steps below show how to bisect ∠*JKL* using a compass and a straightedge.

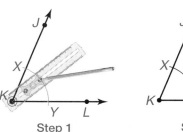
Step 1
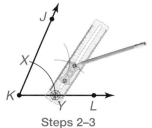
Steps 2–3
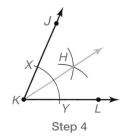
Step 4

Step 1   Place the compass at point *K* and draw an arc that intersects both sides of the angle. Label the intersections *X* and *Y*.

Steps 2–3   With the compass at point *X*, draw an arc in the interior of ∠*JKL*. Using this setting, place the compass at point *Y* and draw another arc.

Step 4   Label the intersection of these arcs *H*. Then draw $\overrightarrow{KH}$. $\overrightarrow{KH}$ is the *bisector* of ∠*JKL*.

    a. Describe what is true about ∠*JKH* and ∠*HKL*.

    b. Draw several angles and then bisect them using the steps above.

In y o u r
own
    words

If two pentagons are similar, are they congruent? Explain why or why not. If two pentagons are congruent, are they similar? Explain why or why not.

23. Maps are designed to be similar to the layout of a city's streets. This map shows a section of London.

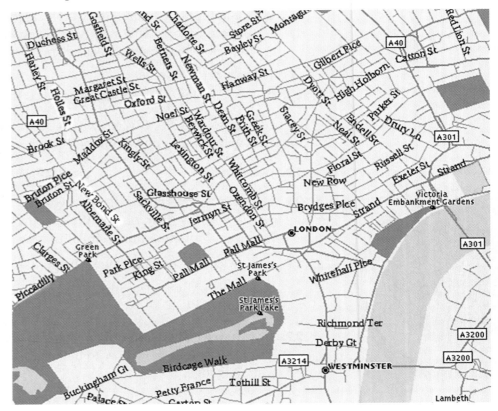

a. The scale of the map is given at the right. How many inches on the map are the same as 1,000 ft in London? Measure to the nearest $\frac{1}{16}$ inch.

1,000 ft

b. What is the distance on the map along Oxford St. between Holles St. and Newman St.?

c. What is the real distance (in feet) along Oxford St. between Holles St. and Newman St.?

d. What is the distance on the map along New Bond St. between Bruton Pl. and Piccadilly?

e. What is the real distance along New Bond St. between Bruton Pl. and Piccadilly?

24. You have two polygons that you know are similar.

a. What would you measure to determine whether the two similar polygons are also congruent?

b. What would you need to know about your measurements to be sure the polygons are congruent? Explain.

25. Preview Mariko proposed this conjecture: "If you are given two similar rectangles with side lengths that share a common ratio of 1 to 2, the ratio of the areas is also 1 to 2. That is, the area of the larger rectangle is twice the area of the smaller rectangle."

Is Mariko correct? If she is, explain how you know. If she isn't, give a counterexample for which the conjecture isn't true.

26. In Investigation 1, you examined rules for testing two figures to determine whether they are congruent. For each pair of figures in this exercise, describe a test you could use to tell whether they are similar.

a. two circles

b. two cubes

c. two cylinders

27. In Investigation 4, you discovered that *similar polygons* have corresponding sides that share a common ratio and corresponding angles that are congruent. For some special polygons, though, you can find easier tests for similarity. For each pair of special polygons below, find a shortcut for testing whether they are similar.

a. two rectangles

b. two squares

Mixed Review

Evaluate each expression.

28. $^-3 \cdot 20$

29. $^-4 \cdot {}^-5.5$

30. $^-2 \cdot {}^-5 \cdot {}^-8$

31. $\frac{1}{8} \div \frac{3}{8}$

32. $\frac{6}{11} \div \frac{2}{11}$

33. $\frac{4}{7} \div 2$

34. $5\frac{1}{16} - 1\frac{1}{8}$

35. $\frac{3}{24} + \frac{2}{3}$

36. $\frac{4}{31} - \frac{5}{62}$

Express each answer in scientific notation. Before adding or subtracting, be sure to change one of the numbers so that both have the same exponent.

37. $6 \times 10^8 + 3 \times 10^7$

38. $3.6 \times 10^4 - 4.5 \times 10^3$

39. Arnaldo wrote $5b + 6$ to represent the total number of blocks in 5 bags plus 6 extra blocks. If he has 66 blocks altogether, how many blocks are in each bag?

Algebra Simplify each expression.

40. $5p + (6p - 12)$

41. $2x - (3x + 1) + (2x + 8)$

42. $(4x + 1) - (3 - 6x) + 2$

43. $3g + 2(h - 1) - (4g - 6)$

44. $(20a + 60b) - 3(a - 2b) + 5$

45. $2d(d - 1) - 3d(2 - 4d)$

Geometry Use the Pythagorean Theorem or the distance formula to find the length of each segment.

46. $(6, 0)$ to $(0, 8)$

47. $(3, {}^-1)$ to $({}^-2, 5)$

48. $(8, 3)$ to $(3, 8)$

49. $({}^-2, {}^-3)$ to $({}^-1.5, {}^-4)$

50. Look at the pattern of squares.

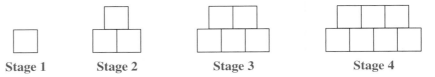

Stage 1   Stage 2   Stage 3   Stage 4

  a. How many squares do you add to go from Stage s to Stage $s + 1$?

  b. How many squares are needed for Stage 5? Stage 7?

  c. How are the new squares placed?

  d. How many squares in all are at each stage shown?

  e. What stage uses 99 squares?

  f. Write an expression for the number of squares in Stage s.

7.2 Polygon Similarity and Congruence

How often do you see polygons? Traffic signs, books and posters, boxes, doors and windows—they all have polygonal shapes.

The most common polygon is probably the rectangle, but triangles are the simplest polygons because they have the least possible number of sides. Two-sided figures like those below don't close, so they aren't polygons.

Every polygon can be divided into triangles. Because of this fact, your knowledge of triangles can help you study other polygons.

 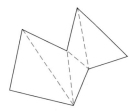 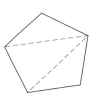

MATERIALS

- linkage strips and fasteners
- protractor

Explore

Use linkage strips to form an equilateral triangle.

To make sure the sides are the same length, start at one vertex and count the *gaps* between the holes until you reach the next vertex. Each gap is 1 unit. All three sides of your triangle should have the same number of units.

Measure each angle of your triangle. (It may help to trace inside the figure on a sheet of paper first.)

Compare your triangle to the others made in your class. Are all the triangles congruent? Are they all similar? How do you know?

Investigation Sides, Sides, Sides

If two figures have corresponding sides of the same length—for example, if two pentagons each have sides of length 2 cm, 3 cm, 4 cm, 5 cm, and 6 cm—can you conclude that the figures are congruent?

MATERIALS

linkage strips and fasteners

Problem Set A

Use linkage strips to create each figure described below. Trace the insides of the polygons with a pencil, and label each side with its length. Write the problem numbers inside the figures so you can refer to them later.

1. Make a quadrilateral with sides of length 4, 5, 6, and 7 units. Keep the sides in that order.

2. Make a pentagon with sides of length 4, 5, 5, 6, and 6 units. Keep the sides in that order.

3. Make a quadrilateral with all sides 5 units long.

4. Make a pentagon with sides of length 6, 5, 6, 5, and 6 units. Keep the sides in that order.

5. Compare your figures to those made by others in your class. Are any of the figures congruent?

All the polygons you created in Problem Set A had more than three sides. If everyone in class now creates triangles, do you think the figures will be congruent?

MATERIALS

linkage strips and fasteners

Problem Set B

Use linkage strips to create each triangle. Trace inside the triangles, and write the problem numbers inside the figures.

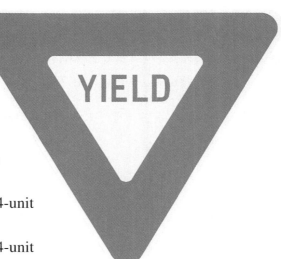

1. Make a triangle with sides of length 3, 4, and 5 units.

2. Make a triangle with sides of length 4, 5, and 6 units.

3. Make a triangle with two 4-unit sides and one 5-unit side.

4. Make a triangle with two 4-unit sides and one 6-unit side.

5. Compare your triangles to those made by others in your class. Did everyone make congruent triangles?

Share & Summarize

1. You can test whether any two polygons are congruent by measuring the sides and angles of both figures. Make a conjecture about what might be an easier test if the polygons are triangles.

2. Give an example to show your test won't work for other polygons.

Investigation And More Sides

You know that you can test whether two polygons are congruent by measuring all their sides and angles. For triangles, though, you need to measure only their sides. This simpler test is called the *side-side-side congruence test*, or SSS for short. Congruent figures must have corresponding sides of equal length. For similar figures, rather than having equal length, corresponding side lengths must share a common ratio.

When you compared figures for similarity, you could have measured all the sides and angles as you did for congruence. In this investigation, you will explore whether there is a test for similarity like the SSS congruence test.

MATERIALS
linkage strips and fasteners

Problem Set C

With your partner, you will make triangles and quadrilaterals. For each problem, each of you should create one of the figures. See if you and your partner can create figures that are *definitely not similar.*

Trace the insides of your figures. If you can create nonsimilar figures, explain how you know they are not similar. If you can create *only* similar figures, tell how you know they are similar.

1. Make one triangle with sides of length 3, 2, and 2 units, and one with sides of length 6, 4, and 4 units.

2. Make one quadrilateral with sides of length 8, 6, 8, and 6 units, and one with sides of length 4, 3, 4, and 3 units. Keep the sides in the order given.

3. Make one triangle with all sides 2 units long, and one with all sides 6 units long.

4. Make one quadrilateral with all sides 2 units long, and one with all sides 6 units long.

5. Make one triangle with sides of length 3, 4, and 5 units, and one with sides of length 4, 5, and 6 units.

Remember

Until it is proven, you can't be sure a conjecture is true.

MATERIALS

• metric ruler
• protractor

Just the facts

Sailors have long used imaginary triangles—formed by their vessel, the horizon, and celestial bodies—to determine their position on the seas.

Think & Discuss

For each pair of figures below, can you tell whether the two figures are similar by knowing just the lengths of their sides? If so, make a conjecture of a way to test for similarity. If not, give a counterexample showing that this information is not enough.

1. two triangles   **2.** two quadrilaterals

Problem Set D

Some of the triangles below are similar to each other. Work with a partner to group all the triangles so that those in each group are similar.

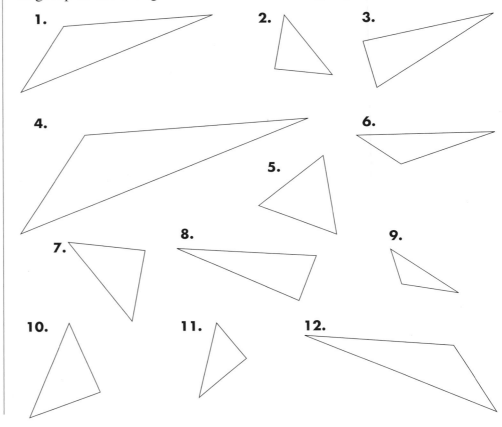

Share & Summarize

When it comes to congruence and similarity, how are triangles different from other polygons?

Investigation ▶3 Angles, Angles, Angles

As you saw in Investigations 1 and 2, you can simply measure the sides of two triangles to test whether they are congruent or similar. With other polygons, you need to know about their angles as well. What if the only information you have about two polygons is the measures of their angles?

MATERIALS
• ruler
• protractor

Problem Set E

Work with a partner to draw a figure that fits each description. Write the problem numbers inside the figures.

1. a triangle with angles 30°, 60°, and 90°

2. a quadrilateral with all angles measuring 90°

3. a triangle with two 60° angles

4. a quadrilateral with three 80° angles

5. a triangle with two 45° angles

6. a triangle with one 110° angle and one 25° angle

7. a triangle with one 90° angle

8. Compare your figures to those made by others in your class. For which problems were all of the figures similar?

Share & Summarize

1. Can you tell whether two triangles are similar by knowing just their angle measures? If so, make a conjecture of a test you can use. If not, give a counterexample showing that this information is not enough.

2. Will your test work on quadrilaterals? If so, explain. If not, give a counterexample.

3. Can you tell whether two triangles are *congruent* by knowing just their angle measures? If so, make a conjecture of a test you can use. If not, give a counterexample.

 **Building Towers**

Architects need to know a lot about geometry and the properties of figures. Some of what you have learned in this lesson may help you build a model as an architect would.

MATERIALS
- toothpicks
- mini-marshmallows (or other connectors)
- rulers

The Challenge

Work with your group for 10 minutes to build the tallest freestanding structure you can. A *freestanding structure* is one that does not lean against anything; it stands on its own. After 10 minutes, measure your structure from the table to its highest point.

Evaluating Your Work

1. How tall is your structure?

2. Look around the room at the different structures. What strategies did the creators of the tallest structures use?

3. What are the common features of the structures that have trouble standing up? What are the common features of the structures that are stronger?

The SSS congruence test you learned about works only for triangles because triangles are the only polygons that are *rigid*. That is, if you build a triangle with three unbendable sides, you can't press on the sides or the angles of the triangle to make a different shape. However, you can change the shapes of other polygons in an infinite number of ways by pressing on the sides. The vertices act like hinges.

Pushing on the sides of a triangle will not change the triangle's shape, but pushing on the sides of other polygons will change their shapes.

4. Does the idea of rigidity help explain which structures seem stronger? If so, explain.

5. If you had the chance to create another structure, what building strategies would you try?

Try It Again

Try your building strategies with another set of materials for another 10 minutes. Your goal this time is to build a freestanding structure *taller* than your first attempt.

6. How tall is your second structure?

7. What strategies did you use this time?

What Did You Learn?

8. Why do you think buildings have triangular supports in their walls?

9. Which of these structures is likely to remain standing the longest? Why?

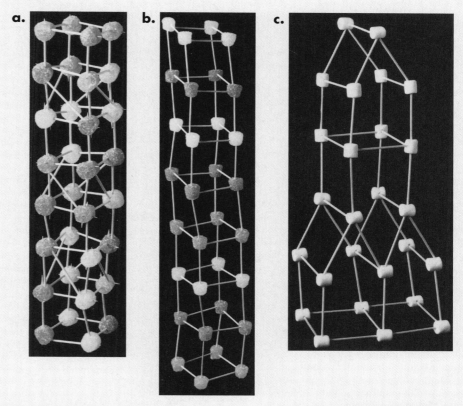

a. b. c.

10. The foreman of a crew that was knocking down a building considered telling the crew to first take out the front and rear walls, leaving a structure like this. Is this a wise decision? Why?

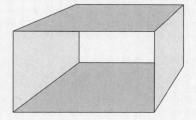

On Your Own Exercises

Practice & Apply

In Exercises 1–3, decide whether the figures described *must be* congruent, *could be* congruent, or *are definitely not* congruent.

1. two quadrilaterals with all sides 4 cm long

2. two triangles with all sides 4 cm long

3. two squares with all sides 4 cm long

The triangles in each pair below are congruent. Find the values of the variables.

4.

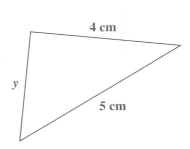

5.

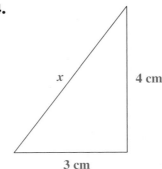

6.

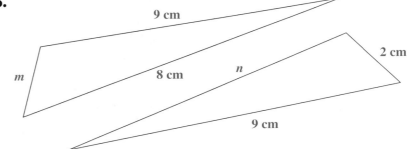

7. The side lengths of six triangles are given. Tell which triangles are similar to a triangle with sides of length 2 cm, 4 cm, and 5 cm.

   a. 1 in., 7 in., 4 in.

   b. 5 cm, 4 cm, 2 cm

   c. 8 in., 4 in., 10 in.

   d. 400 cm, 500 cm, 200 cm

   e. 20 ft, 30 ft, 40 ft

   f. 1 cm, 2 cm, 2.5 cm

impactmath.com/self_check_quiz

In Exercises 8–17, decide whether the figures described *must be* similar, *could be* similar, or *are definitely not* similar.

8. two squares

9. two rectangles

10. two triangles

11. a square with all sides 2 cm and a rectangle with length 2 cm and width 1 cm

12. a triangle with sides of length 2 cm, 3 cm, and 4 cm and a triangle with sides of length 2 ft, 3 ft, and 4 ft

13. a triangle with all sides 2 cm long and another triangle with all sides 2 cm long

14. two quadrilaterals with all right angles

15. two triangles, each with a right angle

16. two triangles with three 60° angles

17. two triangles with two 45° angles

18. Sally thinks that triangles with two corresponding sides of equal lengths and an equal angle must be congruent. To test her conjecture, first draw as many triangles as you can that meet the descriptions in Parts a–c.

a. Two sides have lengths 4 and 7; the angle between them is 115°.

b. Two sides have lengths 5 and 8; the angle between them is 35°.

c. Two sides have lengths 4 and 5; one of the angles not between the two sides is 45°.

d. Do you agree with Sally's conjecture? That is, if you know that two corresponding sides and a corresponding angle of two triangles are congruent, must the triangles be congruent? Explain.

19. Suppose you have two circles. Decide whether they *must be* similar, *could be* similar, or *are definitely not* similar.

20. Consider Triangle T.

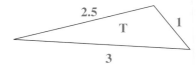

- **a.** Give the side lengths of three different triangles that are similar to Triangle T.

- **b.** Is a triangle with sides of length 7.5, 3, and 9 units similar to Triangle T? Explain.

- **c.** Is a triangle with sides of length 4, 12, and 8 units similar to Triangle T? Explain.

- **d.** If a triangle is similar to Triangle T and its shortest side has a length of s units, how long are the other two sides?

- **e. Challenge** If a triangle is similar to Triangle T and its *longest* side has a length of s units, how long are the other two sides?

21. You can combine two congruent equilateral triangles to form a quadrilateral.

Suppose you form a quadrilateral with another pair of equilateral triangles similar to those above. Decide which of the following is true about the quadrilateral you make and the one above: the two figures *must be* similar, *could be* similar, or *are definitely not* similar.

22. Draw a large triangle. It should take up most of a sheet of paper. Then follow these directions. (The example is to help you follow the directions. Your triangle may look different.)

Mark the midpoint of each side of your triangle.

Connect the midpoints to form four smaller triangles, one of them upside down. Shade that center triangle.

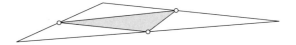

a. You now have four small triangles. How do they compare to the large triangle you began with? (Think about similarity and congruence.)

b. How do the four small triangles compare to each other?

Now, for each triangle that is not shaded, find the midpoints of the sides. In each of these triangles, connect the midpoints. Then shade the new upside-down triangles.

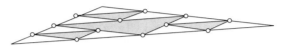

c. You now have 13 triangles inside the original large one. Are all the triangles similar? Are any of them congruent?

Mixed Review

Evaluate without using a calculator.

23. $49.073 - 56.2$ **24.** $0.98 - 2.71$ **25.** $400.06 - 2.2$

26. $\frac{3}{7} + \frac{4}{9}$ **27.** $\frac{2}{3} - \frac{1}{7}$ **28.** $\frac{32}{35} + \frac{3}{70}$

Factor each expression.

29. $3p + 3$ **30.** $\frac{a}{7} - \frac{1}{7}$ **31.** $4a^2b - 16a^3$

Supply each missing exponent.

32. $6 \times 10^? = 600$ **33.** $0.2 \times 10^? = 0.002$ **34.** $1.7 \times 10^? = 17$

35. Suppose you use a photocopying machine to enlarge a picture.

a. Your picture is 10 cm wide and you enlarge it 110%. How wide is the copy of the picture?

b. You use the same setting to enlarge your copy. How wide is the final picture?

36. Suppose you use a photocopying machine to reduce a picture.

a. Your picture is 10 cm wide and you reduce it to 85% of its original size. How wide is the copy of the picture?

b. You use the same setting to reduce your copy. How wide is the final picture?

Area and Perimeter of Similar Figures

You have learned several tests for identifying similar figures, and you know a few ways to create similar figures. How do you think the areas and perimeters of similar figures compare? In this lesson, you will find out.

Think & Discuss

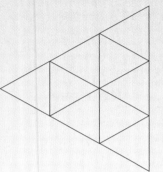

You can combine nine copies of an equilateral triangle to form a larger triangle.

Is the large triangle similar to the small triangle? How do you know?

If the area of the small triangle is 1 square unit, what is the area of the large triangle?

Suppose you wrap a string around the perimeter of the small triangle. How many times longer must a string be if you want to wrap it around the perimeter of the large triangle?

Investigation Dilating Triangles and Rectangles

You have seen that if two polygons are similar, the lengths of their corresponding sides share a common ratio. That means you can multiply the side lengths of one figure by some number to get the side lengths of the other figure.

VOCABULARY
scale factor
dilation

Given similar figures A and B, the **scale factor** from Figure A to Figure B is the number by which you multiply the side lengths of Figure A to get the side lengths of Figure B. The scale factor from the small triangle to the large triangle shown above is 3. Figure B is said to be a **dilation** of Figure A. Dilating a figure creates another figure that is similar, but not necessarily congruent, to the original.

Problem Set A

The sides of this small triangle are 1 unit long. You will build larger equilateral triangles from copies of this small triangle.

1. Trace and cut out several copies of the small triangle.

a. Use your triangles to build larger triangles with the side lengths listed in the table. Make a sketch of each large triangle you build.

b. Copy and complete the table.

Side Length of Large Triangle	Number of Small Triangles in Large Triangle	Scale Factor (small to large)
1		
2		
3		
4		

2. Make a new table with three columns. In the first column, list the scale factors you found in Problem 1. You will complete the other two columns in Parts a and b.

a. First, find the perimeter of each large triangle you created. In the second column of your table, record the ratios of the perimeter of the small triangle to the perimeter of the large triangle.

Write each ratio using the smallest possible whole numbers. For example, the triangle at right has a perimeter of 9 units, so the ratio of the perimeters is 3:9, which is equivalent to 1:3.

b. Now find the area of each large triangle, using the area of the small triangle as your unit. For example, the area of the large triangle above is 9 small triangles. In the third column of your table, give the ratios of the areas.

3. How are the perimeters of the small and the large triangles related?

4. How are the areas of the small and the large triangles related?

5. You've worked with dilating the small triangle to make the large triangle. In fact, each pair of similar figures with different sizes has *two* scale factors associated with it. Imagine dilating each of the large triangles down to the size of the small triangle.

a. For each, what would the scale factor from large to small be?

b. How is the large-to-small scale factor for a pair related to the small-to-large scale factor?

Problem Set B

Each large rectangle below is divided into smaller rectangles that are congruent to each other and similar to the large rectangle.

For Problems 1–5, find the scale factor from the small rectangle to the large rectangle. Then calculate the perimeter and the area of each size rectangle. Record your answers in two tables, with these heads:

Problem Number	Scale Factor (small to large)	Perimeter of Small Rectangle	Perimeter of Large Rectangle

Problem Number	Scale Factor (small to large)	Area of Small Rectangle	Area of Large Rectangle

1. 1 unit, 1 unit

2. 2 units, 2 units

3. 1 unit, 3 units

4. 2 units, 1 unit

5. 3 units, 2 units

Share & Summarize

1. A rectangle with perimeter *p* and area *A* is enlarged by a factor of *f*. Make a conjecture about how to calculate the perimeter and the area of the new rectangle.

2. A triangle with perimeter *p* and area *A* is enlarged by a factor of *f*. Make a conjecture about the perimeter and the area of the new triangle.

3. Use your observations of similar rectangles and similar triangles to make a conjecture about how you can calculate the perimeters and the areas of two similar polygons.

Investigation 2 ▶ Dilation and Formulas

The box below contains possible conjectures for the relationship between the perimeters and between the areas of similar polygons.

> For two similar polygons A and B such that the scale factor from Polygon A to Polygon B is r,
>
> - the perimeter of Polygon B is r times the perimeter of Polygon A.
> - the area of Polygon B is r^2 times the area of Polygon A.

Your investigation with whole-number scale factors supports these conjectures—but you can't try every possible scale factor to verify them. However, you can use the perimeter and area formulas for a polygon to prove that the conjectures are true for *any* scale factor and that polygon.

EXAMPLE

Malik thought of a way to verify the perimeter conjecture for all rectangles and a scale factor of 3.

When I dilate a rectangle by a whole number, the perimeter is dilated by the same number. But what if I don't know anything about the rectangle?

For length L and width W, the rectangle's perimeter is $L + W + L + W$. I just add the side lengths.

If it's dilated by 3, the new length is $3L$ and the new width is $3W$ — so the new perimeter is $3L + 3W + 3L + 3W$. The distributive property lets me factor out a 3, giving $3(L + W + L + W)$, which is 3 times the original perimeter.

Problem Set C

1. Following the steps below, you will use an explanation like Malik's to show that if you dilate this square by a factor of $\frac{1}{2}$, the perimeter of the new square will be $\frac{1}{2}$ the original perimeter.

   a. Malik first wrote the perimeter of the rectangle as a sum of the four side lengths. Write the sum that gives the perimeter of the original square.

   b. Dilate the square by a factor of $\frac{1}{2}$, and draw the new square. How long is each side, in terms of *s*?

   c. Write the sum that gives the perimeter of the new square.

   d. Is the new perimeter half the original perimeter? Explain how you know from the expressions you wrote in Parts a and c.

2. Following the steps below, you will show that if you scale this triangle by 100, the new perimeter will be 100 times the original perimeter.

   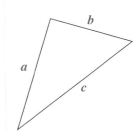

   a. Write an expression for the triangle's perimeter.

   b. If you dilate the triangle by a factor of 100, how long will each side be?

   c. Write an expression for the perimeter of the dilated triangle.

   d. Is the new perimeter 100 times the original? Explain.

3. **Prove It!** Suppose a pentagon has sides of length *a*, *b*, *c*, *d*, and *e*.

   a. Show that if you dilate the pentagon by 0.323, the new perimeter will be 0.323 times the original perimeter.

   b. Show that if you dilate the pentagon by a factor of *r*, the new perimeter will be *r* times the original perimeter.

   c. Explain why a similar argument would work for a polygon with *any* number of sides, not just for five-sided polygons.

Model trains such as this one are built to specific scales. A common scale is 1:87, which means that 87 feet of actual track would be represented by 1 foot of track on the model.

Now you will consider what happens to the *area* of a dilated figure.

EXAMPLE

A square with side length s has an area of s^2.

If you dilate the square by 3, the new sides have length $3s$.

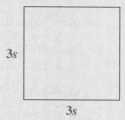

The area of the new square is $(3s)^2 = 9s^2$, which is nine times the original area.

In general, if you dilate an $s \times s$ square by a factor of r, the new square has sides of length rs.

The area of the new square is $(rs)^2 = r^2s^2$, which is r^2 times the original area.

Problem Set D

1. In this problem, you will show that the areas of similar parallelograms are related by the square of the scale factor between them. This parallelogram has length l and height h.

   a. What is the parallelogram's area?

   b. If you dilate the parallelogram by a factor of 2, what happens to the height? Explain your thinking.

   c. If you dilate the parallelogram by a factor of 2, what are the new length and height?

   d. What is the area of the dilated parallelogram?

e. If you dilate the original parallelogram by a factor of $\frac{1}{3}$, what are the new length, height, and area?

f. If you dilate the original parallelogram by a factor of r, what are the new length and height?

g. Prove It! What is the area of the dilated parallelogram in Part f? How does it relate to the area of the original parallelogram?

2. In this problem, you will show that if you dilate *any* triangle, the two areas are related by the square of the scale factor. Start with a triangle with sides x and y, base z, and height h.

a. What is the triangle's area?

b. If you dilate the triangle by a factor of 3.2, what are the new base and height?

c. What is the area of the dilated triangle?

d. If you dilate the triangle by a factor of r, what are the new base and height?

e. Prove It! What is the area of the dilated triangle in Part e? How does it relate to the area of the original triangle?

Share & Summarize

1. If you dilate a polygon by a factor of r, what happens to the polygon's perimeter? Does it depend on r? Do you *know* this is true for all polygons, or is this a conjecture?

2. If you dilate a polygon by a factor of r, what happens to the polygon's area? Does it depend on r? Do you *know* this is true for all polygons, or is this a conjecture?

Investigation 3 Similarity in More Complex Figures

Do the relationships you found for similar parallelograms and triangles hold for more complex polygons? Do they hold for figures with curved sides? You will explore these questions in this investigation.

Problem Set E

The circles and the polygons in Problems 1 and 2 are similar figures. For each pair, do Parts a and b. You may want to record your answers in a table.

a. Find the scale factor from the small figure to the large figure.

b. Find the perimeters and the areas of both figures.

1.

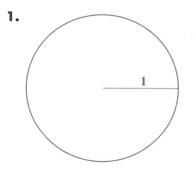 

Remember

A circle with radius r has area πr^2 and circumference (perimeter) $2\pi r$. A trapezoid with bases b_1 and b_2 and height h has area $\frac{1}{2}(b_1 + b_2)h$.

2.

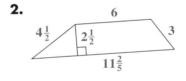 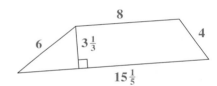

3. How are the perimeters of two similar figures related? Do the shapes of the figures affect this relationship?

4. How are the areas of two similar figures related? Do the shapes of the figures affect this relationship?

By now you should have a good idea of the relationships between the areas and the perimeters of *any* dilated figures. To help you see that those relationships can be extended beyond polygons and circles, think about how you learned to find areas and perimeters of irregular figures.

You can approximate the area of an irregular figure by placing it on a grid and counting the squares inside.

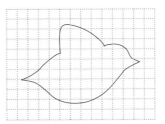

If you dilate the figure *along with the grid,* you get something that looks like this.

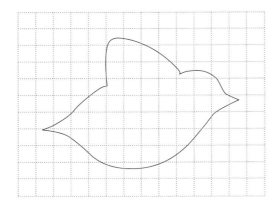

You can approximate the perimeter of an irregular figure by using line segments.

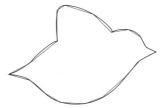

When you dilate the figure, the line segments look like this.

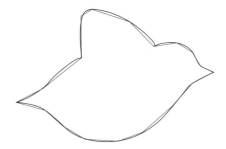

Problem Set F

1. Write a few sentences to explain why dilating the irregular figure on the opposite page by a factor of r dilates its area by a factor of r^2.

2. Write a few sentences to explain why dilating the irregular figure by a factor of r also dilates its perimeter by a factor of r.

3. The figures below are similar.

Perimeter: 5 units
Area: 1 square unit

Perimeter: 15 units
Area: ? square units

a. What is the scale factor from the small figure to the large figure?

b. What is the area of the large figure?

4. The figures below are similar.

Perimeter: ? units
Area: 4 square units

Perimeter: 24 units
Area: 16 square units

a. What is the scale factor from the small figure to the large figure?

b. What is the perimeter of the small figure?

Just the facts

The arms of certain starfish, when broken off, will regenerate into new starfish genetically identical to the original organisms. This process is known as *cloning*.

▶ MATERIALS
ruler

Share & Summarize

These four figures are all similar. Figure A has an area of 3 square units. One of Figures B, C, and D has an area of 12 square units. Which figure is it? How can you tell?

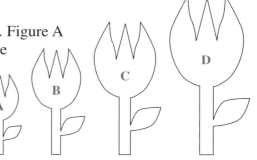

On Your Own Exercises

Practice  **Apply**

In Exercises 1–4, a figure is divided into smaller figures that are congruent to each other and similar to the large figure. Do Parts a–c for each exercise.

    a. Find the scale factor from the small figure to the large figure.

    b. Find the perimeter and area of each figure (small and large).

    c. Test whether the results support your conjectures from the Share & Summarize on page 488.

Remember

The area of a triangle with base *b* and height *h* is $\frac{1}{2}bh$. The area of a parallelogram with base *b* and height *h* is *bh*.

1.

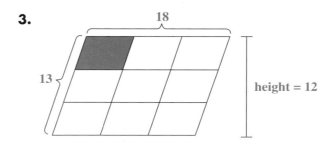

2.

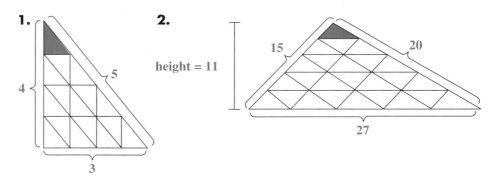

3.

4.

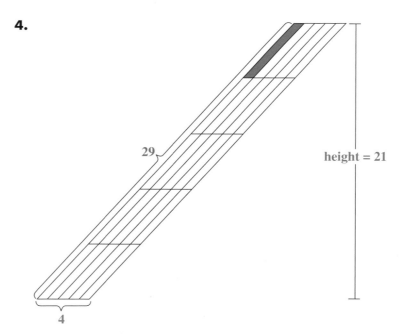

impactmath.com/self_check_quiz

5. This rectangle has sides x and y.

a. What is the rectangle's area?

b. If you dilate the rectangle by a factor of 10, what is the area of the dilated rectangle?

c. Prove It! Show that if you dilate the rectangle by a factor of r, the area of the new rectangle is r^2 times the area of the original.

6. A triangle has perimeter 10 cm and area 4 cm^2. Could you make a similar figure with perimeter 30 cm and area 12 cm^2? Why or why not?

7. A rectangle has perimeter 30 cm and area 25 cm^2. Could you make a similar figure with perimeter 6 cm and area 1 cm^2? Why or why not?

8. Here are two similar figures.

Area = 32 square units   Area = 18 square units

a. What is the scale factor from the large figure to the small figure? How did you find it?

b. What is the scale factor from the small figure to the large figure?

c. Suppose you wrapped a string around the small figure and now want a piece of string to wrap around the large figure. In comparison to the length of string needed for the small figure, how much string will you need for the large figure?

9. Here are two similar figures.

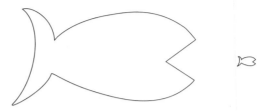

Perimeter: 27 m   Perimeter: 2.7 m

    a. What is the scale factor from the large figure to the small figure?

    b. What is the scale factor from the small figure to the large figure? How did you find it?

    c. If the area of the small figure is n square meters, what is the area of the large figure?

10. A figure has perimeter 30 cm and area 50 cm^2. Could you make a similar figure with perimeter 60 cm and area 100 cm^2? Why or why not?

11. The area of this triangle is 2 square units. Draw a triangle similar to it that has an area of 18 square units.

12. **Measurement** Understanding the relationships in Investigation 1 can help you with metric conversions.

    a. What is the scale factor from a square with a side length of 1 centimeter to a square with a side length of 1 meter?

    b. How many square centimeters fit in a square meter?

    c. What is the scale factor from a cube with a side length of 1 centimeter to a cube with a side length of 1 meter?

    d. How many cubic centimeters fit in a cubic meter?

13. **Prove It!** The area of a regular hexagon with side lengths s is given by $\frac{3}{2}\sqrt{3}s^2$.

Suppose you dilate this hexagon by a factor of f. Prove that the area of the new hexagon is f^2 times the area of the original hexagon.

In your
own
words

Suppose you have two similar figures, one with a perimeter of twice the other. Explain why the area of one is not twice the area of the other.

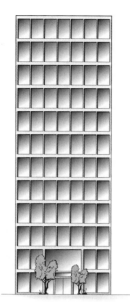

14. Architecture At left is a scale drawing of Zeitner Tower, a steel-framed 12-story building. The scale factor from the drawing to the real building is 1:480.

a. What is the height of the real Zeitner Tower?

b. At right is a sketch of the first floor of the building, using the same scale factor. What is the area of the sketch? What is the area of the first floor of Zeitner Tower?

c. Suppose you wanted to fill Zeitner Tower with popcorn. What volume of popcorn would you need? Explain.

15. The grid lines on this map of a lake are at 1-centimeter intervals. Use the map's scale to approximate the area of the real lake's surface.

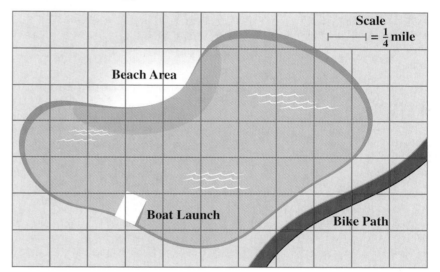

16. Life Science Scientists often work with objects that are too small to see with the unaided eye. A microscope, in a way, creates a scale model of small objects. For example, this photograph of a *Volvox* colony is magnified 150 times.

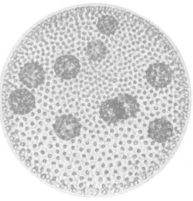

a. Measure the approximate diameter of this entire colony as magnified, in millimeters.

b. What is the approximate diameter of the real *Volvox* colony?

Rewrite each expression as a fraction.

17. $(^-10)^{-5}$

18. $(^-4)^{-2} \cdot (^-4)^{-2}$

19. $3^{-6} \div 3^{-2}$

Rewrite each expression as addition, and calculate the sum.

20. $^-10 - ^-9$

21. $1.5 - ^-0.14$

22. $^-3.33 - ^-1.68$

23. Draw a factor tree to find all the prime factors of 128.

24. Copy and complete the table.

Fraction	Decimal	Percent
$\frac{1}{2}$	0.5	50%
		85%
$\frac{4}{5}$		
	1	

25. Order these fractions from least to greatest.

$$\frac{1}{2} \qquad -\frac{1}{2} \qquad \frac{1}{8} \qquad \frac{1}{4} \qquad -\frac{1}{3} \qquad \frac{1}{9} \qquad -\frac{1}{7}$$

26. Point A is located at (5, 2).

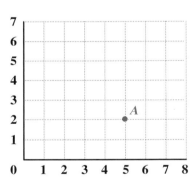

a. Point (3, 1) is located 3 units from Point A along grid lines. Name *all* other points that are located 3 units from Point A along grid lines. (Be careful—some of the points may not be visible on the grid shown.)

b. Point B is located at (3, 5). How many units long is the *shortest* path from Point A to Point B along grid lines?

c. How many *different* shortest paths are there from Point A to Point B along grid lines?

27. Two cylinders have the same radius. The height of one cylinder is three times the height of the other cylinder. How do the volumes of the cylinders compare?

7.4

Volume and Surface Area of Similar Figures

You have dilated two-dimensional drawings. You can also dilate three-dimensional objects. Model trains are scale models of real trains, a globe is a scale model of Earth, and some dollhouses are scale models of real houses.

To see how scaling works in three dimensions, it helps to start with simple block structures.

MATERIALS
cubes

Explore

Build this block structure.

Which of the block structures below has edges that are twice as long as those of your structure? You may want to use your blocks to help you answer the question.

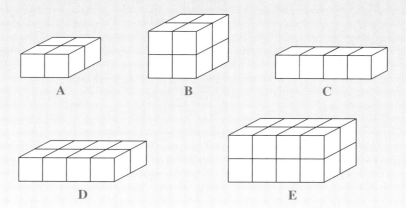

Does the structure you identified have twice the volume of your structure?

Investigation Dilating Block Structures

As with two-dimensional figures, the scale factor between three-dimensional figures is the number by which you multiply the lengths of the original figure to get the corresponding lengths of the new figure. The figure you identified in the Explore activity is similar to the original structure, dilated by a factor of 2.

MATERIALS
- cubes
- graph paper

Problem Set A

For each structure, work with a partner to build a new structure, using a scale factor of 2. Draw the top-count view of each structure you build. (Remember: The *top-count view* shows a top view of the structure and the number of blocks in each stack.)

1.   **2.**   **3.**

MATERIALS
- cubes
- graph paper

Problem Set B

For each structure, work with a partner to build a new, similar structure, dilated by the given scale factor. Draw a top-count view of each structure you build.

1. scale factor 3

2. scale factor 3

3. scale factor $\frac{1}{2}$
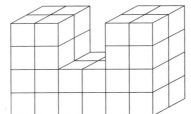

4. scale factor $\frac{2}{3}$
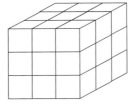

Share & Summarize

Write a letter to a friend explaining the steps you took to solve Problem 2 of Problem Set A.

Investigation Volume of Similar Block Structures

In Lesson 7.3, you found that if you create a polygon that is dilated by a factor of c from an original polygon, the perimeter of the new polygon will be c times the original perimeter. You also found that the area will be c^2 times the original area. Now you will dilate three-dimensional objects and compare the volumes of the dilated objects to the original volumes.

MATERIALS

cubes

Remember

A single block has a volume of 1 cubic unit.

Problem Set C

Work in a group on these problems.

1. For each block structure, find the volume of the original. Then create a structure that has edges twice as long as those of the original structure, and find the new volume. Record your results in a table like the one shown.

   a.   b.   c.

Part	Volume of Original Structure	Volume of Scaled Structure
a		
b		
c		

2. You dilated each structure in Problem 1 by a factor of 2. For each structure, how does the volume of the dilated structure appear to be related to the volume of the original structure?

Problem Set D

Now you will explore what happens to volume for several scale factors. Your original structure for these problems will be a single block, which has a volume of 1 cubic unit.

Working with your group, complete the table below by following these instructions:

- If you are given the *scale factor*, try to build a new block structure similar to the original structure and scaled by that factor. Find the new volume.

- If you are given the *volume*, try to build a block structure similar to the original structure and with that volume. Find the scale factor.

For some entries, you won't be able to *build* the structure, but you will be able to reason about what the volume would be if you *could* build it.

Scale Factor	Volume of Dilated Structure
2	
	27
1	
10	
	$\frac{1}{8}$
$\frac{1}{5}$	
r	

Share & Summarize

Make a conjecture describing how the volume changes when you dilate a block structure. Be sure to say how the change in volume depends on the scale factor.

Investigation Surface Area of Similar Objects

In Investigation 2, you explored the relationship between the volumes of two similar objects. Another measurement for three-dimensional objects is surface area.

Nets allow you to use only two dimensions to show all the surfaces of a three-dimensional object at once. Nets make it easy to explore the relationship between the surface areas of two similar objects.

MATERIALS
- metric ruler
- scissors
- compass
- tape

Remember

A *net* is a flat object that can be folded to form a closed, three-dimensional solid.

Problem Set E

1. Here is a net for a square prism.

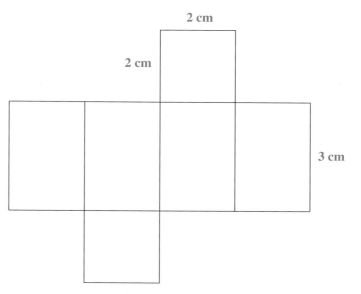

a. Carefully draw and label a net similar to this one, using a scale factor of 2.

b. Trace the net above. Cut out both nets, and fold them to form solids. Are the two prisms similar?

c. Compare the surface areas of your two prisms. You might find it easier to work with the unfolded nets. How many times the surface area of the scaled prism is the surface area of the original prism? Show how you found your answer.

Problems 2 and 3 show a net for a prism and a cylinder. For each problem, do Parts a–c.

> **a.** Draw and label a net similar to the given net, using the given scale factor.
>
> **b.** Find the surface areas of the original object and the scaled object.
>
> **c.** Compare the surface areas. The new surface area is how many times the original surface area?

2. scale factor 3

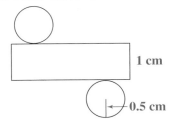

1 cm

0.5 cm

3. scale factor 4

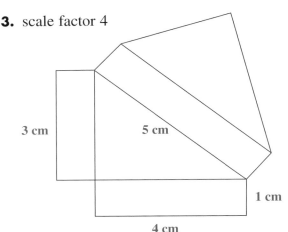

3 cm

5 cm

1 cm

4 cm

4. How does the surface area of a scaled object appear to be related to the surface area of the original object? Be sure to mention scale factors in your answer.

In Lesson 7.3, you saw that the relationships you found between perimeters of scaled polygons and areas of scaled polygons are true for all figures, not just polygons. It's also true that the volumes and surface areas of *all* similar objects have the same relationships as the ones you've probably observed.

That is, if three-dimensional Figures A and B are similar and the scale factor from Figure A to Figure B is *r*, then

- the volume of Figure B is r^3 times the volume of Figure A.
- the surface area of Figure B is r^2 times the surface area of Figure A.

Remember

The surface area of a cylinder with radius *r* and height *h* is $2\pi r^2 + 2\pi rh$ (the area of the top and bottom plus the area of the sides).

Problem Set F

1. The figures below are similar.

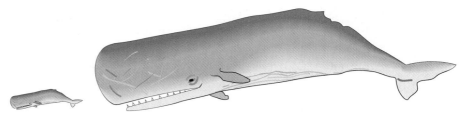

Surface area: 8 square units
Volume: 2 cubic units

Surface area: 200 square units
Volume: ? cubic units

a. Find the scale factor of the small figure to the large figure.

b. Find the volume of the large figure.

2. The figures below are similar.

a. Find the scale factor of the small figure to the large figure.

b. Find the surface area of the small figure.

Surface area: ? square units
Volume: 20 cubic units

Surface area: 450 square units
Volume: 540 cubic units

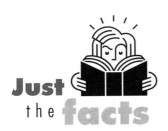
3. The figures below are similar.

Surface area (hull): **40 square units**
Volume (hull): **10 cubic units**

Surface area (hull): **? square units**
Volume (hull): **? cubic units**

a. The small boat's hull is half the length of the large boat's hull. Find the scale factor of the small figure to the large figure.

b. Find the surface area and volume of the large boat's hull.

Share & Summarize

Suppose Figures A and B are similar figures such that the scale factor from Figure A to Figure B is n.

1. If the surface area of Figure A is s, what is the surface area of Figure B?

2. If the volume of Figure A is v, what is the volume of Figure B?

Investigation 4 Giants

In Jonathan Swift's 1726 book *Gulliver's Travels,* Gulliver visits a place called Brobdingnag. There he meets people who are 12 times his size in every dimension. Could the Brobdingnagians really have existed? If the body of a giant were similar to ours, would it be able to support its own weight?

Think & Discuss

In Chapter 2, you used blocks to model human beings so you could think about people's surface areas more easily. If you use a single block to model Gulliver, how many blocks would you need to model a giant 12 times the size in every dimension?

If Gulliver weighed 175 lb, how much would the giant weigh (based on his volume)?

Problem Set G

Your bones must be able to support more than your own weight. If they didn't, you couldn't carry a heavy object, gain weight, or land from a jump without breaking your bones. The maximum weight your bones can support depends on their thickness, or *cross-sectional area*.

1. The sketch shows a cross section of Gulliver's femur (thigh bone), which is very close to a circle. What is the approximate area of the circular cross section?

2. For mammals, the weight a bone can support is related to the area of the bone's cross section. If its cross-sectional area is n in.^{2}, a femur can support $1{,}563n$ lb. How much weight can Gulliver's femur support?

0.6 in.

3. Giants in Brobdingnag are 12 times the size of Gulliver in every dimension. What is the area of the circular cross section of a giant's femur?

4. Assume that the giant's bones have the same strength as human bones. Using the fact that a bone with cross-sectional area n in.^{2} can support $1{,}563n$ lb, find the maximum weight the giant's femur can support.

5. When you walk, one leg must support all your weight as you take a step. Could the giant's leg bone support his weight? Explain how you know.

6. For a leg bone to support the giant's weight (as a maximum), how much cross-sectional area must it have? What would its radius be? Explain.

Just the facts

Jonathan Swift's masterpiece—the full title of which is *Travels into Several Remote Nations of the World, by Lemuel Gulliver*—is written in the form of a journal kept by Gulliver, a ship's physician.

Share & Summarize

1. Explain why there cannot really be giants the same shape as people with bones the same strength as human bones. What do you think is the maximum size a person could grow to and still stand up?

2. Giants that can support their own weight cannot be similar to humans. What *would* giants who can support their weight have to look like? Explain.

On Your Own Exercises

Practice & Apply

1. Bryn dilated Structure A by some factor to get Structure B. What scale factor did Bryn use? How do you know?

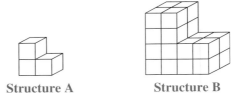

Structure A   Structure B

Determine whether the block structures in each pair below are similar. If they are, find the scale factor. If they aren't, explain how you know.

2.   3.

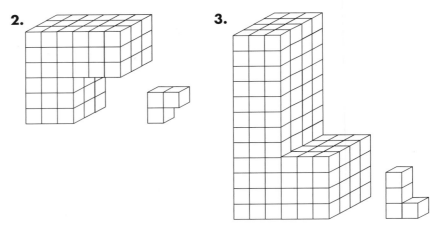

4. Suppose you dilate this block structure by a factor of $\frac{1}{2}$. What would the volume of the new structure be? Explain.

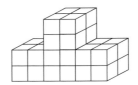

impactmath.com/self_check_quiz

5. Using cardboard, Enrique made a box 30 cm long, 30 cm wide, and 12 cm deep. The box has no top. Enrique then wanted to make a new box, *similar* to the first one, that would hold exactly twice as much.

   a. What is the volume of Enrique's original box?

   b. If its sides were twice as long as those of the original box, how much would the new box hold?

   c. Would the sides of the new box that holds twice as much be twice as long as the sides of the original box?

   d. Challenge Find the approximate dimensions of a box that is similar to the original box and holds twice as much.

6. Suppose you dilated a cylinder with height 3 cm and radius 6 cm by a factor of $\frac{1}{3}$.

   a. What is the volume of the original cylinder?

   b. What is the volume of the scaled cylinder?

   c. Compare the volumes. Tell how the change in volume seems to be related to the scale factor.

7. This is a sketch of a net for a pyramid.

   a. Sketch and label a net for a similar pyramid, dilated by a scale factor of $\frac{1}{2}$.

   b. Find the surface area of the original pyramid.

   c. Find the surface area of the dilated pyramid.

   d. Compare the surface areas by filling in the blank: *The new surface area is _____ times the original surface area.*

13 cm   12 cm   13 cm
10 cm
10 cm
13 cm
13 cm

8. Suppose you dilated a cylinder with height 3 cm and radius 6 cm by a factor of $\frac{1}{3}$. What would be the surface area of the dilated cylinder? Find two ways to get the answer.

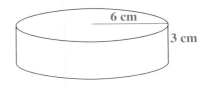

9. Challenge The volume of a sphere with radius r is $\frac{4}{3}\pi r^3$. Hillary dilated a sphere with a volume of $\frac{4\pi}{3}$ cubic units by some factor to get a sphere with radius 3.

a. What scale factor did Hillary use? Find two ways to get the answer.

b. The surface area of a sphere with radius r is $4\pi r^2$. Find the surface area of Hillary's original sphere. Then calculate the surface area of the dilated sphere in two ways.

10. In Investigation 3, you solved several problems about surface areas, volumes, and scale factors of similar figures.

a. Suppose the surface area of a figure is 60 ft^2, and its volume is 40 ft^3. A figure that is similar to this figure has a volume of 5 ft^3. What is the scale factor from the original figure to the new figure? What is the surface area of the new figure?

b. Now make up your own problem. Give the surface area and volume of one figure, and give the surface area or the volume of another, similar figure. Then ask about the volume or surface area (whichever piece of information you did not give) and the scale factor. Include a solution with your problem.

Remember

Gulliver weighed 175 pounds. His thigh bone had a radius of 0.6 in.

11. Literature Brobdingnag wasn't the only land Gulliver visited in his travels. In the land of Lilliput, he encountered people who were as small to him as the Brobdingnagians were large! That means they were $\frac{1}{12}$ Gulliver's size in every dimension.

a. What is the cross-sectional area of a Lilliputian's thigh bone?

b. If the cross-sectional area of a femur is n in.^{2}, it can support about 1,563n lb. How much weight can a Lilliputian's femur support?

c. The amount of weight you found in Part b might not seem like much. However, a human being's femur can support a maximum of 10 times a person's body weight. How much would a Lilliputian $\frac{1}{12}$ the size of Gulliver in each dimension weigh? How many more times a Lilliputian's body weight can his leg support?

12. Life Science Bones make up about 14% of a human being's body weight. About how much would Gulliver's bones weigh?

13. Some breakfast cereals are packaged in boxes of different sizes. The chart gives the dimensions of three box sizes.

Breakfast Cereals

Box Size	Length	Width	Depth
small	27.2 cm	19.0 cm	6.0 cm
medium	30.5 cm	20.8 cm	7.0 cm
large	33.8 cm	24.0 cm	7.5 cm

Are any of the boxes similar to each other? Explain how you know.

14. Latifa says that if you enlarge a block structure by a scale factor of 2, the volume and the surface area of the block structure also increase by a factor of 2. Is she correct? If so, explain why. If not, give a counterexample of a block structure for which this will not happen.

15. Consider cubes of different sizes.

a. Is a cube with edges of length 1 unit similar to a cube with edges of length 10 units? If so, what is the scale factor from the small cube to the large cube? If not, why not?

b. Is a cube with edges of length 3 units similar to a cube with edges of length 5 units? If so, what is the scale factor from the small cube to the large cube? If not, why not?

c. Is a cube with edges of length n units similar to a cube with edges of length m units? If so, give the scale factor from the $n \times n \times n$ cube to the $m \times m \times m$ cube. If not, explain why not.

16. **Prove It!** In Investigation 2, you learned that when a block structure is dilated by a factor of n, its volume is dilated by a factor of n^3.

a. Prove that this statement is true for a $2 \times 2 \times 2$ cube. That is, show that if you dilate such a cube by a factor of n, the volume of the new cube is n^3 times the volume of the original cube. Hint: Use the edge length of the dilated cube to find the new volume.

b. Prove that this statement is true for a rectangular prism with edges of length 1, 3, and 4 units. That is, show that if you dilate such a prism by a factor of n, the volume of the new rectangular prism is n^3 times the volume of the original.

17. **Astronomy** When astronomers create models of the planets in our solar system, they often represent them as spheres. The planets are not perfect spheres, but they are close enough that they can be *modeled* with spheres. Because all spheres are similar to one another, the planets can be thought of as being similar to one another.

The table lists the approximate radii of five planets.

Size of Planets

Planet	Radius
Jupiter	71,400 km
Neptune	24,764 km
Earth	6,378 km
Mercury	2,439 km
Pluto	1,150 km

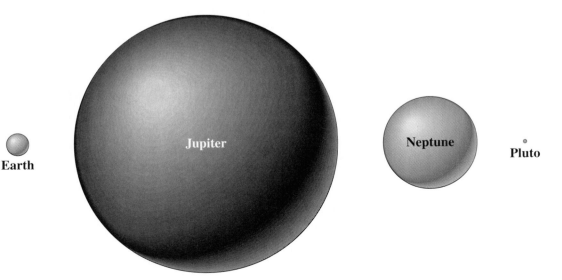

Mercury   Earth   Jupiter   Neptune   Pluto

Round your answers to Parts a–d to the nearest whole number.

    a. How many planets the size of Earth would fit inside Jupiter? Show how you found your answer.

    b. How many planets the size of Earth would fit inside Neptune?

    c. How many planets the size of Mercury would fit inside Earth?

    d. How many planets the size of Pluto would fit inside Earth?

18. Prove It! In Investigation 3, you learned that if you dilate any three-dimensional figure by a factor of n, its surface area is dilated by a factor of n^2.

    a. Prove that this statement is true for a cylinder with radius 3 cm and height 2 cm. That is, show that if you dilate such a cylinder by a factor of n, the surface area of your dilated figure is n^2 the surface area of the original.

    b. Challenge Prove that this statement is true for *any* cylinder dilated by a factor of 2. That is, show that if you dilate a cylinder with radius r and height h by a factor of 2, the surface area of the dilated cylinder is 2^2 times, or 4 times, the surface area of the original.

19. Geography Just as a map of a city is similar to the city itself, a globe is approximately similar to planet Earth. Geoffrey's classroom has a globe with a circumference of approximately 100 cm. Earth's circumference is approximately 40,074 km.

    a. What is Earth's circumference in centimeters?

    b. What is the scale factor of the globe to planet Earth?

    c. As much as 382,478,000 km² of Earth's surface is covered by water. How many square centimeters are in 382,478,000 km²?

    d. On the globe, how many square centimeters represent water?

In your **own words**

Suppose you are given two similar Figures A and B such that the scale factor from Figure A to Figure B is *n*. Tell how to calculate
- the surface area of A if you know the surface area of B
- the surface area of B if you know the surface area of A
- the volume of A if you know the volume of B
- the volume of B if you know the volume of A

20. Literature Gulliver has an ice cube tray. A cube from his tray is close to being a cube with edges of length 1.5 inches. One of the Brobdingnagians has an ice cube tray similar to Gulliver's, with each dimension 12 times as large.

a. What is the volume of an ice cube made in Gulliver's tray?

b. What is the surface area of an ice cube made in Gulliver's tray?

c. What is the volume of an ice cube made in the Brobdingnagian's tray? (Remember, it will be 12 times the size of Gulliver's ice cube in every dimension.)

d. What is the surface area of one of the Brobdingnagian's ice cubes?

e. Suppose Gulliver used the amount of water in one of the Brobdingnagian's ice cubes to make ice cubes from his own tray. How many ice cubes could he make?

f. Which would melt faster, one of the Brobdingnagian's ice cubes or Gulliver's ice cubes from Part e? Explain your answer.

Mixed Review

Fill in each blank to make a true statement.

21. $\frac{1}{9} +$ _____ $= 1$ **22.** $\frac{9}{7} -$ _____ $= 1$ **23.** $\frac{1}{27} \times$ _____ $= \frac{1}{81}$

Find each percentage.

24. 35% of 1   **25.** 1% of 49.5   **26.** 8% of 200

Find the value of m in each equation.

27. $2.2m = 8.8$ **28.** $2m + 6 = 10$ **29.** $5 - 2.5m = 0$

Write each expression without using multiplication or addition signs.

30. $2r + 1.2r + r$ **31.** $0.1s \cdot r \cdot r \cdot r$ **32.** $0.1st + 2.7st$

Find the value of the variable in each figure.

33. area = 36

34. perimeter = 22

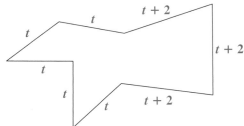

35. perimeter = 26

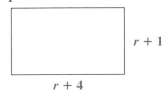

36. The table shows the total number of high school graduates and the number of graduates among 17-year-olds in the United States for several school years.

High School Graduates in the U.S.

School Year	Total High School Graduates	Graduates per 100 17-year-olds
1899–1900	95,000	6.4
1909–1910	156,000	8.8
1919–1920	311,000	16.8
1929–1930	667,000	29.0
1939–1940	1,221,000	50.8
1949–1950	1,200,000	59.0
1959–1960	1,858,000	69.5
1969–1970	2,889,000	76.9
1979–1980	3,043,000	71.4
1989–1990	2,587,000	74.2
1999–2000	2,809,000	69.9

Source: nces.ed.gov

a. What percentage of 17-year-olds in the 1969–1970 school year were high school graduates?

b. What was the change in the percentage of 17-year-olds who were high school graduates from the 1899–1900 school year to the 1999–2000 school year?

c. By how much did the *total number* of high school graduates increase from the 1899–1900 school year to the 1999–2000 school year? What percentage increase is this?

d. The 1900 U.S. Census stated the U.S. population as 76,212,168 people. What percentage of these people graduated from high school in the 1899–1900 school year?

e. The 2000 U.S. Census stated the U.S. population as 281,421,906 people. What percentage of these people graduated from high school in the 1999–2000 school year?

f. Refer to the table of data and Parts d and e. Compare the number of high school graduates in the 1899–1900 school year to the number in the 1999–2000 school year in two ways.

Chapter Summary

In this chapter, you examined two ways in which figures can be considered the same: *congruence* and *similarity.*

You looked at characteristics of congruent and similar figures. For example, congruent figures must be exactly the same shape and size; similar figures can be different sizes but must be the same shape.

In congruent figures, *corresponding sides* and *corresponding angles* must be congruent. In similar figures, corresponding sides must have lengths that share a common *ratio,* and corresponding angles must be congruent.

You discovered tests that allow you to decide whether two triangles are similar or congruent without finding the measurements of both the angles *and* the sides.

You learned that when you *dilate* a figure by a *scale factor* of n, the perimeter is dilated by n and the area is dilated by n^2. When you dilate a three-dimensional object by n, the surface area is dilated by n^2 and the volume is dilated by n^3. You even proved these facts for certain types of figures and objects.

Strategies and Applications

The questions in this section will help you review and apply the important ideas and strategies developed in this chapter.

Understanding congruence and similarity

1. Consider the difference between similarity and congruence.

   a. Can similar figures also be congruent? Do similar figures *have* to be congruent?

   b. Can congruent figures also be similar? Do congruent figures *have* to be similar?

2. Explain how you can tell whether two angles are congruent.

3. Suppose you know that two triangles are similar.

   a. What do you know about their side lengths?

   b. What do you know about their angles?

Testing figures for congruence and similarity

4. Consider the tests you know for congruent and similar triangles.

  a. Describe a congruence test involving only the sides of triangles.

  b. Describe a similarity test involving only the sides of triangles.

  c. Describe a similarity test involving only the angles of triangles.

  d. Compare the three tests you described.

5. How can you tell whether two polygons are similar? How can you tell whether two polygons are congruent?

Understanding how dilating affects measurements

6. The pasture is bordered by a road and a pond. The pasture covers 12 mi^2 of level ground. The map is drawn on a quarter-inch grid.

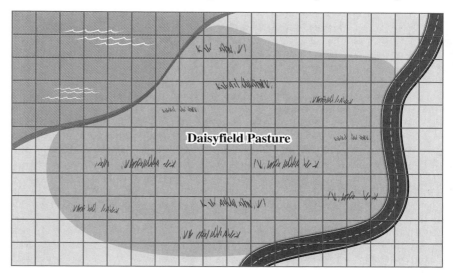

Daisyfield Pasture

  a. Daisyfield Pasture covers about 103 grid squares. Estimate the scale factor from the map to the actual pasture.

  b. The perimeter of the pasture on the map is 10.5 inches. About how much fencing would be needed to enclose the pasture?

7. A dollmaker used his own house as a model for a large dollhouse. The outside of the dollmaker's house has a surface area of 2,628 yd^2. The scale factor from the real house to the dollhouse is 0.25.

  a. What is the surface area of the dollhouse?

  b. If the original house has a volume of 6,160 yd^3, what is the volume of the dollhouse?

8. Prove It! The area of a regular pentagon with sides of length s is about $1.72s^2$.

a. What is the perimeter of the pentagon shown at right?

b. Suppose you dilate the pentagon by a factor of n. What is the length of each side in the new pentagon? Use this length to prove that the perimeter of the new pentagon is n times the perimeter of the original pentagon.

c. Prove that the area of the new pentagon is n^2 times the area of the original pentagon.

Demonstrating Skills

Tell whether the figures in each pair are congruent, similar, or neither.

9.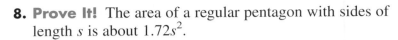

10.

11.

12.

13.

14.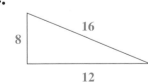

15.

16.
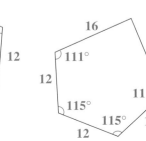

In your
own
words

Describe how to construct an angle that is congruent to another angle. Then draw an angle and construct an angle that is congruent to it.

Decide whether the ratios in each set are equivalent.

17. 3:5 and 9:15

18. $\frac{2}{3}$ and $\frac{3}{2}$

19. 6:18, 3:9, $\frac{2}{6}$, and 1:3

Each pair of triangles are similar. Find the values of *x* and *y*.

20.

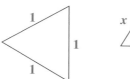 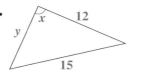

21.

22.

 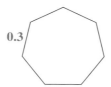

In Questions 23 and 24, do Parts a and b.

a. Find the scale factor from the small figure to the large figure.

b. Find the area of the large figure.

23.

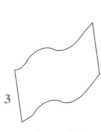

Area: 0.145
square unit

24.

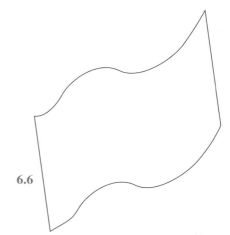

Area: 36
square units

CHAPTER 8

Ratio and Proportion

Gearing Up Understanding *gear ratios* can help you ride your bike more efficiently.

A bike chain goes around a front chain ring (which is connected to the pedal) and a rear chain cog (which turns the back wheel). Changing gears moves the chain to a different rear cog or a different front ring.

The gear ratio for a particular gear tells you how many times the rear wheel rotates each time you rotate the pedals once. You find the gear ratio by counting teeth.

$$\text{gear ratio} = \frac{\text{number of teeth on front chain ring}}{\text{number of teeth on rear chain cog}}$$

For example, if the chain is on a front ring with 54 teeth and a rear cog with 27 teeth, the gear ratio is $\frac{54}{27}$, or $\frac{2}{1}$. This means that the back wheel rotates twice every time the pedals rotate once. If the chain is moved to a rear cog with 11 teeth, the gear ratio is now $\frac{54}{11}$, or about $\frac{5}{1}$. The back wheel now turns five times every time the pedals rotate once.

Think About It If you want to travel as far as possible with the least amount of pedaling, would you want to be in a gear with a high gear ratio or a low gear ratio?

Family Letter

Dear Student and Family Members,

In the next chapter, our class will study ratio and proportion. We will begin with the idea of mixing different strengths of dye to explore ratios. Using this model makes the idea of ratio and proportion easy to see and understand. For example, Mixture A will be darker because the ratio of dye to water is 3:1, and it is only 2:1 in Mixture B.

Mixture A   **Mixture B**

We will also learn how to scale ratios. For example, to make a larger batch of Mixture A, keep the ratio the same but increase the number of cans. This can be done by multiplying both parts of the ratio by the same number.

Percentages are a kind of ratio. Some people may refer to 4 out of 5 athletes while others may describe the same group as 80% of the athletes. Often, we are not interested in the actual amounts of two quantities but only in what percent of the whole each of these quantities represents. Percentages let us compare things on a common scale. Can you think of situations in which you have seen or used percents?

We will use proportions to find missing quantities and to estimate large quantities that would be difficult or impossible to count. For instance, we can estimate the total number of people affected by a flu epidemic by counting the number in a small sample and using the proportion to estimate the total.

Vocabulary Along the way, we'll be learning about these two new vocabulary terms:

> **proportion**   **unit rate**

What can you do at home?

Encourage your student to point out different instances where ratios are used in his or her life, such as finding the cost of 5 cans of beans if 2 cans cost 70 cents. Other examples might include a ratio of adults to students on a field trip or a label that shows your favorite drink consists of 10% juice.

Comparing with Ratios and Rates

TastySnacks Inc. is introducing Lite Crunchers, a reduced-fat version of its best-selling Crunchers popcorn.

The tables list nutrition information for both products.

Original Crunchers Serving size: 35 g	
Total Fat	6 g
Saturated fat	5 g
Cholesterol	0 mg
Sodium	200 mg
Total Carbohydrate	15 g
Dietary fiber	1 g
Sugars	0 g
Protein	2 g

Lite Crunchers Serving size: 35 g	
Total Fat	3 g
Saturated fat	2 g
Cholesterol	0 mg
Sodium	160 mg
Total Carbohydrate	25 g
Dietary fiber	2 g
Sugars	0 g
Protein	3 g

Many comparison statements can be made from this information. Below are some comparisons involving differences, ratios, rates, and percents.

Difference Comparisons

• A serving of Lite Crunchers has 22 more grams of carbohydrate than grams of protein.

• Original Crunchers has 40 more milligrams of sodium per serving than Lite Crunchers.

Rate Comparisons

• Lite Crunchers contains 3 g of protein per serving.

• Original Crunchers contains 200 mg of sodium per serving.

Ratio Comparisons

• The ratio of saturated fat grams to total fat grams in Original Crunchers is 5 to 6.

• The ratio of fiber grams in Lite Crunchers to fiber grams in Original Crunchers is 2:1.

Percent Comparisons

• Almost 67% of the fat in Lite Crunchers is saturated fat.

• Original Crunchers contains 25% more sodium than Lite Crunchers.

Explore

Work with a partner to write more comparison statements about the nutrition data. Try to write a difference comparison, a ratio comparison, a rate comparison, and a percent comparison.

The TastySnacks advertising department is designing a bag for the new popcorn. They want the bag to include a statement comparing the amount of fat in Lite Crunchers to the amount in Original Crunchers. Write a statement you think would entice people to try Lite Crunchers.

Investigation 1 Thinking about Ratios

Remember

Ratios may be written in several ways. Here are three ways to express the ratio "four to one."

4 to 1   4:1 $\frac{4}{1}$

Now that you have had some practice thinking about different types of comparisons, turn your attention to ratios and rates. You already know a lot about them from previous work in this course, earlier grades, and your own life. You will now use that knowledge to reason about ratio and rate comparisons.

Problem Set A

1. Consider this keyboard.

   a. What is the ratio of white keys to black keys on the keyboard?

   b. This pattern of keys is repeated on larger keyboards. How many black keys would you expect to find on a keyboard with 42 white keys?

   c. What is the ratio of black keys to *all* keys on this keyboard?

   d. How many black keys would you expect to find on a keyboard with 72 keys in all?

2. The square tiles on Efrain's kitchen floor are laid in this pattern.

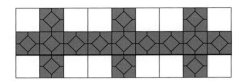

   a. What is the ratio of white tiles to purple tiles in this pattern?

   b. The entire kitchen floor contains 1,000 purple tiles. How many white tiles does it have?

   c. What is the ratio of white tiles to all tiles in this pattern?

   d. If a floor with this pattern has 2,880 tiles in all, how many white tiles does it have?

3. Lucetta participated in a walkathon to raise money for diabetes research. The graph shows the progress of her walk.

   a. At what rate is Lucetta walking?

   b. How many miles would you expect Lucetta to walk in 4.5 hours?

4. Mercedes made this bead necklace at summer camp.

   a. What is the ratio of spherical beads to cube-shaped beads on the necklace?

   b. Mercedes wants to make a longer necklace with beads in the same pattern. If she plans to use 20 spherical beads, how many cube-shaped beads will she need?

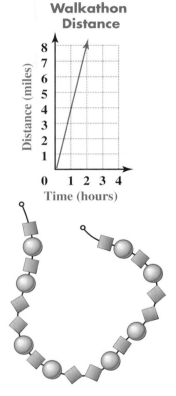

   c. What is the ratio of cube-shaped beads to all beads on this necklace?

   d. Mercedes wants to make a bracelet in the same pattern. If she uses 10 beads in all, how many cube-shaped beads will she need?

5. The two numbers on each card in this set are in the same ratio.

a. Find the missing numbers. Explain how you found your answers.

b. What ratio expresses the relationship between the top number and the bottom number on each card?

c. Draw three more cards that belong in this set.

6. The two numbers on each card in this set are in the same ratio.

a. Fill in the missing numbers.

b. Draw three more cards that belong in this set.

Problem Set B

The lists show the 10 most popular first names in the United States given to children born during the 1990s.

Boys	Girls
1. Michael	1. Ashley
2. Christopher	2. Jessica
3. Matthew	3. Sarah
4. Joshua	4. Brittany
5. Nicholas	5. Emily
6. Jacob	6. Kaitlyn
7. Andrew	7. Samantha
8. Daniel	8. Megan
9. Brandon	9. Brianna
10. Tyler	10. Katherine

Source: *The World Almanac and Book of Facts*

1. What is the ratio of names that start with *J* to names that don't?

2. What is the ratio of students in your class with one of these first names to students whose names are not on these lists?

3. What is the ratio of students in your class whose first *or* middle names are on these lists to the total number of students?

4. Now you will use the lists to make some other ratio comparisons.

    a. Write two ratio statements based on the information on these lists. For example, one possible statement is "The ratio of girls' names that have only the letter *a* as a vowel to all the girls' names is 1 to 5."

    b. Try to write two different comparisons that involve the same ratio.

Share & Summarize

1. Work with a partner to write at least five ratio statements about this quilt, which has white, blue, and purple squares.

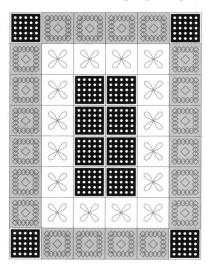

2. Write a ratio statement about the quilt that involves each given ratio.

    a. 1:4

    b. 1:3

    c. 3:5

Investigation 2 Comparing and Scaling Ratios

You have practiced writing ratios to express comparisons between quantities. Reasoning about ratios will help you solve the problems in this investigation.

Problem Set C

Researchers at First-Rate Rags are developing a shade of blue for a new line of jeans. They are experimenting with various shades by mixing containers of blue dye and water.

1. Below are two mixtures the researchers tested. Each blue can represents a container of dye, and each white can represents a container of water. All containers contain the same amount of liquid. Which mixture is darker blue? Explain how you decided.

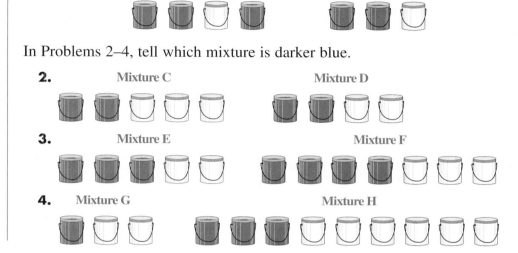

In Problems 2–4, tell which mixture is darker blue.

2.

3.

4.

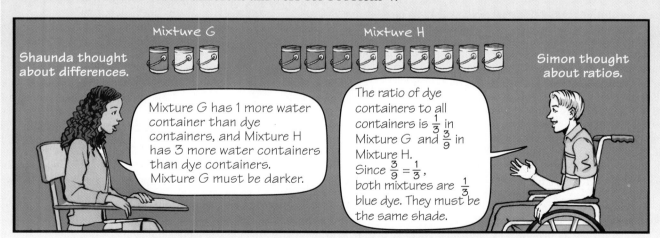

Think & Discuss

Shaunda and Simon found different answers for Problem 4.

Mixture G

Mixture H

Shaunda thought about differences.

Mixture G has 1 more water container than dye containers, and Mixture H has 3 more water containers than dye containers. Mixture G must be darker.

The ratio of dye containers to all containers is $\frac{1}{3}$ in Mixture G and $\frac{3}{9}$ in Mixture H. Since $\frac{3}{9} = \frac{1}{3}$, both mixtures are $\frac{1}{3}$ blue dye. They must be the same shade.

Simon thought about ratios.

Whose reasoning is correct? How would you use the correct student's method to solve Problem 1?

Problem Set D

The researchers at First-Rate Rags have found some shades they like. Now they want to make larger batches of dye.

1. The researchers call this mixture Blossom Blue. Draw a picture to show how Blossom Blue can be created using 12 containers in all.

2. This mixture is called Sky Blue. Draw a set of containers that could be used to create a larger batch of Sky Blue.

3. This mixture is called Robin's Egg. Draw a set of containers that could be used to create a larger batch of Robin's Egg.

4. Order the three mixtures above from darkest to lightest. Use ratios to explain your ordering.

Share & Summarize

1. Describe a strategy for determining which of two shades of blue is darker. Use your strategy to determine which mixture below is darker.

2. Draw a set of containers that would create a larger batch of dye the same shade as Mixture X. Explain how you know the shade would be the same.

Investigation Using Ratio Tables

In the last investigation, you figured out how to make larger batches of dye that would be the same shade as a given mixture. There are many ways to think about problems like these.

The mixture at right is Sky Blue. Jin Lee, Zach, and Maya tried to make a batch of Sky Blue using nine containers.

Here's how they each reasoned about the problem.

I'll draw copies of the 3 containers until I have 9 in all. First I draw the mixture, that's 3 containers. Then I draw it again to get 6 containers. I draw it once more to get 9 containers and I'm done.

Jin Lee

The original mixture has 1 more container of dye than water. The new mixture must also have 1 more container of dye than water. So I'll draw 9 containers, 5 blue and 4 clear.

Zach

In the original mixture, $\frac{2}{3}$ of the containers are blue. If I use 9 containers, $\frac{2}{3}$ of them must be blue. $\frac{2}{3}$ of 9 is 6, so 6 containers must be blue and 3 must be clear.

Maya

Think & Discuss

Did all the students reason correctly? Explain any mistakes they made.

Did any of the students use reasoning similar to your own?

When Jin Lee solved the problem above, she made copies of the original mixture until she had the correct number of containers. Jin Lee could have used a ratio table to record her work. A *ratio table*—a tool for recording many equal ratios—can help you think about how to find equal ratios.

This ratio table shows Jin Lee's thinking.

Blue Containers	2	4	6
Total Containers	3	6	9

All columns of a ratio table contain numbers in the same ratio. The equivalent ratios in this table are $\frac{2}{3}$, $\frac{4}{6}$, and $\frac{6}{9}$.

Problem Set E

1. Complete this ratio table to show the number of blue containers and the total number of containers for various batches of this shade.

Blue		3	4	8	25		200
Total	2		8			74	

2. The ratio table in Problem 1 compares the number of blue containers to the total number of containers. Complete the next ratio table to compare the number of blue containers to the number of clear containers in this mixture.

Blue	2		6	10		90	n
Clear		1.5	3		15		

3. The school band is holding a car wash to raise money for new uniforms. Ms. Chang, the band director, wants to order pizza for everyone. After the car wash last year, 20 people ate 8 pizzas.

   a. Complete this ratio table based on last year's information.

People		10	15	20		
Pizzas	2			8		

   b. How many people will two pizzas feed?

   c. If Ms. Chang is planning to feed 25 people, how many pizzas will she need?

4. Jayvyn and Rosario are planning a party, and they want to figure out how many pints of ice cream to order. At Pepa's party the month before, 16 people ate 12 pints of ice cream. Jayvyn and Rosario want the same ratio of ice cream to people at their party.

   a. Complete this ratio table based on the information about Pepa's party.

People		12	16	20	24
Pints			12		

   b. How many people will 3 pints of ice cream serve?

   c. Jayvyn said that if they extend the table, it will show that 45 pints are needed to serve 60 people. Is Jayvyn correct? How do you know?

Share & Summarize

This mixture is called Sea Blue.

1. Describe how you could find a mixture of Sea Blue that uses 9 containers in all.

2. Make a ratio table to show the number of blue containers and the total number of containers you would need to make different-sized batches of Sea Blue.

Investigation 4 ▶ Comparison Shopping

In the blue-dye problems, you can *compare* ratios to find the darkest mixture, and you can *scale* ratios to make larger and smaller batches of a given shade. Comparing and scaling ratios and rates is useful in many real-life situations.

In this investigation, you will see how these skills can help you get the most for your money.

Problem Set F

1. Abby is inviting several friends to camp in her back yard, and she wants to serve bagels for breakfast. She knows that Bagel Barn charges $3 for half a dozen bagels, and Ben's Bagels charges $1 for three bagels. At which store will Abby get more for her money? Explain how you found your answer.

2. Abby is considering serving muffins instead of bagels. Mollie's Muffins charges $6.25 per dozen. The East Side Bakery advertises "Two muffins for 99¢." Where will Abby get more for her money? Explain.

There are many ways to solve Problem 1 of Problem Set F. Here's how Simon thought about it.

Shaunda thought about the problem by scaling ratios.

Think & Discuss

Did you solve the bagel problem using a method similar to either Simon's or Shaunda's?

Solve Problem 2 of Problem Set F using Simon's or Shaunda's method. Use a different method from the one you used to solve it the first time.

VOCABULARY
unit rate

Simon's method involves finding *unit rates*. In a **unit rate,** one quantity is compared to 1 unit of another quantity. Simon found the prices for one bagel and compared them. Here are some other examples of unit rates.

$1.99 per lb	65 miles per hour	$15 for each CD
24 students per teacher	3 tsp in a tbsp	4 quarts in 1 gallon

Unit rates that involve prices—such as 50¢ per bagel and $1.99 per pound—are sometimes called *unit prices*. Supermarket shelves often have tags displaying unit prices. These tags can help consumers make more informed decisions about their purchases.

Problem Set G

Use unit rates to help you answer Problems 1–3.

1. Camisha has a long-distance plan that charges the same amount for each minute of a call, but the rate depends on where she calls. A 12-minute call to Honolulu costs $4.44. A 17-minute call to Hong Kong costs $7.14. Which long-distance rate is higher: the rate to Honolulu or the rate to Hong Kong? Explain how you found your answer.

2. At FreshStuff Produce, Anthony paid $3.52 for four mangoes. Maria bought six mangoes at FruitMart for $5.52. Which store offers the better price for mangoes? Explain how you found your answer.

3. At Xavier's Music Store, blank CDs are sold in four packages.

   Which package costs the least per CD? Explain how you found your answer.

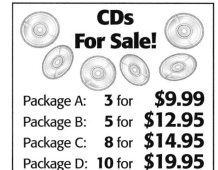

CDs For Sale!		
Package A:	3 for	**$9.99**
Package B:	5 for	**$12.95**
Package C:	8 for	**$14.95**
Package D:	10 for	**$19.95**

4. Coach Mico found several ads for soccer balls in the sports section of the local paper. Use the technique of scaling ratios to find which store offers the best buy. Explain how you made your decision.

Big Kicks — only 10 balls $55

Sport Town — 5 balls $32

SOCCER Warehouse — 20 balls $115

Share & Summarize

1. Explain how unit prices can help you make wise purchasing decisions. Give an example if it helps you to explain your thinking.

2. Suppose a package of 8 pencils costs 92¢ and a package of 12 costs $1.45. Which method would you prefer to use to find which package costs the least per pencil? Explain.

On Your Own Exercises

1. **Sports** Mount McKinley is located in Denali National Park, Alaska. Between 1980 and 1992, 10,470 climbers attempted to reach its summit, a height of 20,320 feet. Of these climbers, 5,271 were successful. The five comparisons below are based on this information.

   i. More than half the people who attempted to reach the summit were successful.

   ii. The ratio of climbers who successfully reached the summit to those who failed is 5,271 to 5,199.

   iii. Of the climbers who attempted to reach the summit, 72 more succeeded than failed.

   iv. The ratio of the total number of climbers to the number of successful climbers is about 2 to 1.

   v. Almost 50% of the people who attempted to climb Denali failed.

   a. Explain why each statement is true.

   b. Suppose that Expert Expeditions arranges group climbing trips on Mount McKinley. Their guides help less-experienced climbers reach the summit. The company is designing an advertising brochure and would like to include one of the statements above. Which do you think they should use? Explain your reasoning.

2. Emilio is laying out the pages of a newspaper. He divides one of the pages into 24 sections and decides which sections will be devoted to headlines, to text, to pictures, and to advertisements.

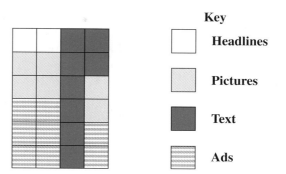

Key

Headlines

Pictures

Text

Ads

Describe something in the layout that has each given ratio.

a. 1 to 3   **b.** 1:4   **c.** 3:4

3. Decide whether Mixture A or Mixture B is darker, and explain why.

Mixture A   Mixture B

4. First-Rate Rags has finally agreed on a shade of blue for their jeans: Ultimate Denim. A sample of the dye is shown below.

How many containers of blue dye and water are needed for a batch of dye that has a total of 75 containers?

For each pair of mixtures, decide which is a *lighter* shade of blue. Then draw or describe a larger batch that will make the same shade as the lighter mixture's shade.

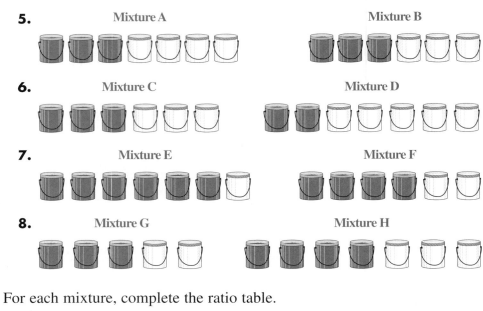

5. Mixture A   Mixture B

6. Mixture C   Mixture D

7. Mixture E   Mixture F

8. Mixture G   Mixture H

For each mixture, complete the ratio table.

9.

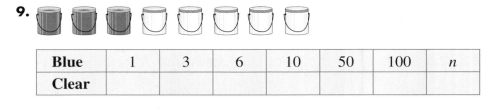

Blue	1	3	6	10	50	100	n
Clear							

10.

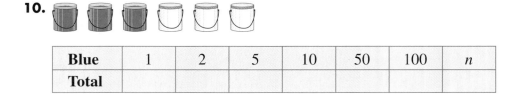

Blue	1	2	5	10	50	100	n
Total							

In Exercises 11–13, is the given table a ratio table? If so, tell the ratio. If not, explain why not.

11.

3	6	9	12	15
5	10	15	20	25

12.

1	4	7	10	13
2	5	8	11	14

13.

3	30	300	3,000	30,000
9	90	900	9,000	90,000

In Exercises 14–17, tell whether the given rate is a unit rate or a non-unit rate.

14. 25 heartbeats in 20 seconds   **15.** 72 heartbeats per minute

16. 30 mph   **17.** 200 miles per 5-hour period

18. To serve 7 people, Winema needed 2 pizzas. What was the per-person rate?

19. At basketball camp, Belicia made 13 baskets for every 25 shots she attempted. What was Belicia's success rate per attempt?

20. Jogging burns about 500 calories every 5 miles. What is the rate of calorie consumption per mile?

In Exercises 21–23, determine the unit price for each offer. Then tell which offer is best.

21. 2 pens for $3, 10 pens for $16, or 5 pens for $7

22. 5 lb of potatoes for $2.99, 10 lb of potatoes for $4.99, or 15 lb of potatoes for $6.99

23. 3 toy racing cars for $7, 5 toy racing cars for $11, 2 toy racing cars for $4.50

Connect & Extend

24. **Preview** Explore the ratios of x- and y-coordinates for some lines.

  a. Graph the line that goes through these points: (6, 4), (12, 8), and (⁻9, ⁻6).

  b. What is the ratio of x to y?

  c. Choose some points with an $x:y$ ratio of $\frac{2}{5}$. Graph the points on the same axes as in Part a. Draw a line through your points.

  d. Now choose some points with an $x:y$ ratio of $\frac{3}{4}$ and graph them. Draw a line through your points.

  e. What do you notice about the three lines you graphed?

  f. Calculate the slope of each line. How are the slopes related to the ratios?

25. Architecture In ancient Greece, artists and architects believed there was a particular rectangular shape that looked very pleasing to the eye. For rectangles of this shape, the ratio of the long side to the short side is roughly 1.6:1. This ratio is very close to what is known as the *Golden Ratio*.

a. Try drawing three or four rectangles of different sizes that look like the most "ideal" rectangles to you. For each rectangle, measure the side lengths and find the ratio of the long side to the short side.

b. Which of your rectangles has the ratio closest to the Golden Ratio?

26. Consider this mixture.

a. Draw a mixture that is darker than this one but uses fewer containers. (Do not make *all* the containers blue!)

b. Explain why you think your mixture is darker.

27. Measurement Susan always forgets how to convert miles to kilometers and back again. However, she remembers that her car's speedometer shows both miles and kilometers. She knows that traveling 50 miles per hour is the same as traveling 80 kilometers per hour. So, in one hour she would travel 50 miles, or 80 kilometers.

a. From this information, find the ratio of miles to kilometers in simplest terms.

b. To cover 100 km in an hour, how fast would Susan have to go in miles per hour? Explain how you found your answer.

28. Denay's class sold posters to raise money. Denay wanted to create a ratio table to find how much money her class would make for different numbers of posters sold. She knew they would raise $25 for every 60 posters sold.

a. Describe how Denay can use that one piece of information to make a ratio table. Assume the relationship is proportional.

b. How much money would Denay's class make for selling 105 posters?

c. Could Denay's class raise exactly $31? If so, how many posters would they need to sell? If not, why not?

29. This table is not a ratio table.

x	20	30	40	50
y	0.5	1	1.5	2

a. Graph the values in the table, and describe the shape of the graph. Draw a line or a smooth curve through the points.

b. Make a table of ratios equivalent to 20:0.5. Include at least four entries in addition to the ratio 20:0.5. Call the values corresponding to the first quantity in the ratio x and the other values y.

c. Graph the values in your new ratio table. Draw a line or a smooth curve through the points.

d. How are the two graphs alike? How are they different?

e. Make another ratio table of x and y values using a ratio close to 20:0.5, such as 18:0.5 or 21:0.5. Graph the values in your table, and draw a line or a smooth curve through the points.

f. Challenge What characteristic do the graphs in Parts c and e share but the graph in Part a does not? Explain why this is so.

30. Kate wants to make a batch of mixed nuts for a party she is planning. The local health food store sells nuts by weight, so Kate can measure out exactly how much she wants. The table shows what Kate wants to buy and how much it will cost:

Cost of Nuts

Nut	Amount	Price
Almonds	10 oz	$3.00
Cashews	6 oz	$3.00
Filberts	12 oz	$3.25
Peanuts	16 oz	$2.00

a. On one coordinate grid, make a graph for each type of nut giving the relationship between the amount and the price. You might want to create ratio tables to help you. You will have four graphs when you're done.

b. Kate paid the same for the almonds as the cashews, but bought different amounts of each. Without calculating, how can you decide which nut gave her more for her money? Which nut is that?

c. Find the price per ounce of each type of nut. Explain how these prices are shown in your graph.

31. Jack is in the grocery store comparing two sizes of his favorite toothpaste. He can buy 3 oz for $1.49 or 4 oz for $1.97. Without calculating unit prices, how can Jack decide whether one size is a better buy than the other?

Mixed Review

Evaluate.

32. $\frac{3}{4} + \frac{6}{7}$

33. $\frac{9}{11} - \frac{2}{5}$

34. $\frac{6}{11} - \frac{3}{4}$

35. $\frac{2}{3} \cdot \frac{9}{20}$

36. $\frac{13}{124} \div \frac{26}{3}$

37. $\frac{12}{17} \div \frac{6}{34}$

Rewrite each expression as a single base raised to a single exponent.

38. $6^2 \cdot 6^2$

39. $n^2 \cdot n^{-4}$

40. $2^3 \div 2^7 \cdot 2^2$

Rewrite each expression without using parentheses.

41. $(6^2)^3$

42. $(3n)^4$

43. $(^-2t)^3$

Evaluate each expression for $a = 1$, $b = 2$, and $c = 3$.

44. $a^c \cdot b^c$

45. $c^0 - (b \cdot c^b)^{-a}$

46. $^-a^c \div b^{-c}$

47. Consider this pattern of squares.

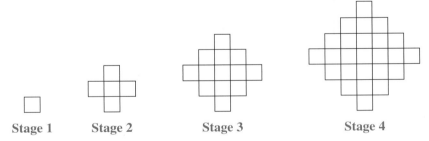

Stage 1   Stage 2   Stage 3   Stage 4

a. For the stages shown, how many squares are added to go from each stage to the next?

b. How many squares are added to go from Stage s to Stage $s + 1$?

c. What is the total number of squares at each stage shown?

d. How many squares will be needed for Stage 5? For Stage 7?

e. Will any stage in this pattern have 100 squares? Explain.

48. Measurement Use the equivalents for various length measurements listed below to answer the following questions.

1 angstrom $= 10^{-10}$ meter
1 meter $= 3.28084$ feet
1 kilometer $= 0.62137$ mile
1 furlong $= 201.168$ meters
1 light-year $= 5,880,000,000,000$ miles

a. An angstrom is a unit used to record the sizes of atoms and other very small things. How many angstroms are in 1 meter?

b. The furlong was originally a measure of the length of a furrow in a field in medieval England. How many feet are in 1 furlong?

c. A light-year is the distance light travels in 1 year. How many kilometers are in 1 light-year?

d. A light-minute is the distance light travels in 1 minute. The sun is 8.3 light-minutes from Earth. About how many miles is the sun from Earth?

e. A light-second is the distance light travels in 1 second. The moon is 1.3 light-seconds from Earth. About how many miles is the moon from Earth?

49. The maximum speeds of several animals are listed in the chart.

a. How many times faster than a chicken is a coyote?

b. How many times faster than an elephant is a cheetah?

c. How many times faster than a garden snail is a giant tortoise?

d. What is the median speed of those listed?

e. What is the mean speed of those listed?

Animal	Speed (kph)
Cheetah	103
Lion	81
Quarterhorse	76
Elk	72
Coyote	69
Zebra	64
Jackal	56
Giraffe	51
Wart hog	48
Human being	45
Elephant	40
Black mamba snake	32
Squirrel	19
Chicken	14
Giant tortoise	0.27
Three-toed sloth	0.24
Garden snail	0.05

Source: QPB Science Encyclopedia © Helicon Publishing Limited

Just the facts

When you look at a star that is 7 light-years from Earth, you are seeing the star as it was 7 years ago because that's how long it took its light to reach our planet!

Remember

Proportional means that as one variable doubles the other doubles, as one variable triples the other triples, and so on.

The seventh graders at Summerville Middle School are selling calendars to raise money for a class trip. The amount of money they raise is *proportional* to the number of calendars they sell. As with any proportional relationship, the graph of the relationship between calendars sold and dollars raised is a line through the origin.

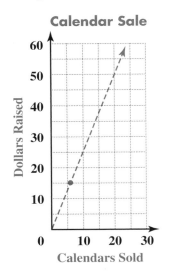

Calendar Sale

Think & Discuss

The point (6, 15) is on the graph. What does this tell you about the calendar sale?

Identify two more points on the graph. Find the ratio of dollars raised to calendars sold for all three points. How do the ratios compare?

You know that the slope of a line is rise divided by run, or this ratio:

$$\frac{\text{rise}}{\text{run}}$$

Slopes are often written as decimals. Decimals are one more way to write ratios.

Find the slope of the line. What does it tell you about the calendar sale? How does it relate to the ratios you found?

Give the coordinates of two other points that are on the line, but that are beyond the boundaries of the graph. Explain how you found the points.

Investigation Reasoning about Proportional Relationships

In the Think & Discuss, you found that the ratio of dollars raised to calendars sold is the same for any point on the graph. For example, (6, 15) and (10, 25) are both on the graph, and

$$\frac{15}{6} = \frac{25}{10}$$

This gives you another way to think about proportional relationships: A *proportional relationship* is a relationship in which all pairs of corresponding values have the same ratio.

Problem Set A

1. Consider these triangles.

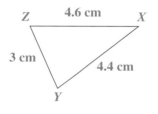

   a. Are the side lengths of Triangle *ABC* proportional to those of Triangle *XYZ*? Explain how you know.

   b. Are the triangles similar? Explain.

2. For each rectangle, find the ratio of the length to the width. Is the relationship between the lengths and widths of these rectangles proportional? Explain.

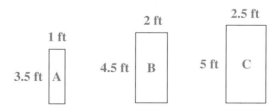

3. On the cards below, the top numbers are proportional to the bottom numbers. Find another card that belongs in this set. Explain how you know your card belongs.

4. Jeff says, "I started with Mixture A. I created Mixture B by adding 1 blue container and 1 clear container to Mixture A. Then I made Mixture C by adding 1 blue container and 1 clear container to Mixture B. Since I added the same amount of blue and the same amount of water each time, the number of blue containers is proportional to the number of clear containers."

Is Jeff correct? Explain.

Mixture A	Mixture B	Mixture C

5. Chris created this tile pattern.

  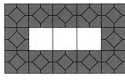 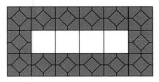

Stage 1	Stage 2	Stage 3	Stage 4

Remember

To be a pattern, the arrangement of tiles must change in a *predictable* way from stage to stage.

a. Describe how the pattern of white and purple tiles changes from one stage to the next.

b. For the stages shown, is the number of white tiles proportional to the number of purple tiles? Explain.

c. Starting with Stage 1 above, draw the next two stages for a tile pattern in which the number of white tiles *is* proportional to the number of purple tiles.

Share & Summarize

1. Describe at least two ways to determine whether a relationship is proportional.

2. Describe two quantities in your daily life that are proportional to each other.

3. Describe two quantities in your daily life that are not proportional to each other.

4. Use the word *proportional* to explain how you can tell whether two triangles are similar.

Investigation ▶2 Solving Problems Involving Equal Ratios

You know that in a proportional relationship, all pairs of corresponding values are in the same ratio. You may find it helpful to think about this idea as you solve the next set of problems.

Problem Set B

1. When the seventh grade calendar sale began, Mr. Diaz, a math teacher at Summerville, bought the first 6 calendars. From his purchase alone, the class raised $15.00. By the end of the first day of the sale, they had raised $67.50. How many calendars had been sold by the end of the first day?

2. The rectangles below are similar. Find the value of x.

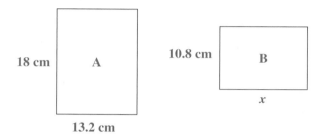

3. Ms. Rosen manages a vegetable market. She uses this graph to find the cost of different amounts of potatoes quickly.

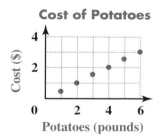

A customer wants to buy 15 lb of potatoes, but the graph doesn't extend that far. How much will the potatoes cost?

There are many ways to solve the problems in Problem Set B.

EXAMPLE

Luis, Kate, and Darnell each thought about Problem 1 in a different way, but all three found the same answer.

Luis used unit rates.

I found out how much they would raise if they sold 1 calendar. I divided $67.50 by that amount.

Kate wrote and scaled a ratio.

They raised $15 for 6 calendars, so I wrote the ratio $\frac{15}{6}$. I multiplied both parts of it by the same number to get an equal ratio in the form of "67.50 to something."

Darnell wrote and solved an equation.

I let n be how many calendars they must sell to raise $67.50. I know
$$\frac{15}{6} = \frac{67.50}{n},$$
and I solved this equation to find the answer.

Problem Set C

1. Complete each student's method and find the number of calendars sold the first day. You should find the same answer with each method.

2. The last time Jack rented skis at Budget Mountain, he paid $19.95 for 3 hours. Today he used the skis for 4 hours 45 minutes. If the rental cost is proportional to time, what will his rental charge be?

3. First-Rate Rags wants to make a large batch of the shade of dye below. The final batch must have 136 containers in all (dye and water). How many containers of blue dye will they need?

4. Eric and his parents went to Switzerland during his summer break. Before they left, Eric exchanged some money for Swiss francs. In exchange for $50 U.S., he received 64 Swiss francs. He returned with 10 Swiss francs. If the exchange rate is still the same, how many dollars will he receive for his francs?

Share & Summarize

Choose one of the problems in Problem Set C, and solve it using a different method from the method you used before. Explain each step clearly enough that someone from another class could understand what you did.

Investigation Solving Proportions

You have solved many problems involving equal ratios. An equation that states that two ratios are equal, such as

$$2:3 = 10:15 \qquad \text{or} \qquad \frac{n}{25} = \frac{3}{15}$$

V O C A B U L A R Y
proportion

is called a **proportion.** One of the solution methods you used in Problem Set C involved writing and solving a proportion. This method is useful for solving lots of different problems.

EXAMPLE

Malik saw a photograph of Vincent van Gogh's famous painting *Starry Night* in a book about art history. The book mentioned that the actual painting is 73.7 cm high, but it didn't give the width.

73.7 cm

x cm

The Museum of Modern Art, New York/Art Resource, NY

Malik decided to measure the photograph and calculate the painting's length. The photograph measured 6.9 cm high and 8.6 cm wide.

Since the photograph is a scaled version of the painting, the dimensions of the painting are multiplied by the same number to get the dimensions of the photograph.

That means that the photograph's $\frac{height}{length}$ ratio is a scaled height-to-length ratio for the painting. Since the two ratios are equivalent, Malik set up this proportion:

$$\frac{73.7}{x} = \frac{6.9}{8.6}$$

To find *x*, he solved his equation.

$\frac{73.7}{x} = \frac{6.9}{8.6}$

$\frac{73.7}{x} = 0.8$ Simplify.

$73.7 = 0.8x$ Multiply both sides by *x*.

$92.1 \approx x$ Divide both sides by 0.8.

Malik concluded that the real painting's length must be about 92.1 cm.

Problem Set D

1. Maya suggested to Malik that he could have used a different proportion. She wrote

$$\frac{x}{73.7} = \frac{8.6}{6.9}$$

   a. Explain why Maya's proportion is also correct.

   b. Solve Maya's proportion. Do you think it is easier or more difficult to solve than Neeraj's proportion?

2. Consider this problem:

   Áaron went to the bank and exchanged 50 Canadian dollars for 32.28 U.S. dollars. Later that day, Nailah went to the same bank to exchange 72 Canadian dollars for U.S. dollars. How many U.S. dollars did she receive?

   a. Set up two proportions you could use to solve this problem.

   b. Solve one of your proportions. How much money in U.S. dollars did Nailah receive?

3. Augustin earns $5 per hour mowing lawns. He wants to buy two CDs that cost a total of $32.50. Set up and solve a proportion to determine how many hours he needs to work to earn enough money.

By now you probably have several strategies for solving problems involving proportional relationships and equal ratios. Now you will have a chance to practice these strategies.

Problem Set E

Solve these problems using any method you like.

1. This ratio table shows the number of cans of paint needed to cover various lengths of fence.

Cans of Paint	2	4	6	8	10	12
Feet of Fence	30	60	90	120	150	180

   a. Find the number of cans needed to paint 75 ft of fence.

   b. Find the number of cans needed to paint 140 ft of fence.

2. The Holmes family has two adults and three children. The family uses an average of 1,400 gallons of water per week. Last month they had two houseguests who stayed for a week. Assuming that the water usage is proportional to the number of people in the house, about how much water did the household use during the guests' visit?

3. The week before Thanksgiving, turkeys go on sale for $1.29 per pound at the EatMore grocery store. How much will a 14-lb turkey cost?

4. Alvin is using a trail map to plan a hiking trip. The scale indicates that $\frac{1}{2}$ inch on the map represents $1\frac{1}{2}$ miles. Alvin chooses a trail that is about $5\frac{1}{2}$ inches long on the map. How long is the actual trail?

5. Viviana loves mixed nuts. The local farmer's market sells a mix for $2.79 per pound. How many pounds can she buy for $10.85?

6. Jack is making his famous waffles for Sunday brunch. He knows that 5 waffles usually feed 2 people. If he wants to serve 11 people, how many waffles should he make?

7. At Camp Poison Oak, there are 2 counselors for every 15 campers. The camp directors expect 75 campers next summer. How many counselors will they need?

Share & Summarize

Did you use the same strategy for solving all the problems in this investigation? If so, explain it. If not, choose one of the strategies you used and explain it.

Investigation 4 ▶ Applications Involving Similarity

Did you ever wonder how people measure the width of the Grand Canyon or the height of a giant Sequoia tree? Clearly, measurements like these cannot be made with a ruler or yardstick. Measurements of very large or very small objects or distances are often made indirectly, by using proportions.

Explore

Zach and Jin Lee are on the decorating committee for their town's holiday parade. They plan to hang balloons on the streetlights along the parade route. First they want to determine how tall the streetlights are so they can find ladders of the right size.

It is late afternoon, and Zach notices that everything is casting long shadows. This gives him an idea.

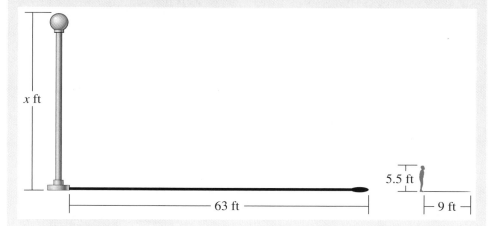

Zach asks Jin Lee to measure his shadow, and she finds that it's about 9 ft long. Using a tape measure, Zach finds that the streetlight's shadow is about 63 ft long.

Zach says, "I know that my height is 5 ft 6 in. Since the sunlight is coming at me and the pole from the same angle, I have two similar triangles. I can figure out how tall the streetlight is!"

• What are Zach's similar triangles? Make a sketch if it helps you think about this.

• How do you know Zach's triangles are similar?

• How can you use the triangles to find the height of the streetlight?

• How tall is the streetlight?

Problem Set F

Jin Lee likes Zach's idea of using shadows to find heights. She decides to show her grandfather how to use the method to find the height of the red-wood tree in his backyard.

She measures the length of the tree's shadow and finds it is 24.5 feet long. Then she holds a 12-inch ruler perpendicular to the ground and finds that it casts a 4.75-inch shadow.

1. Draw and label a simple sketch of the situation.

2. Describe two similar triangles Jin Lee can use to find the tree's height.

3. Set up a proportion Jin Lee could solve to find the tree's height.

4. How tall is the tree?

5. When Jin Lee found the tree's height, she thought all the measurements should be in the same unit, so she converted everything to inches. Her grandfather later asked her, "Could you have left inches as inches and feet as feet, and still found the correct answer?" Explain why the answer to his question is yes.

6. When Jin Lee showed the problem to Zach, he set up this proportion—

$$\frac{1 \text{ ft}}{n \text{ ft}} = \frac{4.75 \text{ in.}}{24.5 \text{ ft}}$$

—and found that the tree was only about 5 ft tall! What did he do wrong?

1. When setting up a proportion involving two triangles, why is it important for the triangles to be similar?

2. The rope on the flagpole at school has frayed and snapped, and a dog ran off with part of it. The principal wants to buy a new rope, but she doesn't know how tall the pole is.

   Write a paragraph explaining to the principal the method of indirect measurement used in this investigation as completely as you can, so she could use it to find the height of the pole.

Lab Investigation ▶ Estimating Heights of Tall Objects

MATERIALS

- protractor
- string or thread
- weight (paper clip or cube)
- tape (optional)
- tape measure, yardstick, or meterstick

In this investigation you will estimate the height of a tall object without actually measuring it. First, you will need to measure the angle of elevation.

The *angle of elevation* is the angle made by the imaginary horizontal segment extending from your eye to an object and the imaginary segment from your eye to the top of the object.

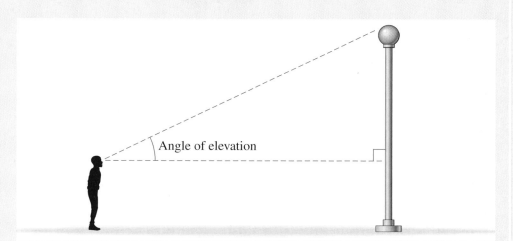

Angle of elevation

First assemble your measurement instrument. Tie a piece of string to a protractor at the middle of the straight side. Attach a weight to the string. Now you are ready to measure the angle of elevation.

Select a position at some distance from the object your class will be measuring. Line up the straight side of your protractor with an imaginary line from your eye to the top of the object. On the drawing below, the top of the object (from the person's point of view) is labeled Point *E*. Hold the protractor close to your eye.

Collect Your Data

1. Look at Angle *ABC* in the drawing. It is formed by the imaginary line through the top of the object and your eye, and by the string. Read the measure of this angle from your protractor, and record it.

2. Now measure the distance from the object to your position. You might need to mark where you're standing, or have a partner measure the distance while you remain standing where you are.

Sketch the Situation

3. Make a sketch similar to the one on the opposite page that represents your situation. Include the same labels for the important points, and label any lengths or angle measures you know.

4. You measured Angle *ABC* with your device. Use that measurement to calculate the measure of Angle *DBE,* which is the angle of elevation.

5. Use your protractor to draw a triangle similar to Triangle *BDE*. Measure the sides of your triangle, and label your drawing with the side lengths.

Calculate the Height

You can use the triangle you drew for Question 5 to find the object's height.

6. Set up and solve a proportion to calculate the length of Side *DE*.

7. Notice in your diagram that the length of Side *DE* is not the height of the object. What is the last thing you need to do to find the height of the object? What is the object's height?

8. Compare the height you found with the heights your classmates determined. How close are they? What might account for any differences?

What Did You Learn?

9. Describe in your own words how you can use the angle of elevation to find the height of a tall object. Be sure to explain how similar triangles help.

Remember

In *similar triangles,* corresponding angles have the same measures.

On Your Own Exercises

1. **Ecology** One of the new energy-efficient cars will travel many miles on a gallon of gas by using a combination of electricity and gasoline for fuel. The table shows estimates of how far the car will travel on various amounts of gas.

Gallons of Gas	0.5	1	1.5	2	2.5	3
Miles	30	60	90	120	150	180

Are the miles traveled proportional to the gallons of gas? How do you know? If so, describe how they are related and write the ratio.

2. Is the number of blue dye containers in these two mixtures proportional to the total number of containers? Explain how you know.

Mixture A   Mixture B

3. The Summerville Co-op sells two types of trail mix. Here are the ingredients.

Mountain Trail Mix	**Hiker's Trail Mix**
8 oz toasted oats	6 oz toasted oats
7 oz nuts	5 oz nuts
5 oz raisins	4 oz raisins

   a. Is the amount of nuts in each mix proportional to the total ounces of mix? Why or why not?

   b. Is the amount of toasted oats in each mix proportional to the total ounces of mix? Why or why not?

4. Set up a proportion for the following situation, and solve it using any method you choose:

   Carla's grandmother wants to sell a gold ring she no longer wears. The jeweler offered her $390 per ounce for her gold, which came to $97.50 for the ring. How much does the ring weigh?

5. Marta is following this recipe for maple oatmeal bread.

Just as she's preparing to mix the ingredients, she realizes her brother used most of the maple syrup for his breakfast. Marta has only $\frac{1}{4}$ cup of syrup, so she decides to make a smaller batch of bread. How much of each ingredient should she use?

Maple Oatmeal Bread

1 cup quick-cooking oats
3 cups bread flour
$\frac{1}{3}$ cup maple syrup
1 tablespoon cooking oil
$1\frac{1}{4}$ cups water
3 tablespoons yeast
1 teaspoon salt

6. Many schools have a recommended student:teacher ratio. At South High, the ratio is 17:1. Next year, South High expects enrollment to increase by 136 students. How many new teachers will they have to hire to maintain their student:teacher ratio?

7. One seventh grade homeroom class has 20 students, 12 of whom are girls. Another homeroom has 25 students, 14 of whom are girls.

a. Do these two classes have the same proportion of girls? Explain how you know.

b. How many girls would have to be in a class of 25 students for it to have the same ratio of girls to students as the smaller class?

8. A farmer wants to cut down three pine trees to use for a fence he is building. To decide which trees to cut, he wants to estimate their heights. He holds a 9-inch stick perpendicular to and touching the ground, and measures its shadow to be about 6 inches. If he wants to cut down trees that are about 40 feet tall, how long are the tree shadows he should look for?

9. A surveyor needs to find the distance across a lake, so she makes several measurements and prepares this drawing. The two triangles are similar. What is the distance across the lake?

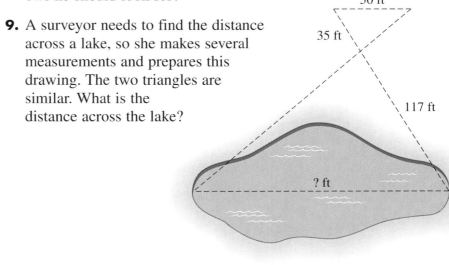

50 ft

35 ft

117 ft

? ft

10. Captain Hornblower is out at sea and spots a lighthouse in the distance. He wants to know how far he is from the lighthouse.

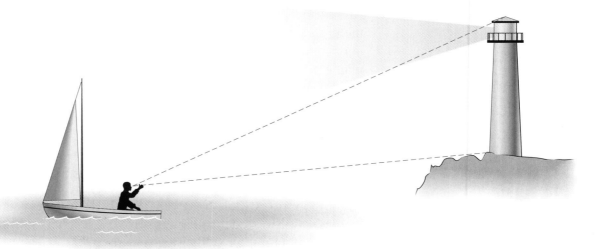

He holds up his thumb at arm's length. Then he brings his hand closer to his eye, until his thumb just covers the image of the lighthouse. His thumb is about 2.5 inches long. He measures the distance from his eye to the base of his thumb and finds that it's about 19 inches. His charts indicate that the lighthouse is Otter Point Lighthouse, which is 70 feet tall.

a. Make a sketch of this situation. Describe the two similar triangles that can be used to find the distance to the lighthouse.

b. Set up a proportion you could solve to find the approximate distance to the lighthouse.

c. Approximately how far is the boat from the lighthouse?

Connect & Extend

11. Christine was born on her father's 25th birthday. On the day Christine turned 25, her father turned 50, and they threw a big party. Then Christine wondered, "Dad's twice as old as I am. Does that mean our ages are proportional?" Answer Christine's question.

12. This recipe makes one and a half dozen peanut butter cookies. Write a new recipe that will make more cookies and has ingredients in the proper proportions.

Peanut Butter Cookies
$\frac{1}{2}$ cup peanut butter
1 cup flour
$\frac{1}{2}$ cup butter
$\frac{3}{4}$ cup brown sugar
$\frac{1}{4}$ cup white sugar
2 egg whites
$\frac{1}{2}$ tsp baking soda
$\frac{1}{3}$ tsp salt

For each tile pattern, decide whether the numbers of tiles of any two colors are proportional to each other. That is, consider three ratios—blue:purple, purple:white, and blue:white—for each stage of the tile patterns. Explain your answers.

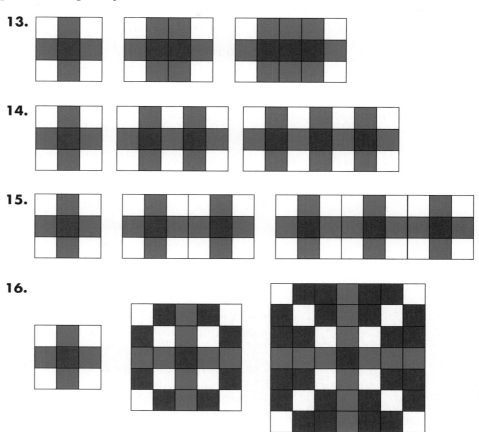

13.

14.

15.

16.

17. Renee is trying to set up a proportion to determine how much gas she will need for a 1,000-mile car trip. She averages 30 miles per gallon of gas. Here is what she wrote:

$$\frac{1,000 \text{ miles}}{G \text{ gallons}} = \frac{1 \text{ gallon}}{30 \text{ miles}}$$

a. Explain what is wrong with Renee's "proportion."

b. Set up a correct proportion, and solve it using any method you choose.

c. Find another proportion that Renee could use to solve this problem. Solve your proportion to check your answer to Part b.

18. Casey found a photograph of himself that was taken when he was three years old. In the photo, he is standing next to a bookcase that his parents still have. Explain how he could figure out how tall he was when the photo was taken.

19. Economics When you travel to another country, you often have to exchange U.S. dollars for whatever currency that country uses. Exchange rates vary frequently, but for this problem, use these exchange rates:

1 U.S. dollar ($) equals

- 0.6 English pound (£)
- 1.55 Canadian dollars ($)
- 5.6 French francs (ff)
- 119.7 Japanese yen (¥)

a. How many Canadian dollars would you receive for $250 U.S.?

b. How many U.S. dollars would you receive for 375 yen?

c. Suppose you are returning to the United States from a trip to England and you have £200 left. How many U.S. dollars will you receive for your pounds?

d. Suppose you are traveling from France to England and have 1,523 francs left. How many pounds will you receive for your francs?

20. Challenge Exchange rates for currency fluctuate every day, depending on world economies. Shenequa's family took a trip to England one summer. The exchange rates at the start and the end of their trip are shown below.

Start of trip: 1 U.S. dollar equals 0.6 English pound
End of trip: 1 U.S. dollar equals 0.65 English pound

Shenequa's mother converted $1,200 into pounds for the trip. They spent £600 in England. At the end of the trip, she converted the remaining pounds back into U.S. dollars.

Suppose Shenequa's mother had converted only $1,000 before the trip and took the remaining $200 U.S. with them. Assume they still spent £600 in England and converted the rest back to U.S. dollars when they returned. Would they have ended up with more money than they did when she had converted the full $1,200? Explain.

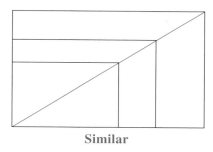
21. Prove It! One way to check whether rectangles are similar is to place them so that one of the corners overlaps, as shown below. Then look at the diagonals of each rectangle.

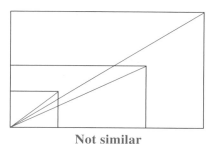

Similar   Not similar

If the diagonals all align, as in the diagram on the left, the rectangles similar. Prove that this is true. (Hint: Consider the triangles that are formed by the diagonals.)

22. Challenge These two right triangles are similar.

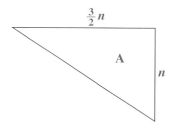

Find the lengths of the legs of both triangles. Show your work.

Mixed Review

Evaluate without using a calculator.

23. $0.012 - 1.3$ **24.** $5.62 - 5.743$

25. $0.09 \cdot 1.1$ **26.** $102.4 \div 2$

27. $0.42 \div 21$ **28.** $0.33 \cdot 11$

Simplify.

29. $k^2 + 2k^2$ **30.** $3t(t^2 - 0.1t)$

31. $3(x - 1) + 5$ **32.** $\frac{1}{5}(2r + \frac{5}{3}p)$

33. $\frac{10y - y^2}{y}$ **34.** $\frac{3(w^3 + tw)}{w}$

Solve each equation.

35. $6.4t - 0.8 = 3.1t - 4.1$

36. $2(1.1 + m) - (2.2 + m) = 10$

37. $\frac{2}{3}\pi = k + \pi$

38. $\frac{7x + 2}{17} = \frac{5x + 12}{65}$

Geometry Tell whether the triangles in each pair are congruent, similar, or neither.

39.

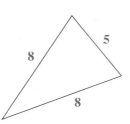

40.

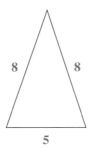
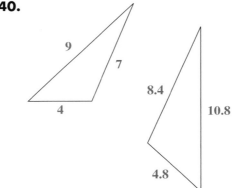

41.

42.

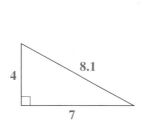
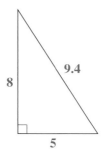

43. Earth Science The table lists the world's 10 largest deserts.

Largest Deserts in the World

Desert	Area (km^2)
Sahara (Africa)	8,800,000
Gobi Desert (Asia)	1,300,000
Australian Desert (Australia)	1,250,000
Arabian Desert (Asia)	850,000
Kalahari Desert (Africa)	580,000
Chihuahuan Desert (North America)	370,000
Takla Makan Desert (Asia)	320,000
Kara Kum (Asia)	310,000
Namib Desert (Africa)	310,000
Thar Desert (Asia)	260,000

Source: *Ultimate Visual Dictionary of Science*. London: Dorling Kindersley Limited

a. What are the mean, median, and mode of the areas listed?

b. How many times the size of the Thar Desert is the Sahara?

c. What percentage of the deserts listed are in Asia?

d. What percentage of the total area of the deserts listed is in Asia?

Percentages and Proportions

Meredith Middle School is putting on a production of *The Music Man.* The eighth grade had 300 tickets to sell, and the seventh grade had 250 tickets to sell. One hour before the show, the eighth grade had sold 225 tickets and the seventh grade had sold 200 tickets.

Think & Discuss

Which grade was closer to the goal of selling all its tickets? Explain your answer.

To figure out which grade was closer to its goal, you may have tried to compare the ratios 225:300 and 200:250. One way to compare ratios easily is to change them to percentages.

Investigation 1 ▶ Percent as a Common Scale

This diagram, called a *percent diagram,* represents the seventh and eighth grade ticket sales. The heights of the left and right bars represent the goal for each grade. The bar in the middle, called the *percent scale,* is a common scale for the two different ratios.

Notice that the three bars are the same height even though they represent different numbers. On the percent scale, the height always represents 100%.

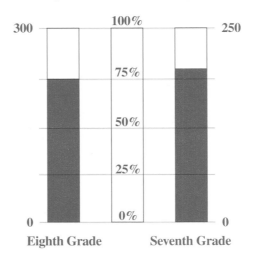

Percents make it possible to represent both grades' ticket sales as some number out of 100. For the eighth grade, 100% represents 300 tickets. For the seventh grade, 100% represents 250 tickets.

Think & Discuss

The eighth grade sold 225 tickets, and the seventh grade sold 200. Why is the bar for the seventh grade taller, when they sold fewer tickets?

Problem Set A

Jefferson Middle School has 600 students, and Memorial Middle School has 450 students. For each of these problems, use this percent diagram to estimate an answer.

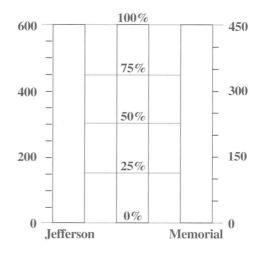

1. A survey of the two schools finds that 300 Jefferson students watch more than 1 hour of television every night, while 270 Memorial students watch more than 1 hour per night.

   a. What percentage of Jefferson students watch more than 1 hour of TV each night?

   b. What percentage of Memorial students watch more than 1 hour of TV each night?

   c. Comparing percentages, do more Jefferson students or more Memorial students watch more than 1 hour of TV each night?

2. Jefferson has 275 girls and Memorial has 250 girls. Which school has a greater percentage of girls? Explain.

3. At each school, 75% of the students play a musical instrument.

   a. About how many students at Jefferson play an instrument?

   b. About how many students at Memorial play an instrument?

4. The math teachers at each school selected 50 students to represent the school at a math competition.

    a. About what percentage of Jefferson students attended the math competition?

    b. About what percentage of Memorial students attended the math competition?

5. Suppose you looked for $P\%$ of 450 and $P\%$ of 600 on a percent diagram. Which would be the greater number? Explain your answer in terms of the diagram.

6. Which represents a greater percentage: t out of 450, or t out of 600? Explain your answer.

M A T E R I A L S

graph paper
(optional)

Problem Set B

Look again at the ticket sales by the seventh and eighth grades at Meredith Middle School. The eighth grade had 300 tickets to sell, and the seventh grade had 250 tickets to sell.

1. At the end of the first week, each grade had sold 40% of their goal.

    a. Draw a percent diagram to represent this situation.

    b. Explain how you can use your diagram to estimate the number of tickets the seventh grade sold, and give your estimate.

    c. Use your diagram to estimate the number of tickets the eighth grade sold.

2. One hour before the show started, the seventh grade had sold 200 tickets and the eighth grade had sold 225. Malik used his calculator to express the ratios of sold tickets to sales goal for both grades as percentages. How could you use your calculator to find $\frac{225}{300}$ as a percentage?

3. Suppose each grade was only 25 tickets short of meeting its goal.

    a. How many tickets would the eighth grade have sold?

    b. How many tickets would the seventh grade have sold?

    c. Calculate the percentages sold for each grade. Comparing percentages, which grade came closer to its goal?

4. By the time the play began, each grade had sold exactly 270 tickets.

    a. What percentage of its goal did the eighth grade sell?

    b. What percentage of its goal did the seventh grade sell?

    c. Draw a percent diagram to represent this situation. Since the seventh grade sold *more* than its goal, you will need to extend the bar for the seventh grade.

5. Determine which ratio in each pair is greater by finding the percentage each represents.

    a. $\frac{56}{69}$ or $\frac{76}{89}$

    b. $\frac{106}{210}$ or $\frac{206}{310}$

    c. $\frac{50}{45}$ or $\frac{210}{205}$

6. Find four ratios equivalent to 75%.

7. Find four ratios equivalent to 115%.

Share & Summarize

Here are two statements about Jefferson and Memorial middle schools.

    a. Jefferson has 275 girls and Memorial has 250 girls, so there are more girls at Jefferson.

    b. Jefferson is almost 46% girls and Memorial is almost 56% girls, so there are more girls at Memorial.

1. Which statement uses a common scale to make a comparison?

2. Which statement do you think is a better answer to the question, "Which school has more girls?" Explain your thinking.

Investigation 2 Proportions Using Percentages

In Lesson 8.2, you used ratios to write proportions and solve problems. In the last investigation, you saw that you can use percentages as a common scale to compare ratios. Percentages can also be used to solve problems.

Problem Set C

Work with your group to solve these problems. Be ready to explain your reasoning.

1. Two department stores, Gracie's and Thimbles, are having holiday sales. The stores sell the same brand of men's slacks. Gracie's usually sells them for $50, but has marked them down to 30% of that price for the sale. At Thimbles, the slacks are marked down to 40% of the usual price of $30. Which store has the better sale price? Explain.

2. Steve and Anna went shopping during a sale at the KC Nickels department store. Anna looked at her receipt and saw that the total before the discount was $72, but she paid only $40. Steve's total before the discount was $38, but he paid only $18. Who saved a greater percentage off the original price? Explain.

3. You are making punch for the class party. The punch will contain juice and soda, with at least 70% juice. You have 4 gallons of juice. What is the greatest amount of punch you could make?

One way to solve a percent problem is to set up and solve a proportion. In most percent problems, you have values for two things and you want to know the value of the third, so your proportion might look like this:

$$\frac{b}{a} = \frac{n}{100}$$

For example, in Problem 3 above, you knew the percentage of juice the punch should contain, 70%, and how much juice you actually had, 4 gallons. You could have set up this proportion:

$$\frac{4}{a} = \frac{70}{100}$$

Think & Discuss

This type of percent diagram is used to help compare two numbers. State in words the problem represented by this percent diagram.

Write a proportion that is illustrated by the diagram and solve it. How did you think about setting up the proportion?

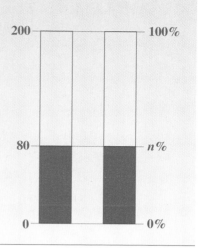

Problem Set D

1. Consider this question: *What percent of 150 is 6?*

 a. Draw a percent diagram to represent the question.

 b. Express the question as a proportion and solve it.

2. Now consider this question: *What is 185% of 20?*

 a. Draw a percent diagram to represent the question.

 b. Express the question as a proportion and solve it.

3. A gardener planted 13% of his tulip bulbs in the border of his garden and the rest in the garden bed. In spring, every one of the bulbs grew into a tulip. He counted 45 tulips in the border. How many bulbs had he planted altogether? Show how you found your answer.

4. Of 75 flights leaving from Hartsfield Airport in Atlanta, 33 went to the West Coast. What percentage of the 75 flights went to the West Coast? Show how you find your answer.

5. The largest land animal, the African bush elephant, may weigh as much as 8 tons. However, that is only about 3.9% of the weight of the largest animal of all, the blue whale. How heavy can a blue whale be? Show how you found your answer.

6. In 2001 Americans spent about $594 billion on recreation, including books, toys, videos, sports, and amusement parks. They spent 171% of that amount on housing. How much money did Americans spend on housing in 2001?

MATERIALS

graph paper
(optional)

Just the facts

The enormous blue whale's primary source of food is plankton—tiny animals and plants that float near the surface of the water.

7. In 1950 the estimated world population was 2.6 billion. In 2003 it was 6.3 billion.

  a. What percentage of the 2003 population is the 1950 population?

  b. What percentage of the 1950 population is the 2003 population?

  c. In 2003 about 292 million people lived in the United States. What percentage of the world's population lived in the United States in 2003?

Share & Summarize

1. Write a proportion that expresses this statement: 32% of 25 is *b*. Solve your proportion for *b*.

2. Write a proportion that expresses this statement: *n*% of *a* is *b*. Explain why your proportion is useful for solving percent problems.

Investigation ▶3 Percent Increase and Decrease

You have used percent diagrams to help you compare two ratios and to calculate percentages. You can use similar diagrams to show how a quantity changes in relation to its original size. These diagrams don't have a percent scale, because one of the two bars represents 100%.

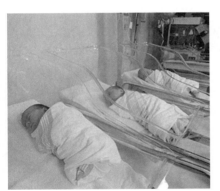

For example, the number of births in Colorado increased from 62,167 in 1999 to 65,438 in 2000. In the diagram, the *percent increase* from 1999 to 2000 is represented by the portion of the "Births 2000" bar above the 100% line.

We can make a visual estimate by comparing the height of the "percent increase" portion of the bar with the portion representing 100%. In this example, the height of the "percent increase" portion is about 5% of the height of the 100% portion.

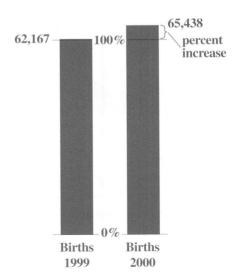

Source: *The World Almanac and Book of Facts*

This diagram shows the change in U.S. military personnel on active duty from 1995 to 2000. The *percent decrease* is represented by the unshaded portion of the bar on the right. In this example, the "percent decrease" portion of the bar looks like about 5% to 10% of the 100% portion.

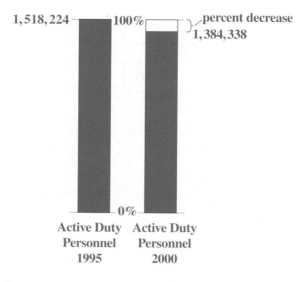

Problem Set E

MATERIALS
graph paper
(optional)

1. As people grow older, they often say that time seems to pass more quickly. Think about whether 2 years seems longer to you now than it did when you were 6 years old.

   a. Make a diagram that shows the percent increase in your age from when you were 6 years old to when you were 8 years old. Draw your diagram so that your age at 6 years old represents 100%. Use your diagram to estimate the percent increase in your age.

   b. Now make a diagram that shows the percent increase in your age from 2 years ago to now. Draw your diagram so that your age 2 years ago represents 100%. Use your diagram to estimate the percent increase in your age.

   c. Does it make sense that time would seem to pass more quickly as you age? Explain.

2. Estimate the percent change from July to August. Specify whether it is percent increase or decrease.

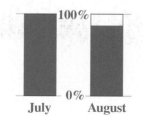

In Chapter 5, you measured your pulse, or heart rate, while resting. If you were to exercise and then measure your heart rate again, it would probably be quite different from your resting heart rate.

To *calculate* the change in your heart rate, you first need to measure your resting heart rate. You will use this rate as a baseline. Then you need to measure your rate after exercise. A comparison of the two rates can be written as a percent increase.

Problem Set F

Find your pulse—either in your neck just below your jaw, or in your wrist.

1. Count the number of pulses in 20 seconds. Use that value to find your resting heart rate in beats per minute.

2. Following your teacher's directions, do some physical activity for 1 minute. As soon as you stop, count the pulses in 20 seconds. Find your heart rate after exercise, in beats per minute.

3. If your resting heart rate in beats per minute is considered 100%, what percentage represents your heart rate after exercise?

4. How might you describe the *change* in your heart rate from a resting to an active state as a percentage?

Think & Discuss

Zoe and Luis both had a resting heart rate of 84 beats per minute. After jumping rope for a short period, they both had a heart rate of 105 beats per minute. Here's how they each calculated the percent increase.

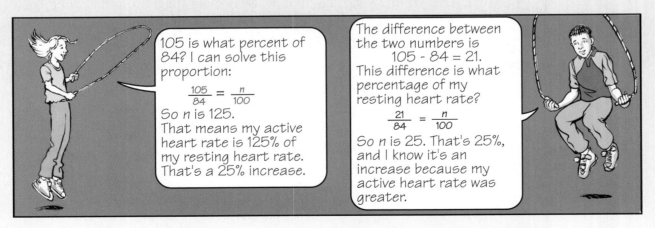

105 is what percent of 84? I can solve this proportion:

$$\frac{105}{84} = \frac{n}{100}$$

So n is 125.
That means my active heart rate is 125% of my resting heart rate. That's a 25% increase.

The difference between the two numbers is 105 - 84 = 21. This difference is what percentage of my resting heart rate?

$$\frac{21}{84} = \frac{n}{100}$$

So n is 25. That's 25%, and I know it's an increase because my active heart rate was greater.

Will these two methods always give you the same answer? Explain how you know.

Problem Set G

Use either Zoe's or Luis's method to solve these problems.

1. In 1985 there were 39,872,520 U.S. households with cable TV. By 2001 the number had increased to 74,148,000 households. What is the percent increase of households with cable TV from 1985 to 2001?

2. Shoes regularly priced at $55 are on sale for $41.25. What is the percent decrease from the original price to the sale price?

3. In 1981, the average hourly wage in the United States was $7.25. In 1991, it was $10.32, and in 2001, it was $14.32.

   a. What is the percent change of the hourly wage from 1981 to 1991? Tell whether it is a percent increase or decrease.

   b. What is the percent increase of hourly wage from 1991 to 2001?

Share & Summarize

The tree grew from 125 cm to 133 cm in one year. What was the percent increase in the height that year? Explain how you solved the problem.

Investigation Percentages of Percentages

To figure a tip for a restaurant bill, some people take a percentage of the total, after taxes. Since the tax is a percentage of the subtotal, these diners are taking a percentage of a percentage.

There are several other situations in which this might happen.

Food and drinks	$20.00
tax (4%)	0.80
total	20.80
tip (15%)	3.12
	$23.92

Think & Discuss

Increase 100 by 20%. Decrease the result by 20%. Is the result the original number, 100? Explain.

Problem Set H

Barry's Bargain Basement and Steve's Super Savings Store are located in the Los Angeles garment district.

BARRY'S BARGAIN BASEMENT SALE
Everything's a bargain!
All items are marked with an arrival date.
1 month old, 10% off the original price!
2 months old, 20% off the original price!
3 months old, 30% off the original price!
4 months or older, 50% off the original price!

STEVE'S SUPER SAVINGS STORE
We won't be undersold!
All items are marked with an arrival date.
If it's been here 1 month, take 10% off!
If it's been here 2 months, take another 20% off the *already reduced* price!
If it's been here 3 months or longer, take another 30% off the *already reduced* price!

1. Rashonda went to Barry's Bargain Basement to buy a pair of jeans. The jeans had been in the store for 1 month. The original price was $40.

   a. What was the sale price?

   b. What percentage of the original price was the sale price?

2. Julie was a smart shopper. The pair of shoes she wanted had an original price of $50 at both Barry's and Steve's. The shoes had been in both stores 2 months.

   a. How much would Julie pay for the shoes at Barry's? At Steve's?

   b. Explain why these sale prices are different.

3. Hernando, a frequent shopper at Barry's, has his eye on a hat with an original price of $20.

   a. The hat has been in the store for 1 month. How much will Hernando save?

   b. Hernando is considering waiting another month to get a better price. How much will he save after 2 months?

   c. Is the amount Hernando would save with the 10% discount double the amount he would save with the 20% discount?

   d. How can you easily determine how much Hernando would save after 3 months?

4. Jasmine found a sweatshirt at Steve's with an original price of $30.

   a. If the sweatshirt has been in the store long enough to earn the 10% discount, how much will Jasmine save?

   b. If the sweatshirt has been there long enough to earn the 20% discount, how much will she save from the original price?

   c. Is the dollar amount of the second discount twice the dollar amount of the first discount? Explain why or why not.

   d. Jasmine calculated the final price this way: "First I take 10% off, and then I take 20% off, so all together I'm taking 30% off. That's $9, so the final price is $21." Is Jasmine correct? Explain.

   e. If Jasmine buys the sweatshirt after it's been in the store 2 months, what is the total percent discount from the original price? Show how you found your answer.

5. An item originally marked at $100 is in both stores for more than 4 months. Where would you get the better buy? Explain.

6. Is the ad for Steve's—which claims they won't be undersold—true when Steve's prices are compared with Barry's prices? Explain.

Problem Set I

With most photocopy machines, you can reduce or enlarge your original by entering a percentage for the copy. If you enter 200%, the copy machine will double the dimensions of your original.

For example, if you copy a 3-inch-by-5-inch photograph at the 200% setting, the copy of the photograph will measure 6 inches by 10 inches.

1. Lydia wanted to enlarge a 2-cm-by-4-cm drawing. She set a photocopy machine for 150% and copied her drawing.

   a. What will the dimensions of the copy of the drawing be?

   b. The drawing still seemed too small to Lydia. She put her copy in the machine and enlarged it 150%. What are the new dimensions?

   c. Lydia likes this size. She now wants to enlarge another 2-cm-by-4-cm drawing to this same size. What percent enlargement should she enter to make the copy in just one stage?

   d. Is the percentage you calculated in Part c the sum of the two 150% enlargements? Why or why not?

2. **Challenge** Suppose you put a print in a copy machine and enlarge it by selecting 180%. What percent reduction would you have to take to return the image to its original size?

Share & Summarize

Suppose you reduce an 8-inch-by-10-inch picture by 90% and then increase the copy by 90%.

1. Will you get the same size picture you started with? Explain.

2. What is the percent change from the original size to the final size?

On Your Own Exercises

Practice  Apply

1. Travel surveys show that 68% of people who travel for pleasure do so by car. Of 338 students from Brown Middle School who traveled for pleasure last year, 237 did so by car. The same year, 134 students from Mayville Middle School traveled for pleasure, and 88 did so by car.

   a. Draw a percent diagram to illustrate this situation.

   b. From which school did a higher percentage of those who traveled for pleasure do so by car?

   c. How do the two schools compare to the survey results?

For each pair of ratios, determine which is greater by finding the percentages they represent.

2. $\frac{13}{12}$ or $\frac{65}{60}$

3. $\frac{23}{28}$ or $\frac{44}{51}$

4. $\frac{328}{467}$ or $\frac{16}{23}$

5. Find four ratios equivalent to 30%.

6. Find four ratios equivalent to 25%.

In Exercises 7–9, decide which is a greater number.

7. $P\%$ of 13   or $P\%$ of 23

8. $P\%$ of 368   or $P\%$ of 712

9. $P\%$ of 3   or $P\%$ of 4.7

In Exercises 10–12, decide which represents a greater percentage.

10. t out of 100   or t out of 200

11. t out of 89   or t out of 91

12. t out of 0.7   or t out of 0.5

13. Consider this question: *5% of what number is 15?*

    a. Draw a percent diagram to represent the question.

    b. Express the statement as a proportion and solve it.

14. 98% of what number is 50?

15. 117% of what number is 3.5?

16. 7.5 is what percent of 9?

17. 20 is what percent of 15?

18. Sam works 12 hours a week. He recently received a 5% raise, and his new weekly pay is $115.92.

    a. What was Sam's weekly pay before the raise?

    b. What was his hourly pay before the raise?

    c. What is his new hourly rate?

19. **Social Studies** In 2000 about 20,530,000 children between the ages of 10 and 14 lived in the United States. That was about 7.3% of the entire U.S. population. Find the 2000 U.S. population.

In Exercises 20 and 21, estimate the percent change. Be sure to indicate whether it is an increase or a decrease.

20. the price of lettuce from one week to the next

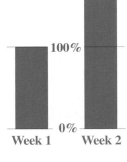

21. the price of a CD player from one month to the next

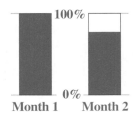

22. Social Studies The table lists U.S. crime rates from 1997 to 2000.

Crime in the United States

Year	Crimes per 100,000 Inhabitants
1997	4,930.0
1998	4,619.3
1999	4,266.8
2000	4,124.0

Reprinted with permission from *The World Almanac and Book of Facts 2003.*

a. What is the percent change in the crime rate from 1997 to 1998? Indicate whether it is a percent increase or decrease.

b. What is the percent change in the crime rate from 1999 to 2000? Indicate whether it is a percent increase or decrease.

c. What is the percent change in the crime rate from 1997 to 2000? Indicate whether it is a percent increase or decrease.

23. A used minivan priced at $20,000 went on sale for 5% off.

a. What is the van's reduced price?

b. A month later the van was discounted 5% off the sale price. What are the dollar amounts of the first discount and the second discount?

c. At the end of the year, the van was discounted 5% off the last sale price. What was the price of the minivan after the third discount?

d. What is the total percent decrease after all three discounts?

24. Carl wanted to enlarge a picture of a tree. On his original, the tree measured 4 cm high. He wanted the copy to measure 7.5 cm high.

a. First he made a copy 120% the size of the original. What was the percent increase in the tree's height?

b. What did the height of the tree become?

c. Carl made a 200% enlargement of his copy. What was the tree's height in the new copy?

d. If Carl wanted to enlarge another 4-cm-high picture to the height you found in Part c, what percentage should he enter into the machine?

e. Carl selected 70% to reduce the picture from Part c. What was the percent decrease in the tree's height?

f. What was the height of the tree in the new copy?

g. What percentage should Carl select to copy the tree in Part f and produce a tree 7.5 cm high?

25. A toy store sells a certain brand of yo-yo. One year the store sold 200 of the yo-yos every month.

  a. The next year, monthly sales of the toy decreased 30%. How many yo-yos per month were sold the second year?

  b. Third-year sales increased 30% per month over the second year's monthly sales. How many yo-yos were sold per month in the third year?

  c. What was the total percent change from the first-year monthly sales to the third-year monthly sales—that is, after the 30% decrease followed by the 30% increase? Was it a percent increase or decrease? Explain.

  d. How much of a percent change is needed now for sales to return to 200 per month? Explain.

26. A family bought a house valued at $150,000. An office park was then constructed next to the house. Five years after the family had bought the house, its value had dropped 30%.

  a. What was the house's value after 5 years?

  b. What percent increase is now needed for the family to be able to sell the house for what they paid for it? Explain.

Connect & Extend

27. Social Studies The bar graphs show information about education levels in the United States in a recent year. The left bar shows the percentages of unemployed people at each level, and the right bar shows the percentages for the employed.

Education Levels in the U.S.

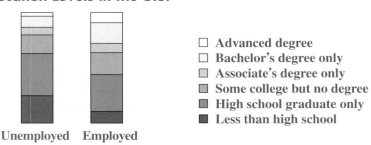

□ Advanced degree
□ Bachelor's degree only
□ Associate's degree only
■ Some college but no degree
■ High school graduate only
■ Less than high school

Unemployed   Employed

Source: U.S. Bureau of the Census. *Statistical Abstract of the United States*

  a. Measure the heights of the bars in centimeters. This height represents 100% of the people classified in each graph.

  b. Now measure the height of the portion for high school graduates who are employed. Use that height to calculate the approximate percentage of employed people who only have a high school diploma.

c. Use a similar procedure to calculate the approximate percentage of unemployed people who stopped their education at high school.

d. Choose two more education levels, and repeat Parts b and c for each of them.

28. **World Cultures** The table shows data about communications in several countries.

Country	Phone Lines per 100 People (2002)	Cellular Phone Subscribers per 100 People (2000)	Number of Personal Computers per 100 People (2002)
Argentina	22	18	8
Australia	54	64	52
Bosnia	12	9	0
China	17	16	2
Iran	20	3	7
Japan	60	62	38
Mexico	15	25	7
South Africa	11	27	7
Sweden	72	89	56
Thailand	10	26	3
United States	66	49	63

Source: International Telecommunications Union.

a. The entries in the third column are "per 100 people." Desta said, "The entry for United States is 63. That means 63% of people had a telephone line in 2002." Nicolás convinced her that wasn't true. Explain why Nicolás was correct.

b. The entries in the second column are "per 100 people." The entry for United States is 49. What would the entry be if the scale were "per 10 people"?

c. Choose one category, and write a paragraph comparing the given countries in terms of that category. You might include information about which country has the highest entry, which has the lowest entry, which have entries close to each other, and how the United States compares to the other countries.

d. Choose one country other than the United States and compare its data to that for the United States.

29. Make up a reasonable word problem that requires finding 65% of 120. Show how to solve your problem.

30. The students at Graves Middle School held a fund-raiser to help purchase new gym equipment. The school newspaper reported that the eighth grade raised the most, $234, while the sixth and seventh grades each raised 30% of the total amount collected. However, the paper didn't report the total amount raised by the three grades!

Use the information given to find the total amount raised and the amounts raised by the sixth and seventh grades.

31. Challenge Suppose $S > 0$ and $S = 2T$.

　　a. Which is greater, 80% of T or 40% of S? Explain your answer.

　　b. Does your answer to Part a hold if $S < 0$? Explain.

32. Susana typed an English paper on her computer. Using a font size of 12, the first page contained 250 words. When Susana changed the font size to 10, the first page held 330 words.

　　a. What is the percent increase in the number of words on the page from the larger font (size 12) to the smaller font (size 10)?

　　b. Susana's paper was long enough to fill several pages. What percent increase in the number of words on three pages would you expect when she changes from font size 12 to font size 10? Explain.

　　c. When Susana printed her paper using a font size of 12, it was exactly five pages long. About how many pages would you expect it to be if she uses a font size of 10?

　　d. Write a formula for calculating the approximate number of pages in a font size of 10 if you know the number of pages in a font size of 12.

33. The table shows the resting and active heart rates of several students. The active rates were taken just after different types of exercises.

　　a. Complete the table.

Name	Resting Heart Rate	Active Heart Rate	Difference (Active − Resting)	Percent Increase
Juanita	91	138		
Kris	92	190		
Nathan	82	157		
Aaliyah	88	176		
Nieve	85	131		
Garrett	79	147		
Ramón	77	119		

b. Which student had the least percent increase? Notice that the student with the least percent increase did not have the least difference in heart rates. Explain why this makes sense.

c. Find the student whose heart rate increased 100%. How does the difference between this student's active and resting heart rates compare with his or her resting heart rate?

d. What is the ratio of Kris's active rate to his resting rate? How is this ratio related to the percent increase in his heart rate? (Hint: Write the ratio as a decimal.)

34. States have different sales taxes. Usually the price of an item is marked without sales tax, and you pay the tax as a percent of the price. In 2003 Massachusetts had a 5% state sales tax, Hawaii had a 4% state sales tax, and California had a 6% state sales tax. In each problem, P is the price *without* tax, and T is the price *with* tax.

a. What is the percent increase from P to T in Massachusetts, Hawaii, and California?

b. What percentage of P is T in each of these three states?

c. For each state write the ratio of T to P as a fraction and as a decimal.

35. Nutrition Food advertisements often contains claims worded to give the best impression.

a. Explain why the claim "20% fewer calories" is, by itself, not very informative.

b. Some reduced-fat milks are labeled "2% milk" or "98% fat free." Below are nutrition data for 1 cup of milk from a carton of whole milk and a carton of 2% milk. How much of a reduction of fat is "98% fat free"?

Whole Milk	**2% Milk**
Calories 150	Calories 140
Total Fat 8 g	Total Fat 5 g
Total Carbohydrates 13 g	Total Carbohydrates 15 g
Protein 8 g	Protein 10 g

c. Explain why advertisers might prefer "98% fat free" over, for example, stating the percent change in the quantity of fat.

36. Karla's Discount Madness store truly doesn't want to keep merchandise in the store for long. Any item that's been in the store for a month is marked down 20%. For each additional month an item stays in the store, Karla's staff marks it down another 20% from the previous month's price.

The "Ever-Whiny Baby" doll didn't generate much interest. It began as a $30 doll. Several months later, Karla's still had lots of the dolls in stock.

a. Complete the table, showing the price of the doll each month.

Doll Price

Month	0	1	2	3	4	5	6	7
Price ($)	30							

b. Graph the data in your table. Connect the points with a smooth, dashed curve.

c. What kind of relationship does this appear to be? Look back through your book if you need help remembering the relationships you've studied.

In your **own words**

Describe several ways percentages are used to compare different quantities.

37. Savings accounts pay *interest* on the money in the account. Each month, a percentage of the money in the account is added to the balance, and that money is included when the next month's interest is calculated.

For example, suppose an account had $100 in it on January 1, and the account pays 12% interest per year. Each month, $\frac{1}{12}$ of that amount, or 1%, will be paid on the amount in the account.

a. After 1 month, 1% of the $100 is added to the original $100. How much money is now in the account?

b. After 2 months, 1% of the amount currently in the account is added. How much money is now in the account?

c. How much money is in the account after 6 months?

d. This method of calculating interest is called *compound interest.* Another method is *simple interest.* With simple interest, only the original amount is used to calculate the interest each month. That is, if the account starts with $100, 1% monthly simple interest would add exactly $1 each month.

How much money would be in this account after 6 months?

Mixed Review

Evaluate.

38. 3^{-2} **39.** $2^2 \cdot 2^3$ **40.** $2^2 \div 2^{-2}$

41. $0.005 \cdot 0.1$ **42.** $10^2 \cdot 0.1$ **43.** $100^2 \cdot 0.3$

44. 0.2^2 **45.** 0.2^{-2} **46.** $\sqrt{169}$

Solve each equation. Be careful: Some of these equations have more than one solution.

47. $t^2 = 49$ **48.** $x^2 - 4 = 5$

49. $2(c - 0.1) = 3c - 1.1$ **50.** $\frac{r}{5} = \frac{5}{r}$

51. $3z - 2 = z$ **52.** $\frac{1}{a^2} = \frac{2}{8}$

53. Mr. Ritter was curious about the copy machine in his office. He wanted to know if pressing the 200% button meant that the area of a picture would be doubled. What would you tell him?

54. Geometry Find the volume and surface area of this cylinder.

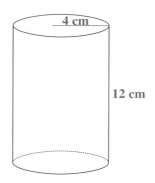

4 cm

12 cm

55. Consider this cylinder.

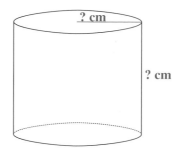

? cm

? cm

a. The base of this cylinder has an area of 30.25π cm^2. Find the cylinder's radius.

b. The rectangular piece that can be used to form the curved side of this cylinder has an area of 110π cm^2. Find the cylinder's height.

56. Use the expression given to complete the table. Then write another expression that gives the same values. Check with the values in the table.

n	0	1	2	3	100
$\dfrac{4n^2 + 1}{2}$					

57. Physical Science The table lists the elements that make up Earth's crust.

Earth's Crust

Element	Mass (%)
Oxygen	49.13
Silicon	26.00
Aluminum	7.45
Iron	4.20
Calcium	3.25
Sodium	2.40
Potassium	2.35
Magnesium	2.35
Hydrogen	1.00
Others	1.87

Source: *Ultimate Visual Dictionary of Science*

a. If you had a 1,000-kg sample of Earth's crust, how many kilograms of it would you expect to be iron?

b. If you had a 500-kg sample of Earth's crust, how many kilograms of it would you expect to be potassium?

c. What percentage of Earth's crust do the three components that are in the greatest abundance make up?

d. Aluminum, iron, calcium, sodium, potassium, and magnesium are classified as metals. What percentage of Earth's crust is composed of metals? (Assume the "Others" category contains no metals.)

e. Compare the amount of calcium in Earth's crust to the amount of hydrogen in two ways.

Interpreting and Applying Proportions

Justin and Cleavon manage a supermarket. They came across this statement from an ad for potato chips:

In a national survey, 60 out of 100 people preferred potato chips over pretzels; only 40 out of 100 preferred pretzels.

Think & Discuss

When they discussed the ad, Justin said, "50% more people prefer potato chips than pretzels."

Cleavon said, "Of the people who were surveyed, there's a 20% difference between those who prefer potato chips and those who prefer pretzels."

Who is right? Could *both* Justin and Cleavon be right? Explain.

Suppose you were an advertiser and overheard Justin's and Cleavon's comments. Whose statement would you rather use in an ad for potato chips?

Investigation Interpreting Comparisons

You probably hear comparisons that use ratios and percentages every day. News reports and advertisements, such as TV commercials, use such comparisons all the time. People who write advertisements try to state comparisons in a way that will make you want to buy their products.

Problem Set A

In a recent election by a seventh grade class with 90 students, Razi received 36 votes and Jorge received 54 votes.

1. Tell whether each of the following statements presents this information accurately. For each accurate statement, explain or give a calculation to show why it is accurate.

   a. The ratio of the number of students who voted for Razi to the number who voted for Jorge is 2:3.

   b. 40% of the voters preferred Razi.

   c. Jorge received 20% more of the total vote than Razi received.

   d. 50% more people voted for Jorge than for Razi.

   e. 18 more people voted for Jorge than for Razi.

   f. Jorge received $\frac{3}{5}$ of the votes.

   g. The number of people who voted for Jorge is 1.5 times the number who voted for Razi.

2. Which of the *accurate* statements in Problem 1 seems to give the best impression of the class's preference for Jorge? Why?

3. Which of the *accurate* statements in Problem 1 seems to minimize the class's preference for Jorge? Explain.

4. Do some of the statements seem more informative than others?

5. Some ratios compare a part of some group to the whole group, like the number of students who voted for Razi, 36, compared to the total number of voters, 90. Such ratios are called *part-to-whole ratios.*

   Other ratios compare a part of a group to another part of the same or a different group, like the number of students who voted for Razi, 36, compared to the number who voted for Jorge, 54. These kinds of ratios are called *part-to-part ratios.*

   a. Which statements in Problem 1 make part-to-whole comparisons?

   b. Which statements in Problem 1 make part-to-part comparisons?

In the next problem set, you will see how different impressions can be created by presenting the same data in different ways.

Problem Set B

The organizers of a basketball tournament had one team slot left to fill, but two teams they wanted to invite. They decided to look at the two teams' win-loss records.

The Wayland Lions had won 24 games and lost 16. The Midway Barkers played more games, and their record was 32 wins and 22 losses.

1. Find the ratio of wins to the total number of games for the Lions. Do the same for the Barkers. Are these part-to-part or part-to-whole ratios?

2. Find the ratio of losses to the total number of games for each team.

3. What percent of all their games did the Lions win? What percent of all their games did the Barkers win?

4. What percent of all their games did the Lions lose? What percent of all their games did the Barkers lose?

Now you will compare the teams using a different method of comparison.

5. Use ratios to compare each team's wins to losses. Are these part-to-part or part-to-whole ratios?

6. For each team, the number of wins is what percent of the number of losses?

7. Use a percentage to report how many more wins than losses each team had.

8. One of the organizers thinks the Lions is definitely the team to invite. What statements—using comparisons and other information—might this organizer make so that the Lions look like the best choice?

9. Another organizer thinks the Barkers is the team to invite. What statements—using comparisons and other information—might this organizer make so that the Barkers look like the best choice?

Share & Summarize

In 2001 California had about 19,720,000 registered motor vehicles. About 9,960,000 were cars. The other 9,760,000 were trucks, buses, and motorcycles. That same year, New York had about 10,710,000 registered motor vehicles: 8,800,000 cars and 1,910,000 other vehicles.

Write at least five comparison statements using these statistics.

Investigation 2 ▶ Predicting Based on Smaller Groups

People sometimes want to estimate large quantities that are difficult or impossible to count—such as the number of animals or plants of a particular endangered species, or the number of people affected by a flu epidemic. If you had to count each individual member of these populations, it would be costly and time-consuming, and probably impossible to get exactly right.

In this investigation, you will learn how proportions can help you to estimate such numbers without actually counting every member. In Chapter 10, you will learn more about this type of estimation.

▶ MATERIALS
bag of beans

Problem Set C

Each student in your group should take 5 beans from your bag and mark them. Put all the marked beans back into the bag with the other beans, and mix them carefully.

Without looking in the bag, take out 20 beans and record the number of marked beans.

1. What is the number of marked beans?

2. Estimate the total number of beans in the bag. How did you make your estimate?

3. What did you need to assume to estimate the total number of beans?

4. Do you think your estimate would be better, worse, or the same if your bag contained thousands of beans? Why? How might you modify the method you used to make the estimate more accurate?

To estimate the numbers of different animals in the wilderness, scientists use a method called *capture-tag-recapture*. This method is similar to the process you used to estimate the number of beans in your bag.

Using this method, scientists capture a certain number of animals and mark them using collars, rings, or other tags. They then release the animals. Some time later, they capture another group of animals and count how many in that group are tagged. They can then solve a proportion to estimate the total number of animals.

The puffin is a seabird that builds its nest in the cliffs of coastal islands.

Problem Set D

1. Suppose biologists tagged 43 blue whales in the Antarctic waters. The next year, they caught 70 blue whales and found that 6 of them had tags. Estimate the total number of whales in the area. Explain your method.

2. Red wolves have been classified as extinct in the wild, but some still live in captivity. Some efforts have been made to restore red wolves to forests in North Carolina and Tennessee.

   Suppose biologists caught 20 wolves from those forests, tagged them, and then freed them. Later they caught 15 wolves and found that 5 had tags. Estimate the number of red wolves in the forests, and explain how you solved the problem.

3. Suppose ornithologists tagged and released 240 bald eagles from across the United States. A couple of months later, they caught 100 birds and found that 3 of them had tags. Estimate the number of bald eagles in the United States.

Surveys don't use the capture-tag-recapture method, but they do use proportions to make estimates of large populations of people.

Problem Set E

The population of Massachusetts is about 6.35 million people. A survey questioned 1,000 people across the country and found that 294 people had two televisions in their homes and 392 people had three or more televisions. The survey also found that 657 people had cable television service.

1. Estimate how many homes in Massachusetts have two televisions. Assume the proportions for Massachusetts are the same as for the sample.

2. Estimate how many homes have three or more televisions.

3. Estimate how many homes in Massachusetts receive cable television.

4. Discuss with your partner the methods you used to solve Problems 1–3. What assumptions did you make and why?

Share & Summarize

What method would you use to estimate the number of people in a large crowd, such as the large gathering at Times Square in New York City on New Year's Eve?

On Your Own Exercises

Practice  Apply

1. There are 20 trees in a small city park, 5 oaks and 15 birches.

   a. Compare the number of oaks to the number of birches using a ratio.

   b. What is the ratio of birches to the total number of trees? What is the ratio of oaks to the total number of trees?

   c. What percentage of all trees in the park are birches? What percentage are oaks?

   d. Now compare the number of birches to the number of oaks by stating the ratio of birches to oaks.

   e. The number of birches is what percentage of the number of oaks?

   f. What percentage more birches are there than oaks?

   g. Write as many comparison statements about the trees in this park as you can.

2. The manager of the Stevenson Middle School cafeteria surveyed 100 students to find which school lunch they liked least. Here are her results.

   Students Rate Lunches

Lunch	Number of "Worst" Ratings
Hamburger	5
Pizza	5
Tuna surprise	75
Macaroni and cheese	15

   a. Represent these results using percentages.

   b. Represent these results using differences in numbers of "Worst" ratings.

3. **Social Studies** The table gives measles-immunization data for several countries.

   a. Write a ratio statement comparing Afghanistan and Cambodia.

   b. Write a percent statement comparing Eritrea and France.

   c. Compare Israel and Nepal using a statement that is not a ratio or percent statement.

   Infants Immunized for Measles

Country	Percent of Infants
Afghanistan	42
Cambodia	72
Eritrea	38
France	82
Israel	94
Nepal	45

   Source: *The State of the World's Children 1999*, UNICEF, New York.

 impactmath.com/self_check_quiz

4. Miki's Bag-It shop carries a line of luggage with the following sizes and prices.

Types of Luggage

Item	Dimensions	Price
22" rolling suitcase	22" × 14" × 7"	$155
26" rolling suitcase	26" × 17" × 9"	$220
29" rolling suitcase	29" × 20" × 10"	$240

a. For each item, find the volume and the price per cubic inch.

b. Use a percentage to write a statement comparing the volumes of the 22" and 26" suitcases. Compare the 26" and 29" suitcases in the same way.

c. Write a statement comparing the prices of the 22" and 26" suitcases. Compare the 26" and 29" suitcases in the same way.

d. Write a statement comparing the prices per cubic inch of the 22" and 26" suitcases. Compare the 26" and 29" suitcases in the same way.

5. Biology To estimate the number of white-tailed deer in Idaho, biologists captured and tagged 500 deer and then released them back into the forest. A year later the biologists captured 280 deer and counted 2 deer with tags. Estimate the actual number of deer in the forest.

6. **Biology** Marine biologists estimated the number of manatees living in the waters off the coast of Florida. They caught and tagged 120 manatees and then let them go. Of the 150 manatees they caught the next year, 9 were marked. Estimate the number of manatees living in these waters.

7. **Social Studies** In March 1999 the *New York Times* and CBS News polled 1,469 registered voters in New York. The poll found that 48% of the people would support Hillary Rodham Clinton if she ran for senator, while only 39% said they would support the mayor of New York City, Rudoph Giuliani. Suppose the number of registered voters in New York at the time was 14 million.

   a. How many registered voters would you expect to support Clinton at the time of the poll? How many would you expect to support Giuliani?

   b. Voter turnout is generally a fraction of the number of people registered to vote. Assuming a 50% turnout, how many actual votes would you expect Clinton to receive if the election was held at the time of the poll? How many would you expect Giuliani to receive?

   c. What percentage of the registered voters would you expect to actually vote for Clinton?

8. There are 864 students and 32 teachers at Scott School. There are 1,428 students and 42 teachers at King School.

   a. Find the ratios of students to teachers at each school. Can you form a proportion with these ratios? Why or why not?

   b. Suppose King School wants to make its student-teacher ratio close to that at Scott School. How many new teachers would King have to hire?

   c. Can you write a part-to-whole ratio using these data? Explain.

9. Annette recorded how she spent her time each day for a week.

Daily Activities

Activity	One Week, in Hours	School Year (39 Weeks), in Days	75 Years (Age 5 to 80), in Years
Sleeping	54.6		
At school	35.0		
Socializing	14.1		
Watching TV	11.2		
Doing homework	9.0		
Talking on phone	8.4		
Reading	7.0		
Eating	7.0		
Bathing, grooming	6.3		
Other activities	15.4		

   a. What assumption would you have to make to complete the table?

   b. Is the assumption in Part a reasonable for the "School Year" column? If not, would it be reasonable for *some* of the activities? If so, which ones?

   c. Is the assumption in Part a reasonable for the "75 Years" column? If not, would it be reasonable for *some* of the activities? If so, which ones?

   d. Complete those entries in the table for which your assumption seems reasonable.

   e. Calculate the percentage of time Annette spent in each activity during the week she made her records.

   f. Make a circle graph of the percentages from Part e.

A healthy adult human has 20 to 30 trillion red blood cells, with an average life span of 120 days. To keep up with this demand, the body replaces about 2 million red blood cells every second.

10. **Life Science** A sample of blood was diluted 100,000 times. The drawing represents a photograph of the red blood cells in 20 cubic millimeters of the diluted blood as taken through a microscope. Each dot represents one red blood cell.

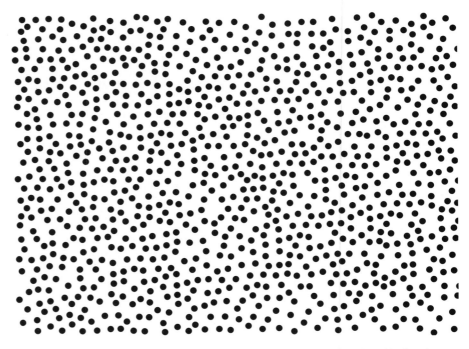

   a. Think of a way to estimate the number of red blood cells in the drawing without counting all of them. Explain your method.

   b. Use your method to estimate how many red blood cells are in 20 mm^3 of the sample of diluted blood.

   c. About how many red blood cells were in 1 mm^3 of this blood sample before it was diluted?

11. **Social Studies** When a political group holds a rally to show support for a cause or to protest, the police will usually estimate the number of people attending. The estimates are often made by counting the number of people in a particular location. That number, and the area of the selected location, can be used to set up a proportion for estimating the total attendance.

   The organizers of such rallies will often make their own estimates—and usually their estimates are much higher.

   Suppose the organizers of an event want their estimate to be as high as possible to show that lots of people support their cause. How might they choose the location where they will count people attending? Explain.

In your **own** **words**

Describe the ways you can compare two quantities. Give an example of a situation for each comparison method you describe.

12. Geography The table shows population and land area for five states.

Population and Area in the U.S.

State	2000 Population	Land Area (mi²)
Alaska	626,932	570,374
California	33,871,648	155,973
Illinois	12,419,293	55,593
Utah	2,233,169	82,168
Vermont	608,827	9,249

Reprinted with permission from *The World Almanac and Book of Facts.*

a. Estimate how many people live in a particular 15 mi² area in California and in a particular 15 mi² area in Vermont.

b. What assumption did you have to make to solve Part a? Is this a reasonable assumption? Explain.

c. Calculate the population density—the number of people per square mile—for each of these states. Which state is most crowded?

Mixed Review

Evaluate for $k = 1$, $m = 2$, and $n = 4$.

13. k^n

14. n^m

15. $\dfrac{k}{m^m}$

16. $\dfrac{m^n}{k}$

17. n^{-m}

18. $\dfrac{k}{m^{-m}}$

19. $n(m^{k-n})$

20. $m^{2m}(m^m)$

21. Irene made a bead necklace by alternating groups of 3 blue beads and 2 red beads. She used a total of 75 beads. How many beads of each color are in her necklace? Explain how you found your answer.

22. You are traveling by train and bus. The train averages 70 mph, and the bus averages 50 mph.

a. Let t and b represent how many hours you ride on the train and the bus, respectively. Write an expression for the total miles you travel on your trip.

b. Suppose the total distance you travel is 600 miles. Use your answer from Part a to express this as an equation.

c. If you travel 600 miles and spend 5 hours on the train, how long was the bus ride?

Chapter Summary

V O C A B U L A R Y
proportion
unit rate

Comparisons can take many forms, including differences, rates, ratios, and percentages. In this chapter, you learned to compare ratios using equivalent ratios and *unit rates*. You also used percentages as a common scale for comparisons.

You combined difference with percentages when you worked with percent change. You also learned how to use *proportions* to solve problems involving ratios and percentages, including problems involving similarity.

Strategies and Applications

The questions in this section will help you review and apply the important ideas and strategies developed in this chapter.

Comparing and scaling ratios and rates

1. The Quick Shop grocery store sells four brands of yogurt.

Brand	Meyer's	Quick Shop	Rockyfarm	Shannon
Price	2 for $1.50	3 for $2	$.75 each	$.80 each
Size	8 oz	6 oz	6 oz	8 oz

a. For each brand, find the ratio of price to ounces.

b. For each brand, find the unit price.

c. Use the unit rates to list the brands from least expensive to most expensive.

d. Explain how you could use the ratios in Part a instead of the unit rates to list the brands by how expensive they are.

2. Every summer, Joe makes his famous Peach Cooler. To make the drink, he mixes 3 quarts of tea with $\frac{1}{2}$ quart of peach juice.

a. Find the ratio of tea to peach juice in Joe's Peach Cooler.

b. Find the ratio of tea to Peach Cooler (that is, to the final drink).

c. If Joe has only 2 quarts of tea, how much peach juice should he add?

d. For a party, Joe wants to make 7 quarts of Peach Cooler. How much tea and peach juice does he need?

e. Explain how you found your answer for Part d.

Writing and solving proportions

3. Quinn's favorite toy store was running a contest. Contestants had to guess the number of marbles in a jar. All the marbles were clear, so Quinn had an idea. She convinced the owner to let her add 20 marbles to the jar and then remove 20 marbles. Not seeing how this could help her, the owner agreed.

Quinn put in 20 colored marbles, mixed all the marbles in the jar thoroughly, and then, without looking, drew out 20 marbles. She had drawn 3 colored marbles and 17 clear marbles.

a. Write a proportion Quinn could use to estimate how many marbles are in the jar.

b. Solve your proportion.

c. How many marbles should Quinn guess are in the jar?

4. The sandhill crane shown here has a wingspan of approximately 6 feet. Measure the crane's wingspan and height in the picture, and use that information to estimate the real crane's height.

Using percentages to make comparisons

5. Social Studies Public elementary and secondary schools in the United States receive money from different levels of government. The table shows National Education Association estimates for several states for the 2001–02 school year. All numbers are in thousands of dollars.

School Funding Sources

State	Total	Federal	State	Local and Intermediate
California	35,054,650	3,108,332	19,920,649	12,025,669
Hawaii	1,364,412	101,842	1,231,799	30,771
Maine	1,520,325	95,181	698,005	727,139
Mississippi	2,502,975	325,505	1,422,642	754,828
New Hampshire	1,365,391	42,742	83,529	1,239,120

a. Which of these states received the most federal money?

b. Which of these states relies most on federal money?

c. Explain why Part b is a different question from Part a.

Understanding percent change

6. In 1999 about 48.5 million people visited the United States. In 2000 there were 50.9 million visitors. In 2001 there were 45.5 million.

a. Describe two ways to find the percent change from 1999 to 2000. Explain how to tell whether the change is an increase or a decrease.

b. Find the percent change from 1999 to 2000 and from 2000 to 2001. Be sure to specify if each change is an increase or a decrease.

Writing and interpreting comparisons

7. A brand of shampoo used to be packaged in 12-ounce bottles and priced at $3.60. When the company switched to 15-ounce bottles, they kept the pricing the same to be more competitive. To get attention on the store shelf, they printed "25% more FREE!" on the new bottles. (That doesn't mean someone can take 25% of the shampoo in a bottle without paying for it!)

a. Explain what is meant by the statement "25% more FREE!"

b. Write as many statements as you can comparing the old size and price to the new size and price.

Demonstrating Skills

8. Examine the tile pattern.

    a. Write the ratio of white tiles to purple tiles.

    b. Write at least five ratios for the tile pattern.

9. Paul wants to tile a large area using this pattern. Make a ratio table of possible numbers of purple and white tiles that he could use.

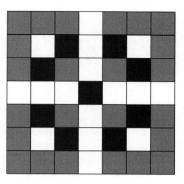

Find the value of the variable in each proportion.

10. $\frac{12}{5} = \frac{x}{9}$

11. $\frac{3.2}{y} = \frac{4}{7}$

12. $\frac{92}{36} = \frac{23}{w}$

13. $\frac{a}{4.7} = \frac{13}{61.1}$

Find the missing measures for these similar figures.

14.

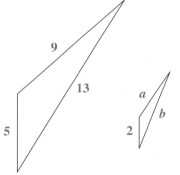

15.

Set up and solve a proportion to answer each question.

16. 17 is 5% of what number?

17. What percent of 450 is 25?

18. What is 32% of 85?

CHAPTER 9

Interpreting Graphs

Real-Life Math

Rain, Rain, Go Away When the student council at Cedar Lane Middle School in Seattle, Washington, started planning its annual car wash benefit, the members needed to agree on a date. They knew that the car wash would be less successful on a rainy day, so they decided to choose a day in a month when it doesn't usually rain very much. They consulted the graph at the right to help them make their decision. Which month do you think would be the best choice?

The students also created other graphs to help them plan for and track the results of their car wash.

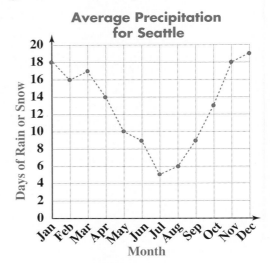

Average Precipitation for Seattle

Days of Rain or Snow / Month

Think About It Can you think of what kinds of information the students may have used to make their other graphs?

Family Letter

Dear Student and Family Members,

In our next chapter, we will be interpreting graphs and understanding the stories behind them. For example, the graph at the right shows the amount of money raised at car washes held by the student council.

Here are some ways we can interpret the graph to learn more of the story:

- About how much more money was raised in 2004 than in 2003?

- Between which two years was there a decrease in the number of dollars that were raised?

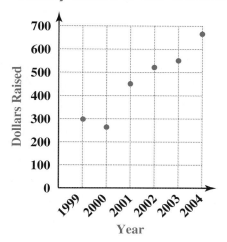

Money Raised at Car Washes

Many graphs represent mathematical relationships, which can also be represented by equations or formulas. We will learn about the shapes of graphs that represent repeating, linear, and quadratic relationships as well as exponential relationships, which show situations where a quantity is repeatedly multiplied by a number greater than 1.

Vocabulary
Along the way, we'll learn about two new vocabulary terms:

line graph   **multiplicative inverse**

What can you do at home?

You can help your student read the stories behind graphs by discussing graphs you find in magazines, newspapers, web sites, or advertisements. Try to find the story behind the graph by asking these questions:

- What is the graph about?

- What does each point represent?

- Is there a comparison between items in the graph?

- Can you use the graph to predict something that is not explicitly shown on the graph?

Finding the story behind the graph can be an enjoyable and imaginative experience that you can share with your student. Have fun!

Graphing Change over Time

The shape of a graph can tell you a lot about the relationship between two variables. When the variable on the horizontal axis is time, the graph shows how the variable on the vertical axis changes as time passes.

Think & Discuss

Kate's mother is a health care worker at a community clinic. As part of her job, she sometimes creates and analyzes graphs. Many of the graphs she makes show how variables change over time.

Match each situation below to the graph you think is most likely to describe how the variable in the situation changes. Explain your choices.

1. a newborn baby's weight during the first few months of life

2. the number of spots on a child with chicken pox from when Kate's mother first sees him until he is well again

3. the monthly number of flu cases reported over several years

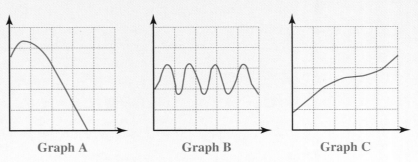

Graph A   Graph B   Graph C

Investigation 1 Growing Up

Each time one of his daughters celebrates a birthday, Mr. Fernandez measures her height and records it on a chart.

Age (years)	Height (inches) Maria	Luisa	Rosi
3			$36\frac{1}{2}$
4		37	$37\frac{1}{2}$
5		$37\frac{3}{4}$	$40\frac{1}{2}$
6	45	42	$43\frac{1}{2}$
7	$46\frac{1}{2}$	44	$44\frac{1}{2}$
8	$48\frac{1}{4}$	$45\frac{1}{4}$	$46\frac{1}{2}$
9	$50\frac{1}{2}$	47	$47\frac{1}{2}$
10	52	$47\frac{3}{4}$	
11	54		
12	$57\frac{1}{2}$		

Problem Set A

Use the information from Mr. Fernandez's chart to answer these questions.

1. Why might some of the height values be missing?

2. How tall was each girl at age 6?

3. Put the girls' names in order, from the girl who was shortest at age 6 to the girl who was tallest.

4. Put the girls' names in order, from the girl who was shortest at age 7 to the girl who was tallest.

5. How much did each girl grow between her sixth and seventh birthdays?

6. Put the girls' names in order, from the girl who grew least between 6 years and 7 years to the girl who grew most.

7. Was the girl who grew most between 6 and 7 years the tallest at 7 years? Explain.

Graphing the data in Mr. Fernandez's growth chart can give you an over-all picture of how the girls' heights changed over the years.

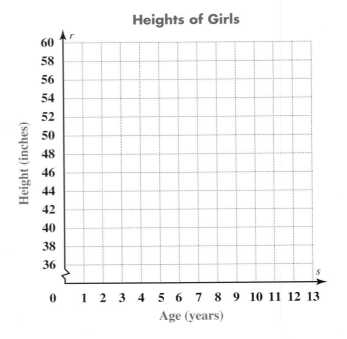

Problem Set B

1. On a grid like the one below, plot points to show how tall Maria was at each birthday.

Heights of Girls

Of course, Maria's height didn't suddenly change on each of her birth-days! The chart and graph show how tall Maria was on each birthday but not how tall she was in between birthdays.

2. You can use your graph to estimate Maria's height at times between her birthdays. Connect the points on your graph to get an idea of how Maria grew during each year.

When the points on a graph are connected by line segments, the graph is called a **line graph.**

3. If Mr. Fernandez had taken measures every month between Maria's birthdays, would the data points between any two birthdays be likely to fall on a straight line? Why or why not?

4. On the same grid you used for Maria, draw line plots for the heights of Luisa and Rosi. You might want to use a different color for each girl.

5. Use your graph to predict Rosi's height at 10 years of age.

6. How can you tell from the graphs which girl grew most between ages 6 and 7?

7. Between which two birthdays did Luisa grow most? Between which two birthdays did she grow least?

8. Describe an easy way to use the graph to find the year in which Rosi grew most.

9. Think about what you found in Problems 6–8. What can you conclude about the shapes of the graphs and the rates at which the girls grow?

Share & Summarize

The graph shows how a plant grew over time, beginning when it first appeared above the ground.

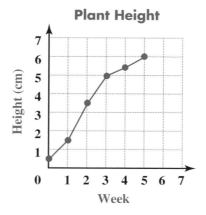

Plant Height

1. How tall was the plant after 1 week?

2. When was the growth rate fastest, and when was it slowest? Explain.

3. Estimate how tall the plant was after 10 days. How did you find your answer?

Investigation ▶ 2 ▶ Filling It Up

In this investigation, you will examine how the water level in containers of various shapes changes as water is poured into them and draw graphs of the data you collect.

MATERIALS

- cup or scoop
- large container of water tinted with food dye
- three transparent jars of different sizes and with straight, vertical sides
- 3 pencils or pens of different colors
- a marker
- metric ruler
- graph paper

Problem Set C

One member of your group should get three different jars for the group. Suppose you use a scoop to pour the same amount of water into each jar. You mark the water levels on the jars, add another scoop of water to each jar, and mark the levels again. You continue scooping and marking until one jar is nearly full.

1. Sketch graphs to show how you think the water level will rise in each jar as you pour water into it. Put the graphs on the same set of axes, and use a different color for each graph. Put height on the vertical axis and number of scoops on the horizontal axis.

Now try it. One person should pour a scoop of water into each jar. Another should mark the water's height on the side of the jar after each scoopful. Put the same number of scoopfuls in each jar, and continue until one jar is nearly full.

2. Measure the heights of the marks you made on each jar. Record the information in a table like this one.

Scoopfuls	Jar 1	Jar 2	Jar 3
0	0 cm		
1			
2			
3			
4			
5			

3. On a new set of axes, draw three graphs, one for each jar, that show the actual heights of the water plotted against the number of scoops. Use the same colors for each jar that you used in Problem 1.

4. Examine the patterns made by your graphs. What do you notice?

5. Compare these graphs with the predictions you drew in Problem 1. Write a paragraph comparing your predictions with the actual results.

Problem Set D

Eva and Tayshaun allowed water from a leaking faucet to fall into the cup shown below. They marked the height of the water on the cup every minute. At the end of 6 minutes, the cup looked like this.

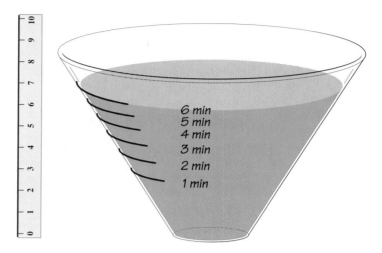

1. Why do the marks get closer together as time passes?

2. Use the ruler at the side of the cup to find the height of the water at each minute. Draw a line graph of the data, and describe its shape.

3. Have you noticed that often when people fill bottles and vases with a narrow top, the water suddenly gushes out the top? Why do you think this happens?

Share & Summarize

1. Ramona owns a cabin high in the mountains. Her main water source is a tank shaped like a cylinder. To fill the tank, Ramona hires a small tanker truck. The tanker's pump sends a steady stream of water into the tank. Make a rough sketch of a graph that shows how the height of the water in the tank will change as it fills.

2. George owns the cabin nearest to Ramona's. He also uses a water tank, but his tank has sides that slant outward slightly. Make a rough sketch of a graph that shows how the height of the water in George's tank will change as the tanker truck fills it with a steady stream of water.

Lab Investigation ▶ Filling Odd-shaped Containers

MATERIALS

- cup or scoop
- large container of water tinted with food dye
- three transparent containers of odd shapes
- a marker
- metric ruler
- graph paper

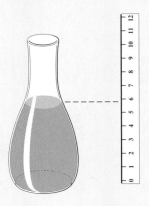

In Investigation 2, you discovered what happens to the height of the water as it fills containers with straight and slanted sides. Some containers don't have such simple shapes.

Make a Prediction

With your group, get three containers with different shapes. As you did in Investigation 2, imagine filling the containers with water, one scoop at a time.

1. For each container, sketch a rough graph that shows how you expect the height of the water to change as you fill the container. Discuss your graphs with your group, and redraw your sketches if necessary. Explain why you think the graphs will look like your sketches.

Try It Out

Now try the experiment. One person should pour a scoop of colored water into each container. Another should mark the height of the water on the side of the container after each scoopful. Continue with each container until they are all filled.

2. Measure the heights of the marks you made on each container. Be sure to measure the vertical height, as shown at left. Record the information in a table.

3. Use the data from your table to draw a graph for each container.

4. Were your sketches roughly the right shapes? If they weren't, try to explain why your predictions weren't accurate.

Design Your Own Container

5. Design an odd-shaped container and draw it.

6. On a different sheet of paper, sketch a graph that shows how you expect the water height to change as you fill your container.

7. Trade graphs with another member of your group. Look at the graph you have been given, and try to imagine what shape container might give that graph. Make a rough sketch of the container you imagine, and then return it to your classmate.

8. Compare your container design to the container your classmate drew. Do the containers look alike? If not, could your graph match both containers?

What Did You Learn?

9. Draw the container that might go with each graph. All four containers hold the same amount of liquid, and it is poured at a constant rate.

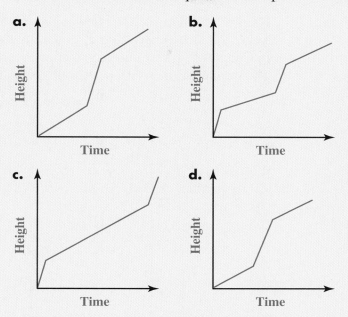

a. Height vs. Time

b. Height vs. Time

c. Height vs. Time

d. Height vs. Time

10. Tanks that hold oil for heating houses are often cylindrical. However, they are usually installed so that the circular bases are on the sides rather than at the top and bottom.

The tank's fuel gauge uses the height of the oil inside the tank to show how full the tank is. Oil tanks are not filled completely to allow for some expansion if the temperature changes. However, if the tank were filled completely, the height of the oil in a full tank would be equal to the diameter of the tank.

a. Sketch a graph that shows how the height of the oil changes as you fill the tank.

b. Challenge When the tank is $\frac{1}{4}$ full, is the height of the oil $\frac{1}{4}$ of the tank's diameter? Explain.

Investigation 3 ▶ Walking About

Graphs are often used to show where an object—a person, an ant, a ball, a car, a planet, or anything else that moves—is located at various times. In Chapter 5, for example, you considered a race between five brothers. The graphs of their distances from the starting line are shown below.

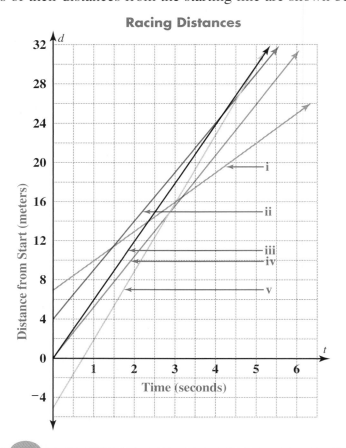

Racing Distances

Think & Discuss

A graph's shape illustrates, or gives information, about a situation.

• The graphs all have different slopes. What does that tell you about the brothers in the race?

• To give the slower boys a chance to win, the brothers started at different places. How is this shown in the graph? Explain.

• Suppose the brothers' dog, King, was waiting at the finish line. When the boys started running, King ran toward them. After a couple of seconds, he turned around and ran back to the finish line. Give a rough description of what King's graph would look like.

Problem Set E

In the following problems, a person walks from one side of a room to the other. You can match graphs with different walks by thinking about the situation.

Discuss each problem below with your partner until you are sure of your answers and feel you could convince someone else you are correct.

1. Zach walks slowly and steadily across the front of his classroom, from the left wall to the right wall. Which graph best shows his distance from the left wall at any time? Explain your reasoning to your partner.

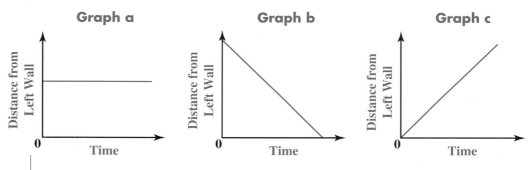

2. Each graph below uses the same scales on the axes. Which shows the fastest walk from left to right, and which shows the slowest? Explain your reasoning to your partner.

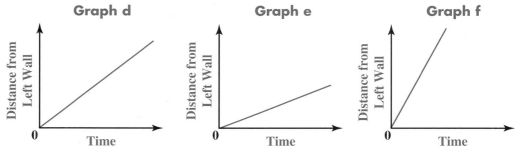

3. Suppose Jin Lee walks in the opposite direction, from the right wall to the left wall. Sketch a graph of Jin Lee's distance from the *left* wall over time.

4. The graphs below all use the same scale.

   a. Choose the graph that best represents the following:

     i. walking slowly from left to right

    ii. walking slowly from right to left

   iii. walking quickly from left to right

    iv. walking quickly from right to left

    v. standing still in the center of the room

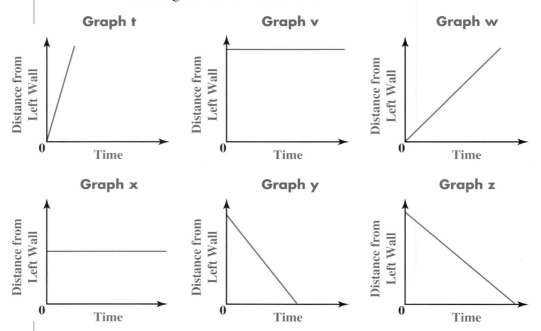

   b. There is one extra graph. Describe the walk it shows.

5. Look at the eight graphs below.

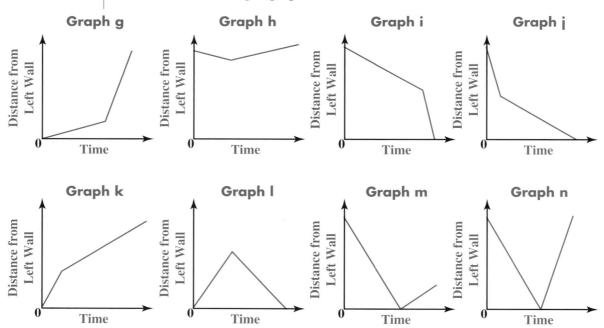

a. Choose the graph that best shows each of these walks:

i. slow to the right, then fast to the right

ii. fast to the right, then slow to the right

iii. slow to the left, then fast to the left

iv. fast to the left, then slow to the left

v. slow to the left, then slow to the right

vi. fast to the right, then fast to the left

vii. fast to the left, then fast to the right

b. There is one extra graph. Describe the walk it shows.

Problem Set F

Sophie starts to walk slowly from the left wall of a large hall and moves faster as she walks to the right.

1. Suppose you mark the floor at fixed intervals of time—for example, every 2 seconds—to show where Sophie was as she walked across the room.

Which of the following statements would best describe the marks? If necessary, you and your partner can experiment to discover what happens.

a. The marks will be equally spaced across the room.

b. The marks will get further apart from left to right.

c. The marks will get closer together from left to right.

2. Which of these graphs best describes Sophie's walk? Explain how the graph links to the description you chose in Problem 1.

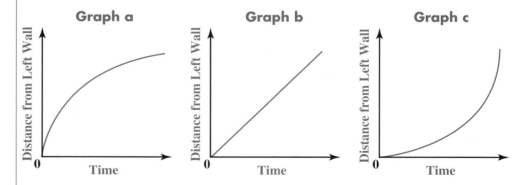

Graph a   **Graph b**   **Graph c**

Distance from Left Wall — Time

3. What kinds of walks are shown in the other graphs of Problem 2?

Problem Set G

Many cities around the world use historical trains for tourist trips. The famous steam train *Puffing Billy* in the hills near Melbourne, Australia, is really a set of several small trains. About three trains run on any day, taking people from Belgrave to Lakeside and back.

The line is mainly a single track, so when trains from each direction meet, one has to wait for the other to pass. Of the eight stations along the route, Menzies Creek and Emerald are the only ones with double tracks.

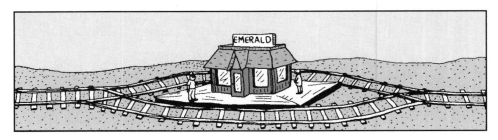

One day Train A left the Belgrave station at 8:54 A.M., Train B at 9:30 A.M., and Train C at 10:00 A.M. Train C met Train A as Train A was making the return trip. The graphs show the trains' movements up to 11:30 A.M. The names of the stations and their distances from Belgrave are shown on the vertical axis.

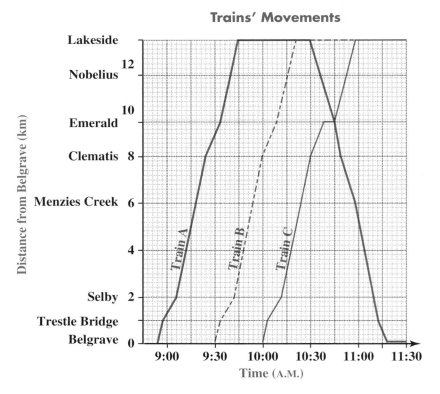

1. When did Train A arrive in Lakeside?

2. For how long did Train A stop at Lakeside?

3. The section from Trestle Bridge to Selby is the steepest climb. How do the graphs shows this? Explain how this relates to the speed of the trains.

4. Which train took the longest to get from Belgrave to Lakeside? Why do you think it needed more time?

5. Along which sections did Train A reach its greatest speed on the return trip to Belgrave?

6. It is downhill from Emerald to Clematis. How does the graph show that Clematis to Emerald is a steep climb?

7. At 11:30 A.M., where was each of the three trains?

MATERIALS
graph paper

Share & Summarize

One winter, Malik conducted a weather experiment. He stuck a meter-stick upright in his backyard on December 1. Every morning when there was snow on the ground, he measured the depth of the snow. He made this graph from his data, counting December 1 as Day 1.

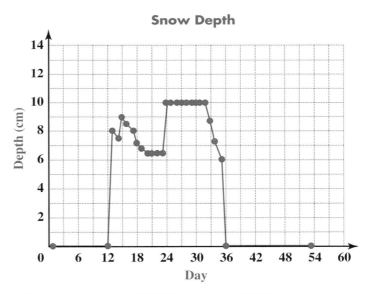

Snow Depth

1. When did the first snow fall? How much fell?

2. During a sudden warm spell, rain washed away all the snow left from the last storm. How is this event shown on the graph? When was the rainfall?

3. That winter, school was canceled because of snow only once, on January 23. Make a rough copy of the graph, and add a section showing what it might look like by January 24.

On Your Own Exercises

Practice & Apply

1. Dikembe and Aissa are twins. The graphs show their weights during the first 20 years of their lives.

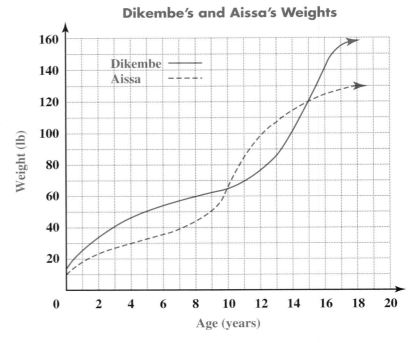

Dikembe's and Aissa's Weights

a. From the graphs, how much weight did each twin gain between 10 and 18 years of age?

b. Between what ages did Dikembe weigh more than Aissa did at those same ages? Explain how you know from the graph.

c. At what ages did the twins weigh the same?

d. When was Aissa gaining weight most rapidly? Explain how you know from her graph.

e. Estimate how much weight per year Aissa was gaining at that time.

f. When was Dikembe's rate of weight gain greatest? Estimate his rate of weight change at that time.

g. Who was putting on weight faster at age 16? How do you know?

    impactmath.com/self_check_quiz

2. The tables give average heights of girls and boys for different ages.

Girls

Age (yr)	2	4	6	8	10	12	14	16
Average Height (in.)	35.6	40.1	45.3	50.0	53.8	58.5	63.1	64.0

Boys

Age (yr)	2	4	6	8	10	12	14	16
Average Height (in.)	34.2	40.6	45.6	50.0	54.2	59.0	64.2	68.4

Source: National Center for Health Statistics

By graphing the data, you can get an idea of when young people usually grow the most and the least. You can easily compare the information for girls and boys.

a. On a grid, plot girls' average heights at different ages. Label the vertical axis from 30 in. to 70 in. Should you connect the points? Explain.

b. In another color, plot boys' average heights at different ages. If it makes sense, connect the points.

c. Use your graph to estimate the average height of girls and of boys at 15 years of age.

d. From your graph, decide at what ages boys and girls are, on average, the same height and when they are not.

e. Write a brief paragraph describing the relationship between age and average height for boys and for girls.

3. The graph shows the height of grass as it grows over time.

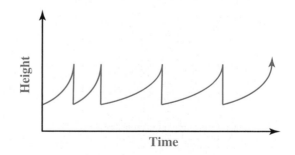

a. When the graph is steep and increasing, what is happening to the grass?

b. When the graph is not so steep, what is happening to the grass?

c. The vertical line segments show the most dramatic change. What might these show?

4. Many science laboratories have beakers with sides that slant inward. Sketch a graph that shows how the water level in the beaker would change if water is poured in at a constant rate.

5. Ramona's cabin has a water tank shaped like a right cylinder. George owns the cabin nearest to Ramona's, and his water tank has sides that slant outward slightly. (See the drawings on page 607.) When they close up their places at the beginning of the winter, they empty their tanks so there isn't any water in them to freeze during the winter.

Assuming the water drains from the tanks at a constant rate, sketch graphs that show how the water height in each tank will change as the tanks drain.

6. Match these descriptions of a person's walk across a room with the graph that best shows the walker's distance from the left wall.

  i. slow to the right, stop, slow to the right

  ii. fast to the right, stop, slow to the right

  iii. fast to the left, stop, fast to the right

  iv. slow to the right, stop, fast to the right

  v. fast to the left, stop, fast to the left

  vi. fast to the left, stop, slow to the left

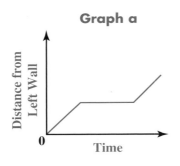

Graph a

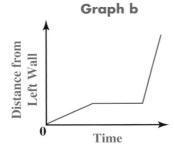

Graph b

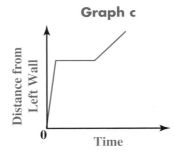

Graph c

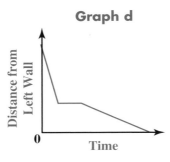

Graph d

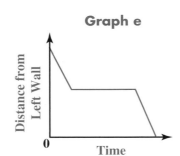

Graph e

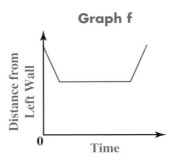

Graph f

Connect & Extend

7. In the movie *Forrest Gump,* the hero decides to run across America.

   a. If Forrest runs an average of 5 miles per hour for 10 hours each day, how many miles per day will he cover?

   b. New York to San Francisco is approximately 4,000 miles. If Forrest maintains his pace, how many days will this journey take?

   c. If Forrest rests for 10 days at the halfway mark, how long would the entire journey take?

   d. Draw a graph of the journey, including the halfway rest. Put time (days) on the horizontal axis and miles on the vertical axis.

   e. On the same graph, draw a line for someone who gave Forrest 20 days' head start and arrived in San Francisco at the same time. What must this person's average speed be if he or she also ran for 10 hours per day?

8. The graph shows approximately how Charles' weight changed the year he was 11.

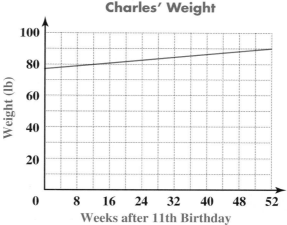

Charles' Weight

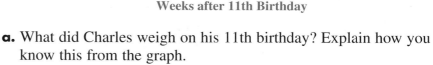

Weeks after 11th Birthday

   a. What did Charles weigh on his 11th birthday? Explain how you know this from the graph.

   b. Use data from the graph to calculate Charles' average weight gain each week.

   c. Write a rule in words and in symbols to describe the relationship between time after Charles' birthday t in weeks and weight w in pounds.

   d. Assuming Charles continues to grow at the same rate, how old would you expect him to be when he weighs 100 lb? 120 lb?

   e. If Charles continues to grow at the same rate, what will his weight be when he is 15 years old? 20 years old?

9. Social Studies The population of the United States has grown steadily over the years. The table shows the population at various intervals. The data are from the U.S. Census, which is taken every 10 years.

U.S. Population

Year	Population (millions)	Percent Increase	Year	Population (millions)	Percent Increase
1790	3.9	—	1900	76.2	
1800	5.3		1910	92.2	
1810	7.2		1920	106.0	
1820	9.6		1930	123.2	
1830	12.9		1940	133.2	
1840	17.1		1950	151.3	
1850	23.2		1960	179.3	
1860	31.4		1970	203.3	
1870	38.6		1980	226.5	
1880	50.2		1990	248.7	
1890	63.0		2000	281.4	

Source: U.S. Bureau of the Census

a. Copy and complete the table, adding the percentage increase from one time interval to the next.

b. Which 10-year interval shows the greatest percentage increase?

c. Notice that the increase from 1860 to 1870 is lower than the previous increases. What historical event might have caused this?

d. The increase from 1930 to 1940 is very low. What historical event might have caused this?

e. Make a graph of the data, with year on the horizontal axis and population on the vertical axis.

f. Use the trend of your graph to predict the U.S. population in the year 2050.

10. Tina Lao lives with her parents, two siblings, and her mother's parents. The Laos have a water tank on their farm property for their household water use. The graph shows how the amount of water in the tank changed over one day.

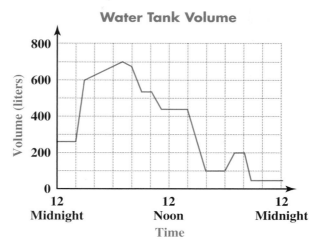

Water Tank Volume

a. Write a few paragraphs about what events might have happened during the day to explain parts of the graph. Be creative!

b. What total volume of water (in liters) was used or lost from the tank during the day? Explain.

c. How many liters per hour were removed between 2 P.M. and 4 P.M.?

11. The cylindrical water tank at Ramona's mountain cabin has a diameter of 2 m and a height of 3 m. Suppose the tanker truck fills the tank at a rate of 0.072 m³ per minute.

a. Find the tank's volume. How long will it take to fill it?

b. Carefully draw a graph (not a sketch) showing the *volume* of water in the tank as time passes, until the tank is full. Be sure to show the scales on both axes.

c. Now draw a graph showing the *height* of the water in the tank as time passes, until the tank is full. Show the scales on both axes. Hint: What is the height when you first begin filling? When is the tank full, and what is the height then?

d. Use your graph from Part c to find the height of the water after 1 hour.

e. Use your graph from Part b to find the volume of the water after 1 hour.

f. Challenge Think of another way to find the volume of the water after 1 hour. How can you use that volume to check your answer to Part d?

12. Two friends are sailing their boats, *Free Time* and *Relaxation,* in a lake. The lake's two piers are at each end of the widest section, a 2-mile stretch. The graphs show the boats' distances from the west pier. The solid line shows *Free Time*'s distance; the dashed line shows *Relaxation*'s.

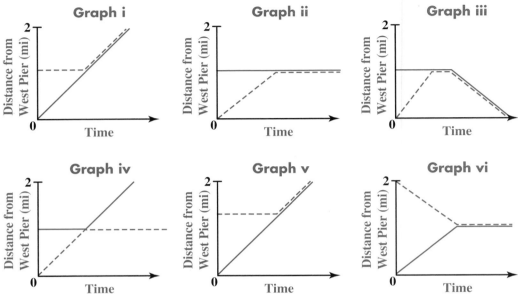

a. Describe what each graph shows. Begin by stating where each boat is at the start.

b. The scales are the same for each graph. Explain what the steepness of a given part of the graph tells you about the boat's speed.

c. How does the slope tell you which direction a boat is sailing?

13. The train schedule shows station arrival times for the three trains on the *Puffing Billy* line between 11:40 A.M. and 1:08 P.M.

Station	Train A	Train B	Train C
Belgrave	leave 11:40 A.M.	12:34 P.M.	1:08 P.M.
Trestle Bridge	11:42 A.M.	12:29 P.M.	1:03 P.M.
Selby	11:50 A.M.	12:26 P.M.	1:00 P.M.
Menzies Creek	arrive 12:00 noon leave 12:17 P.M.	12:14 P.M.	12:48 P.M.
Clematis	12:23 P.M.	12:08 P.M.	12:38 P.M.
Emerald	arrive 12:32 P.M. leave 12:41 P.M.	leave 12:05 P.M. arrive 12:00 noon	12:35 P.M.
Nobelius	12:47 P.M.	11:52 A.M.	12:27 P.M.
Lakeside	12:51 P.M.	leave 11:45 A.M.	leave 12:20 P.M.

a. Draw a graph of the data in the schedule. Put time on the horizontal axis, from 11:30 A.M. to 1:30 P.M. Label the vertical axis with the station names as done in the graph in Problem Set G.

b. Write some questions that could be answered using your graph. Be sure to include answers.

Mixed Review

Evaluate each expression without using a calculator.

14. $0.51 - 1.2$　　　**15.** $3 - 0.006$　　　**16.** $0.015 - 0.0015$

Solve each equation.

17. $3(2a - 1) = 3$　　**18.** $2(3 - 3b) = {}^-7.2$　　**19.** $\frac{3a + 4.5}{1.5a} = 2.5$

Find each percentage.

20. 32% of $100　　　**21.** 80% of $200　　　**22.** 2.5% of $10,000

23. Suppose Laura invests $200 on January 1, 2000, at 8% interest per year. Her original investment and the yearly interest are reinvested each year for 10 years.

a. How much money is in Laura's account at the end of 2000 (to be reinvested for 2001)?

b. Make a graph, on axes like those at right, showing the growth of the account's value for the 10 years the money is invested.

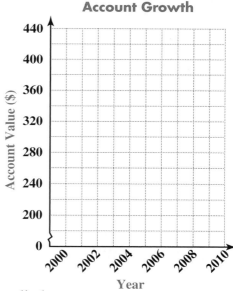

Account Growth

Geometry Find the volume of each cylinder.

24.

25.

26.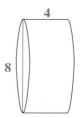

9.2 Graphs and Equations

You have investigated many real-world situations that can be modeled by linear relationships. In this lesson, you will encounter three more common types of relationships. As with linear relationships, their graphs have special shapes and their equations have particular forms.

Think & Discuss

A gas station has a 20,000-gallon underground storage tank. The owners know that the amount of gas they sell varies from day to day. By looking at their sales over several weeks, though, they conclude that they sell approximately 800 gallons per day.

The graph below can be used to keep track of about how much gas is left in the storage tank in the days after it has been filled.

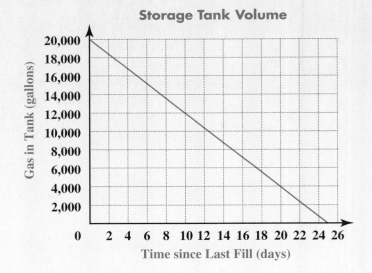

Storage Tank Volume

How can the graph be used to find the amount of gas remaining in the tank for any number of days since it was last filled?

How long does it take to empty the tank after it has been filled?

How is this shown on the graph?

Write an equation for estimating the amount of gas g in the tank d days after it has been filled.

Investigation Graphs for Squares

In the Think & Discuss, you inspected a graph of a linear relationship. You know that linear relationships can be expressed symbolically, using an equation in the form $y = ax + b$.

In this investigation, you will explore a different kind of relationship. As with linear relationships, all relationships of this type can be expressed using equations with a similar form, and their graphs have similar shapes.

Problem Set A

1. Use your calculator to find the value of x^2 for various values of x. Choose 10 values of x from $^-3$ to 3. Record your results in a table.

2. Draw a graph of the values in your table. Put x values on the x-axis and x^2 values on the y-axis. Before you start, decide on a reasonable scale.

3. Do your points seem to form a pattern? If so, describe the pattern.

4. Is it sensible to connect the points on your graph? Why or why not? If it does make sense, connect them by drawing a *smooth curve* through them, rather than drawing line segments between them.

5. Write the equation that your graph shows.

6. Extend the curve you drew so you can use it to estimate 3.2^2. Check your answer with your calculator.

7. Use your graph to estimate the square roots of 7. That is, estimate the values of x for which $x^2 = 7$. Explain what you did to find your answer.

Just the **facts**

The graph of $y = x^2$ is called a *parabola*.

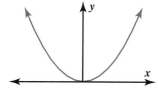

The paths made by the water in the photograph are also parabolas.

Problem Set B

Simon and Shaunda were playing *What's My Rule?*

Output = $n^2 + 3$, where n stands for the input.

1. Write Shaunda's rule in words.

2. Choose two values for *n*, one positive and one negative, and check that you and your partner agree on outputs for them.

3. Draw a graph for Shaunda's rule, using 10 inputs from ⁻4 to 4. If it makes sense, connect the points using a smooth curve.

4. How is your graph like the one you drew in Problem Set A? How is it different?

5. When it was his turn, Simon used the rule *output* = $n^2 + 1$. Discuss with your partner what the graph for this rule might look like. Would it be similar to any graphs you have seen before? If so, which?

Share & Summarize

Describe the shape or pattern of the curve that results when an input variable is squared.

Investigation Graphs for Inverses

VOCABULARY
**multiplicative
 inverse**

The next type of relationship you will look at uses the *multiplicative inverse* of a quantity. The **multiplicative inverse** of a number is the result when 1 is divided by the number. For example:

- The inverse of 4 is $\frac{1}{4}$.
- The inverse of $^-13$ is $-\frac{1}{13}$.
- The inverse of x is $\frac{1}{x}$.

MATERIALS
graph paper

Problem Set C

1. Complete the table, showing the values of $\frac{1}{x}$ (that is, $1 \div x$) for various values of x.

x	$^-4$	$^-2.5$	$^-2$	$^-1$	$^-0.1$	0.25	1	1.5	2	3	4
$\frac{1}{x}$											

2. Explain why you cannot use $x = 0$ as an input.

3. Graph the values in your table, with $\frac{1}{x}$ on the y-axis. Make the x-axis from $^-4$ to 4. Decide on a good scale for the y-axis.

4. Write the equation that your graph shows.

5. Since $\frac{1}{x}$ cannot be evaluated for $x = 0$, the graph of $y = \frac{1}{x}$ cannot cross the y-axis. The parts of the graph on either side of the y-axis can't be connected.

   a. Draw a smooth curve through the points that fall to the left of the y-axis to show the pattern. It may help to plot more points if you are having trouble deciding on the shape of the graph.

   b. Now draw a smooth curve through the points that fall to the right of the y-axis. Again, you may need to plot more points.

6. Use your graph to estimate the multiplicative inverse of 2.8. Check your answer with your calculator.

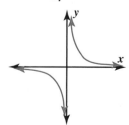

Just the facts

The graph of $y = \frac{1}{x}$ is called a hyperbola.

Problem Set D

Kenneth thought of a new *What's My Rule?* game.

For this game I'll use this rule:
$output = \frac{6}{n}$,
where *n* is the input.

1. Write Kenneth's rule in words.

2. Choose two values for *n*, one positive and one negative, and check that you and your partner agree on outputs for them.

3. Draw a graph for Kenneth's rule, using 10 inputs from ⁻4 to 4. If it makes sense, connect the points using a smooth curve.

4. How is your graph like the one you drew in Problem Set C? How is it different?

5. In another game, Liani used the rule *output* $= \frac{12}{n}$. Discuss with your partner what the graph for her rule might look like. Would it be similar to any graphs you've seen before?

Share & Summarize

1. Why does the graph of $y = \frac{1}{x}$ have two parts?

2. Imagine starting at Point (1, 1) on the graph of $y = \frac{1}{x}$ and moving along the graph so that the *x* values increase. What happens to the *y* values?

Investigation ▶3 Using Graphs to Estimate Solutions

You have learned how to solve equations using backtracking, guess-check-and-improve, and by doing the same thing to both sides. You have also used graphs to estimate solutions to problems.

For example, in Problem Set A, you used your graph of $y = x^2$ to estimate the values of x for which $x^2 = 7$. To make a good estimate from a graph, you need to draw the graph as accurately as possible.

MATERIALS
graph paper

Problem Set E

This little-known formula allows you to estimate the air temperature by measuring the speed of ants crawling around:

$$t = 15s + 3$$

where t is temperature in °C and s is the ants' speed in cm/s.

1. Draw a graph to show the relationship between temperature and speed given by the formula. Put s on the horizontal axis with values from 0 to 3.

2. Andrea timed some ants and estimated their speed to be 2.5 cm/s. Use your graph to estimate the temperature.

3. Early in the morning, Andrea estimated the ants' speed to be about 1.2 cm/s, but by late afternoon it was 2.0 cm/s. Use your graph to estimate the change in temperature during that time.

4. Write an equation you could solve to find an ant's speed when the temperature is 25°C. Use your graph to estimate the solution to your equation.

5. Use your graph to estimate how much the ants' speed would change if the temperature increased from 10°C to 20°C.

Just the **facts**

A temperature of 10°C is equivalent to 50°F. A temperature of 25°C is equivalent to 77°F.

Problem Set F

When an object is dropped, the formula $d = 4.9t^2$ shows how the distance in meters that the object has fallen is related to the time in seconds since it began to fall. Below is a graph of this relationship.

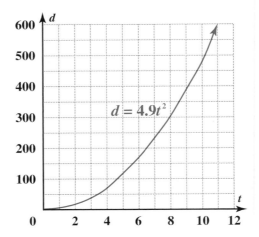

1. A child dropped a toy from a balcony, 60 meters above the ground. Write an equation you could solve to find how long it will take for the toy to hit the ground.

2. Use the graph to estimate a solution of your equation. Try to give your answer accurate to the nearest tenth of a second.

3. Solve your equation. Round your answer to the nearest tenth of a second. Is the answer close to your estimate?

4. Use the graph to estimate the solution of each equation, accurate to the nearest tenth of a second.

   a. $100 = 4.9t^2$ **b.** $490 = 4.9t^2$ **c.** $20 = 4.9t^2$

Problem Set G

Some musical instruments, like pipe organs and other wind instruments, create sound by causing columns of air to vibrate. Short columns of air vibrate quickly and make high-pitched sounds. Long columns of air vibrate slowly and make low-pitched sounds.

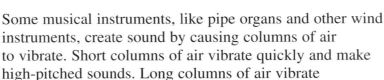

Just t h e **facts**

Flutes, clarinets, and trombones are wind instruments: air is blown into them to create sound. The organ is another wind instrument: pressing keys on the keyboard sends air through pipes, producing music.

Belita was trying to make a musical instrument from a series of pipes of different lengths. She wanted to find a mathematical relationship between the sound and the length of pipe. The *pitch* of a sound tells how high it is. Belita used a device that measures pitch. She blew into pipes of various lengths and recorded the pitches in a table.

Pitch (cycles per second)	64	128	192	261	300	395	438	512
Length of Pipe (centimeters)	80	41	26	19	18	13	12	10

Belita made a graph from her data, adding a smooth curve to fit the points.

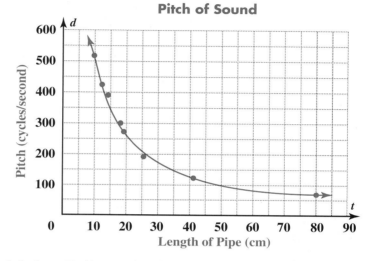

Pitch of Sound

1. Explain how Belita could justify drawing a curve through the points even though she only knew what happened for some pipe lengths.

2. Does the graph look like any of those you have studied in this lesson? If so, which ones?

3. Use the graph to predict the length of pipe Belita needs to produce a pitch of 160 cycles/second.

4. Use the graph to predict the pitch of a pipe 60 cm long.

Share & Summarize

Here is a graph of $y = \frac{x^2}{4} + 2$.

1. Use the graph to estimate the solution of the equation $6 = \frac{x^2}{4} + 2$.

2. Explain how you found your answer.

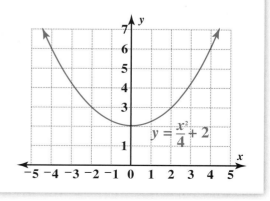

$y = \frac{x^2}{4} + 2$

Investigation 4 Doubling Up

Remember

A quantity grows exponentially if it is repeatedly multiplied by a number greater than 1.

In Chapter 3, you looked at situations that showed *exponential growth*. For example, you may remember the legend of the ruler who gave a subject a grain of rice on the first square on a chessboard. On each square after that, the ruler placed twice as many grains as on the previous square.

In this investigation, you will explore graphs of such relationships.

Think & Discuss

Try folding a large sheet of paper in half, then in half again, and then in half again, for a total of eight folds.

Are you able to fold the paper eight times? Why or why not?

MATERIALS
graph paper

Problem Set H

1. Suppose Kyle takes a sheet of paper and cuts it in half. Kyle then cuts both of those pieces in half. How many pieces would he have now?

2. Copy and complete the table.

Stage	0	1	2	3	4	5
Pieces	1	2				

3. Could the relationship between the stage and the number of pieces be linear? Why or why not? What is the pattern in the table?

4. Graph the data. Should you connect the points? Explain. If it makes sense to connect the points, draw a solid curve through them. Otherwise, draw a dashed curve through them.

5. Write a sentence or two describing the shape of your graph.

6. Consider the pattern in the table you created. You can describe the pattern with the equation $p = 2^t$, where t is the stage number and p is the number of pieces. The numeral 2 in the equation shows that doubling is involved. Use the equation to find the number of pieces after 10 rounds of cuts.

7. How many pieces would there be if Kyle could do 20 stages of cuts?

Problem Set I

In the late 1700s, before Uranus, Neptune, and Pluto had been discovered, Johann Daniel Titius and Johann Elert Bode made a conjecture that there existed a rule that would predict how far a planet was from the sun.

First they numbered the planets. They started with Earth as Planet 1, moving away from the sun as the planet number increased. Since Venus and Mercury are closer to the sun than Earth is, Venus is Planet 0 and Mercury is Planet $^-1$.

The distances from the planets to the sun are often expressed in *astronomical units,* or AU for short. One astronomical unit is the average distance from Earth to the sun. (The average is used because the distance varies during the year.)

1. Copy and complete the table, which shows how Titius and Bode predicted the distances from the sun. When they first invented the rule, they found it worked only if they ignored 3 when numbering the planets. Skip this row for now; you will complete it later.

Distance from the Sun

Planet	Planet Number, n	Predicted Distance, d
Mercury	$^-1$	$0.4 + 0.3 \cdot 2^{-1} = 0.55$ AU
Venus	0	$0.4 + 0.3 \cdot 2^0 = 0.7$ AU
Earth	1	$0.4 + 0.3 \cdot 2^1 =$
Mars	2	$0.4 + 0.3 \cdot 2^2 =$
Jupiter	4	$0.4 + 0.3 \cdot 2^4 =$
Saturn	5	$0.4 + 0.3 \cdot 2^5 =$

2. What is the rule? Write it first in words, and then in symbols.

Kitt Peak National Observatory
Tohono O'Odham Reservation, Arizona

3. After Titius and Bode found their rule, astronomers used it to search for a planet with the number 3. This led to the discovery in 1801 of Ceres, an asteroid in the asteroid belt between Mars and Jupiter. Predict the distance the asteroids are from the sun, and complete Row 3 of your table.

4. In 1781, Uranus was discovered using Titius and Bode's rule while searching for planet number 6. Predict its distance from the sun, and add a line to your table for this planet.

5. Plot the points from the table on a grid, and draw a dashed line through them to make a smooth curve.

6. The table gives the actual average distance from the sun to each planet, as calculated by techniques that bounce radio waves off objects in space and measure the time it takes for the signals to return. On the same set of axes, plot points to represent the actual distances. Does the rule seem to give a result close to reality?

Distance from the Sun

Planet	Planet Number, n	Actual Distance, d
Mercury	-1	0.39 AU
Venus	0	0.72 AU
Earth	1	1.0 AU
Mars	2	1.5 AU
Asteroids	3	2.9 AU
Jupiter	4	5.2 AU
Saturn	5	9.5 AU
Uranus	6	19.2 AU

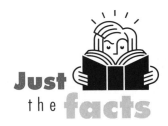

Remember

A dashed line is often used to show the shape of a curve even though the in-between points have no real meaning.

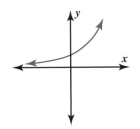

Just the facts

The graph of $y = 2^x$ is called an *exponential curve*.

Share & Summarize

1. How does the graph of the doubling pattern differ from a linear pattern?

2. The equation for a doubling pattern has the form $y = 2^x$. The equation for a linear pattern has the form $y = ax + b$. Use these equations to explain the difference in the graphs.

On Your Own Exercises

Practice & Apply

1. **Geometry** The surface area (in square centimeters) of a cube with an edge length of d cm is given by the formula $A = 6d^2$.

   a. Use the formula to complete the table.

d (cm)	0	0.5	1	1.5	2	2.5	3
A (cm²)							

   b. Plot the points in the table. If it is reasonable to do so, connect them with a smooth curve.

   c. Compare your graph to the graph of $y = x^2$, which you created in Problem Set A.

2. For a game of *What's My Rule?* Takiyah decided to use the rule *output* $= n^2 - 3$.

   a. What shape will the graph of Takiyah's rule have?

   b. Without plotting any points, make a rough sketch of the graph of Takiyah's rule.

3. **Geometry** Melisenda and Jade were drawing rectangles, each with an area of 20 square centimeters.

   H | area = 20 cm²

   B

   a. Write a formula for finding the base B given the height H.

   b. What values make sense for H? Must H be a whole number? Can it be negative?

   c. Use your calculator to find at least 10 possible values for B and H, including decimals or fractions if they make sense.

   d. Draw a graph of the data from your table.

   e. Would it make sense to connect the points on your graph? Explain. If it makes sense, do so.

4. Malik's rule for *What's My Rule?* was *output* $= \frac{20}{n}$. Without plotting points, make a rough sketch of the graph of Malik's rule.

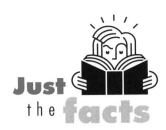

Just the facts

In 1999, the euro was introduced as a new currency to be used by all countries in the European Union.

5. The currency in the United Kingdom is the pound (symbol £). On a particular day in June 2003, the rule for converting pounds to U.S. dollars was $D = 1.65P$, where D was the price in U.S. dollars and P was the price in pounds.

    a. Draw a graph of this relationship. Put P on the horizontal axis, and show amounts in pounds up to £150.

    b. On that day, a British collector on the Internet advertised a set of miniature cars for £125, including postage. Write an equation you could solve to find how much this was in U.S. dollars, and solve your equation. Use your graph to check your solution.

    c. On that same day in June, Megan exchanged $120 she had saved toward a holiday in Wales. Write and solve an equation to find how many pounds she received. Use your graph to check your solution.

    d. What are the advantages of using either the graph or the rule to solve the equations you wrote?

6. Geometry The area A of a circle of radius r can be found using the formula $A = \pi r^2$. This graph shows the relationship.

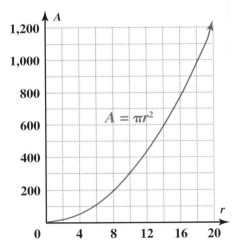

    a. A pastry recipe claims that a certain quantity of dough will roll out to an area of 1,000 square inches. A cook plans to make a large circular pie crust. Write and solve an equation to find what radius he can expect the pie crust to have. Use the graph to check your solution.

    b. A pizza shop sells a 10-inch pizza for $6.50 and a 20-inch pizza for $15.00. Use the graph to help you decide which pizza costs more per square inch of pizza. Explain.

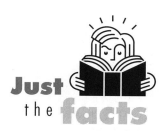
7. Scuba divers usually descend to a desired depth and spend some time exploring there. Although the amount of oxygen needed varies from person to person, the following rule gives an estimate of how long an average diver can stay at a particular depth. T is the maximum time under water in minutes, V is the volume of air available in cubic meters, and D is the depth in meters.

$$T = \frac{120V}{D}$$

a. Michael has 1 m^3 of air in his tank. Write an equation to show the relationship between the time T he can stay under water and the depth D to which he can dive.

b. Find 10 possible pairs of values for T and D for Michael's equation. Draw a graph of the relationship.

c. Use your graph to find how long Michael can stay at a 25-m depth. Check your answer using the equation you wrote in Part a.

d. Use your graph to find how deep Michael can dive if he wishes to stay under water for 15 minutes at the maximum possible depth. Check your answer by using the formula.

8. Social Studies The table shows the United Nations estimates for the world population from 1750 to 2000.

a. Plot these data on a graph. Put the year on the horizontal axis, from 1750 to 2100, and make the scale on the vertical axis go up to 10 billion people. Draw as smooth a curve as you can through your points, extending the curve beyond 2000.

b. Use your graph to estimate the world population in the year you were born and in the year you will turn 18. What is the likely increase in population over that period?

c. Use your graph to estimate the 1980 world population.

d. The actual 1980 world population was about 4,440 million, or 4.44 billion. How close was your estimate?

World Population

Year	Population (millions)
1750	790
1800	980
1850	1,260
1900	1,650
1950	2,520
1970	3,700
1990	5,270
2000	6,100

9. Businesses often set prices for their products. Businesses would like to sell many products at a high price—but if the price is too high, people won't buy the products.

Kate and her friends earn money by selling greeting cards they create on a computer. To help decide how much to charge, they surveyed 100 people about the price they would pay for a card. They organized the information in a table.

Price ($)	0	1	2	3	4	5
Number of People Surveyed Who Would Buy One Card	100	80	60	40	20	0
Income ($)						

a. Copy the table, and complete it with the income the friends could make for each price. For example, if they set the price at $1.00, they could sell 80 cards to these 100 people for a total of $80. Explain how you found your answers.

b. Make a graph showing the income for the various prices. Connect the points in your graph with a smooth, dashed curve. What shape does the curve have?

c. Look closely at the curve you drew. What price do you think the friends should charge to earn the most income?

d. Write an expression that gives the number of cards sold for any price p.

e. If the friends sell n cards for p dollars each, what will their income be?

f. Replace the n in your expression from Part e with the expression you wrote for Part d. Use the distributive property to rewrite your new expression.

g. Consider your rewritten expression in Part f and the graph you made in Part b. How do the expression and the graph compare to the equations and graphs from Investigation 1?

10. **Geometry** Sean and Fiona are planning a vegetable garden. They figure they need 36 square yards to grow all they would like. They want the garden to have a rectangular shape, but they can't decide on its length and width.

a. Find at least 10 sets of lengths and widths the friends could use to get a total area of 36 yd^2. Use your data to make a graph, with length on the horizontal axis. Connect the points to form a smooth curve, if it makes sense to do so.

Sean and Fiona decided to enclose their garden with a fence to protect it from animals. To keep material costs down, they want the rectangle to have the least possible perimeter.

b. Calculate the perimeter of each rectangle you found in Part a. Label each plotted point on your graph with the corresponding perimeter. Do you see a pattern in the labels?

c. What point on the graph do you think would have the least perimeter? What size garden should Sean and Fiona create?

11. A video games arcade has a special deal on school holidays. Instead of paying $1 for each game, you pay $2.40 admission and 40¢ per game.

a. Copy and complete the table.

Number of Games	1	2	4	7	10
Regular Price ($)	1.00				
Holiday Price ($)	2.80				

b. Plot the data in your table on one set of axes. Make two graphs that show the costs of playing different numbers of games at the regular price and at the holiday price. Connect the points using solid or dashed lines, whichever make more sense. Explain why the type of lines you chose makes sense.

c. Find an equation for the cost of playing any number of games at the holiday price. Use your equation to determine how much it would cost to play three games.

d. Suppose you have $6 to play video games on a holiday. Write an equation you could use to determine how many games you could play. Use your graph to estimate the solution of the equation.

e. For which numbers of games are the holiday prices lower than the regular prices? For which numbers of games are they higher?

12. Physical Science David and Angela were experimenting with gravity and falling objects. They dropped stones from four heights and carefully timed how long it took them to hit the ground.

These are their results.

Height (ft)	0	5.2	11.8	14.7	20.1	28.7
Time (s)	0	0.57	0.86	0.96	1.12	1.34

a. Plot the results on a graph with time on the horizontal axis and height on the vertical axis. Connect the points using a smooth curve.

Angela thought the curve looked like the graph of $y = x^2$. Using H for height and t for time, she suggested that an equation in the form $H = Kt^2$ might fit the data.

b. Choose a time other than 0 from the table. Substitute the time and the corresponding height into the equation $H = Kt^2$, and solve the new equation.

c. Repeat Part b for the other times except 0. What might be a good value for K?

d. Rewrite Angela's equation, using your value from Part c for K.

e. Angela and David realized they could use the graph to test their throwing arms. One of them threw a baseball as high as possible. The other watched carefully until the ball reached its highest point, and then timed it until it hit the ground. By looking at the graph, they could figure out how high they had thrown the ball.

These are the times the baseball took to fall. Use the graph to estimate how high the ball rose each time. Check your answers using your equation from Part d.

Time (s)	1.05	1.20	1.43	1.31
Height (ft)				

13. Maya took a sheet of paper about 0.1 millimeter thick and cut it in half. She stacked the pieces, one on top of the other.

a. How thick is Maya's two-sheet stack?

b. Maya cut each of the pieces in half and stacked all the new pieces. How thick is the new stack?

c. Suppose Maya continues to cut all the pieces, stack them, and then measure the stack's height. Write an equation that gives the thickness t after any number of cut-and-stacks c.

d. How thick will the pile be after 10 cut-and-stacks? After 20? Answer first in millimeters, and then convert to meters.

e. How many cut-and-stacks will Maya need to make a thickness of at least 1 kilometer? How thick will the pile be?

f. The moon is about 400,000 km from Earth. Find the number of cut-and-stacks needed for Maya's pile of paper to reach the moon.

14. Cut a long strip about 1 centimeter wide from a sheet of paper with no creases. Fold it in half, end to end, once. When you open it, there will be one crease.

a. Continue to fold the strip in half, end to end. At each stage, open it and count how many creases it has. Complete the table.

Times Folded, f	0	1	2	3	4	5
Creases, c	0	1				

b. Graph this relationship.

c. Write an equation to describe the relationship between the number of creases and the number of folds. Hint: It may help you see the pattern if you add 1 to each of the c values.

Just the facts

The moon's orbit about Earth is elliptical, not circular, so its distance from Earth changes. The nearest it comes to Earth (this point in its orbit is called the *perigee*) is about 363,000 km; the farthest (the *apogee*) is about 405,000 km.

In your
own
words

Explain how a graph can help you solve an equation. Make up a problem of your own to use as an example.

15. Julian and Gina conducted a game with their class of 32 students. The students divided into 16 pairs. One student in each pair tossed the coin, while the other called "heads" or "tails." The 16 winners of the tosses paired off and played again.

This knockout competition continued until only one person was left. That person was declared the winner of the game.

a. How many rounds were played before the game winner could be determined?

b. If the game is played a second time, do you think the winner from the first game is more likely, just as likely, or less likely to win the second game than anyone else in the class? Explain.

c. Suppose the students tried the game with the entire school population of 512 students. How many rounds would it take to find the winner?

d. Supposed they tried it in a city of 1,048,576 citizens. How many rounds would it take to find the winner?

e. About how many rounds would it take to find the winner if they could get the whole world to play? Assume the world population is about 6 billion.

Mixed Review

Evaluate each expression.

16. $\frac{11}{12} - \frac{7}{12} + \frac{2}{3}$

17. $\frac{2}{7} + \frac{1}{8} + \frac{1}{2}$

18. $\frac{2}{9} - \frac{1}{7} + \frac{1}{3}$

19. $\frac{1}{6} + \frac{1}{2} - \frac{1}{12}$

Write each number in scientific notation.

20. 237

21. 637,800

22. 0.057

Evaluate each expression. Write your answers in standard notation.

23. $(4.5 \times 10^{-3}) - (2.5 \times 10^{-5})$

24. $(1.7 \times 10^{-4}) + (9.2 \times 10^{-5})$

Geometry Find each missing side length.

25.

26.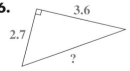

27.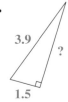

Nutrition Information

Serving Size:
  4 cookies (28 g)
Calories: 120

Amount per Serving
Total Fat 4 g
  Saturated Fat 2 g
Total Carbohydrate 19 g
  Dietary Fiber 0 g
  Sugars 7 g
Protein 1 g

28. Nutrition information for Damon's favorite cookies are shown on the label.

a. Damon usually eats eight cookies when he comes home from school. How many servings is this, according to the package?

b. One gram of fat contains 9 calories. How many calories in one serving of these cookies are from fat?

c. What percentage of the calories in one serving of these cookies are from fat? Show how you found your answer.

d. One gram of protein contains 4 calories. What percentage of the calories in one serving of these cookies are from protein?

e. If Damon decides to consume only 1.5 grams of saturated fat in his afternoon cookie snack, how many cookies can he have?

29. Measurement How many days is 1,000,000 seconds?

30. How many years is 1,000,000 hours?

31. How many decades is 3,500 years?

32. Every day for two weeks, Amato kept track of how many times he heard someone say "you know" during a class discussion. He made a graph of his data.

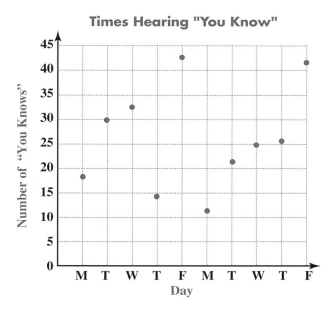

a. How many "you knows" did Amato hear in all?

b. On what day of the week did Amato hear the most "you knows"? What percentage of all the "you knows" were heard on this day?

Repeating Relationships

Think about how many things repeat in your life. Perhaps you have a daily routine. You attend school for five days and then don't for two days. Seasons come and go. In this lesson, you will explore some relationships that repeat.

Explore

Find the missing values in the table for the number the minute hand on a circular clock face would point to for each given time.

Time	Minute-Hand Position
12:00	12
12:15	3
12:30	
12:45	
1:00	
1:15	
1:30	
1:45	

Time	Minute-Hand Position
2:00	
2:15	
2:30	
2:45	
3:00	
3:15	
3:30	
3:45	

You probably noticed that the table shows a repeating pattern. What is the pattern? Explain why it repeats every hour.

Plot the data from the table. Put time on the horizontal axis and minute-hand position on the vertical axis.

Think of some other repeating patterns you could see in a clock, and graphs you could produce from those patterns. Write at least two sets of instructions for producing a clock pattern that repeats, and state how often it repeats.

Investigation 1 Repeating Patterns

In this investigation, you will look at simple repeating patterns to learn something about their graphs.

MATERIALS
graph paper

Problem Set A

Damian started his summer job early, beginning work after school in May. He gets a paycheck every four weeks, which he deposits in his bank account. Each week he takes out $30 for spending money.

Damian wants to figure out how much money he will have saved by the beginning of his family vacation at the end of July. The table shows his predictions of transactions in his bank account for May through July.

Account Balance

Date	Deposit	Withdrawal	Balance
May 4	$400	$30	$570
May 11		$30	$540
May 18		$30	$510
May 25		$30	$480
Jun 1	$400	$30	$850
Jun 8		$30	$820
Jun 15		$30	$790
Jun 22		$30	$760
Jun 29	$400	$30	$1,130
Jul 6		$30	$1,100
Jul 13		$30	$1,070
Jul 20		$30	$1,040

1. What must the balance have been before May 4?

2. Plot the data in the "Balance" column. Put the date on the horizontal axis and the balance on the vertical axis.

3. Think about whether you should connect the points on your graph. Damian's withdrawals and deposits occurred all at once, not in fractional amounts. That means the graph would have to show sudden changes, not gradual changes.

   a. How much money would be in the account on May 5, 6, 7, 8, 9, and 10?

   b. Add points for those days to your graph.

   c. Does it make sense to connect those points? Explain. If it makes sense, do so.

   d. Should you connect the points for May 10 and May 11? Explain. If it makes sense, do so.

4. Revise your graph until you are convinced it shows the relationship for *all* times from May 4 to July 26. Discuss it with your partner.

5. What is Damian's balance on July 15? If the pattern continues, what do you expect it to be on August 3? On August 9?

MATERIALS
graph paper

Problem Set B

When Darnell is nervous, he often paces slowly and steadily across a room. Suppose Darnell started from the left wall of his room and walked across to the right wall. When he reached the right wall, he turned immediately around and paced back to the left wall, then walked again from left to right, and so on.

1. Which graph best shows Darnell's distance from the left wall at any time? Why?

Graph a

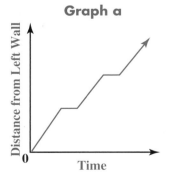

Graph b

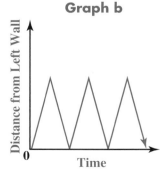

Graph c

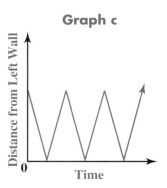

2. The other graphs also describe a person walking in a room. Describe the situation shown by each of the other graphs.

A child is playing on a slide. He climbs up, sits down, and then starts to slide, gaining speed as he descends. At the bottom, he gets up quickly, runs around to the ladder, and climbs up again. He does this over and over.

3. Sketch a graph to show how the height of the child's feet above the ground is related to the time.

4. Would your graph look different if, instead of using the height of the child's feet, you used the height of his head? Sketch this graph on the same set of axes as the graph for Problem 3.

Share & Summarize

1. Give an example of a repeating number pattern involving money.

2. What would the graph of your pattern look like?

Investigation Repeating Patterns in Life

In each repeating relationship you have graphed, the graph showed the same basic shape repeated over and over at fixed intervals of time. The following problems show how one particular kind of repeating pattern can appear in two different situations.

MATERIALS
graph paper

Problem Set C

As a bicycle wheel turns, its tire valve is at different heights from the ground. Imagine the following experiment:

Chalk marks are placed on a mountain bike tire, one mark at every fourth spoke. The bicycle rolls along slowly. Whenever a chalk mark touches the ground, the height of the valve is measured in inches. When the first measurement is taken (Measurement 0), the valve is 8 inches above the ground. The table lists the valve heights from Measurement 0 to Measurement 16.

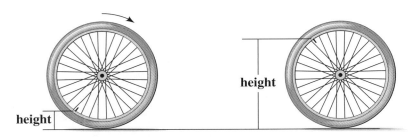

Measurement	0	1	2	3	4	5	6	7	8	9	10	11	12	13	14	15	16
Valve Height (in.)	8	3.5	2.5	8	18	22.5	21.5	15.5	8	3.5	2.5	8	18	22.5	21.5	15.5	8

Just the **facts**

Most mountain bike wheels have 32 spokes. Wheels are also made with 28 spokes (for someone who wants a lighter wheel) and 36 spokes (for someone who wants a more durable wheel).

1. Draw a graph of the data. Connect the points with a smooth curve.

2. Describe the shape of your graph. Use the tire situation to explain why the graph has this shape.

Patterns are very useful for making predictions. For example, the repeating pattern of the tides is so regular that tide times can be predicted for many years ahead. People who live near the sea often use tide tables.

Problem Set D

The graph shows the height of tides over a week in a particular seaport. When the height of the water reaches a high point and just starts to recede, the ocean is said to be at *high tide*. When the height reaches a low point, the ocean is at *low tide*. The same basic shape is repeated, but the pattern is not exactly the same each day.

The graph starts on Monday at 10 A.M. Every vertical grid line represents 2 hours.

Height of Tides

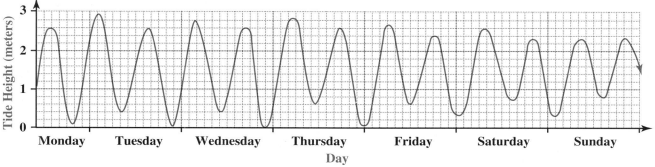

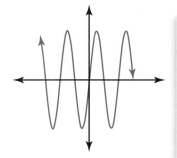

1. Look at how the tide height changed on Tuesday.

  a. When were the high tides? Were they the same height? Explain.

  b. When were the low tides? Were they the same height? Explain.

2. How often does the pattern repeat?

3. For what situations might people need to know the tide patterns?

Share & Summarize

1. Give an example of a repeating pattern you might encounter.

2. How could you make a graph from your pattern? What would the graph of your pattern look like?

Investigation 3 ▶ Repeating Patterns in the Weather

Because certain weather patterns repeat year after year, it is possible to describe the typical weather for a location at a given time of year.

MATERIALS
graph paper

Problem Set E

The table lists the temperatures every 2 hours over a three-day period in a recent month for three cities: Norfolk, Virginia; San Francisco, California; and St. Louis, Missouri.

Source: National Climatic Data Center, OnLine Climate Data, www.ncdc.noaa.gov.

	Temperature (°F)								
	Norfolk, VA			San Francisco, CA			St. Louis, MO		
Time	Day 1	Day 2	Day 3	Day 1	Day 2	Day 3	Day 1	Day 2	Day 3
1 A.M.	37.9	55.0	41.0	45.0	46.9	46.9	63.0	30.0	23.0
3 A.M.	35.1	55.0	41.0	46.0	46.9	45.0	62.1	30.0	21.0
5 A.M.	34.0	55.9	39.0	41.0	46.0	44.1	63.0	28.9	21.0
7 A.M.	35.1	61.0	35.1	41.0	42.1	44.1	66.0	34.0	24.1
9 A.M.	46.0	64.9	37.9	45.0	44.1	48.9	71.1	41.0	28.9
11 A.M.	57.9	66.9	45.0	46.9	46.9	48.9	69.1	43.0	35.1
1 P.M.	60.1	66.0	46.9	50.0	50.0	50.0	60.1	39.9	36.0
3 P.M.	63.0	72.0	45.0	52.0	51.1	55.0	44.1	36.0	37.9
5 P.M.	61.0	70.0	43.0	52.0	51.1	48.9	37.0	33.1	34.0
7 P.M.	55.0	68.0	39.0	51.1	50.0	46.9	34.0	32.0	30.0
9 P.M.	55.9	51.1	34.0	50.0	48.9	45.0	33.1	31.1	28.9
11 P.M.	54.0	45.0	30.9	48.0	48.0	46.9	30.9	30.0	31.1

1. Working with your group, graph the data for each city. Put the times for the entire three days on the horizontal axis, and the temperature on the vertical axis. Use the same scales for each of the three graphs so you will be able to compare the graphs later. Label the temperature axis in a way that will account for the range of temperatures in the data. Connect your plotted points with line segments.

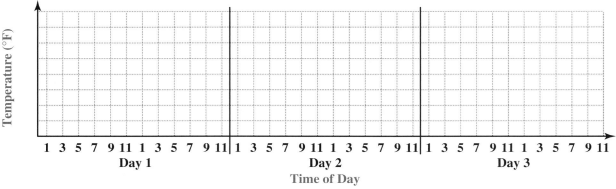

2. Explain how your graphs show the following things:

   a. the daily maximum temperature

   b. the daily minimum temperature

   c. when temperatures are changing most rapidly

3. What do the patterns indicate might happen the next day (Day 4) in San Francisco?

4. How are the three graphs similar? How do they differ?

5. Why do you think the similarities you noted occur?

6. Make a rough sketch of the likely temperature patterns for where you live. Your graph should show an estimate of the maximum daily temperatures over an entire year. Discuss your graph with a partner and explain why it is shaped the way it is.

Share & Summarize

1. Describe some repeating patterns involving weather.

2. What advantage does graphing such data have over just using a table?

3. What do the slopes of the line segments in the graphs you created for Problem Set E show?

On Your Own Exercises

1. Every Tuesday Pablo deposits his allowance of $10 into his bank account. Every Thursday he deposits the $15 he earns from his part-time job. Then, every Friday, he takes out $12 for weekend spending.

   a. Draw a graph that shows the amount in Pablo's account over the first three weeks (21 days) after opening the account and making his first Tuesday deposit (on Day 1).

   b. How often does the graph repeat? How are the repeated sections different from each other?

   c. In which week will Pablo's account first reach $100? Explain.

2. **Sports** Tennis is played on a court 78 feet long, with a net in the center. To warm up before a game, players will stand at either end of the court and hit the ball back and forth. Suppose two players are warming up in this way. Assume the ball travels at a constant speed of 50 ft/s.

   a. Draw a graph that shows the ball's distance from the first player for the 10 s after the player's first hit. Use a grid like this one.

   b. On the same grid, use another color to draw a graph that shows the ball's distance from the *net* over the same period of time.

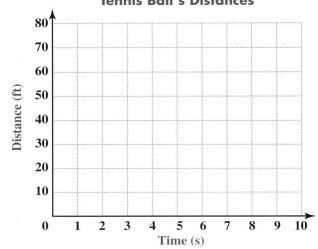

Tennis Ball's Distances

3. Tala and her friend Jing were playing on the swings. Tala noticed that Jing went up, came down, and then rose again in a repeating pattern. Which of these three graphs do you think best represents Jing's height over time? Why?

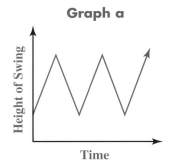

Graph a

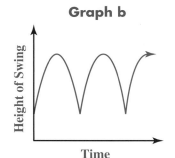

Graph b

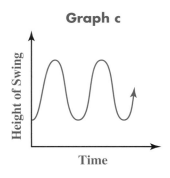

Graph c

4. When the carnival was in town, Maria watched her sister Rosi ride the carousel. As the carousel turned, Rosi moved away from Maria, and then back toward her again. The table shows how the distance between the girls changed as the carousel revolved. Data for the first 24 seconds of the ride are shown.

Time (s)	0	4	8	12	16	20	24	28	32	36	40	44	48
Distance (ft)	13	7	13	20.2	23	20.2	13						

a. Plot the data, with times up to 50 seconds on the horizontal axis.

b. Describe the pattern in your graph. What do you think will happen over the next 24 s?

c. Complete the table, assuming the movement continues in the same pattern.

d. Add the new points to your graph. If it makes sense to do so, draw a smooth curve connecting the points in your graph.

5. Earth Science Rain tends to fall at certain times of the year in many areas, and even at certain times of the day in some. The graph shows the monthly rainfall for Honolulu, Hawaii, from 1993 to 1996. January 1993 is Month 1, and December 1996 is Month 48.

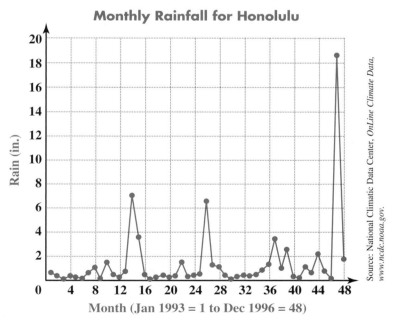

Monthly Rainfall for Honolulu

Month (Jan 1993 = 1 to Dec 1996 = 48)

Source: National Climatic Data Center, *OnLine Climate Data, www.ncdc.noaa.gov.*

a. Describe the pattern in the graph.

b. Are there any months that don't seem to follow the pattern well? Explain.

c. From the pattern in the graph, how often should the people in Honolulu expect a lot of rain? About when would that occur?

6. **Earth Science** Although the temperature in Hawaii doesn't vary as much as it does in other places, the seasons still have an effect. The table gives Honolulu's average maximum temperature for every other month from 1993 to 1997.

Average Maximum Temperature (°F)

Month	1998	1999	2000	2001	2002
January	81.6	79.6	78.6	82.1	80.2
March	82.3	80.7	82.0	81.6	80.4
May	82.7	83.2	85.1	84.9	83.8
July	85.8	85.8	87.4	87.9	87.1
September	87.8	87.2	87.1	88.8	87.9
November	83.5	83.0	83.1	83.1	84.0

Source: Western Regional Climate Center

a. Plot these data, with time on the horizontal axis and temperatures from 75°F to 95°F on the vertical axis. Connect the points with line segments.

b. Describe the pattern in your graph.

c. At what time of year does Honolulu experience its highest temperatures? Its lowest temperatures?

Connect  **Extend**

7. Tony solved three pattern puzzles. Each puzzle had seven numbered cards. To solve each puzzle, Tony had to arrange the cards to show a repeating pattern that involved addition and subtraction.

a. The first puzzle had seven cards, numbered 2, 4, 6, 7, 8, 9, and 11. Tony's solution is shown below. What is the repeating pattern in this solution?

2	7	4	9	6	11	8

b. The cards in the second pattern puzzle were numbered 7, 8, 9, 10, 12, 13, and 14. Tony found a solution that started with 10 and ended with 7. Find Tony's solution, and describe the repeating pattern.

10						7

c. The third puzzle's cards were numbered 0, 1, 2, 3, 4, 5, and 7. Tony's solution started with 1 and ended with 7. Find his solution, and describe the pattern.

1						7

d. Each of the three patterns can be continued to the right and to the left. For example, in the first puzzle, the first five extra numbers to the right would be 13, 10, 15, 12, and 17. The first five extra numbers to the left, from left to right, would be 1, ⁻2, 3, 0, and 5. So the fifth extra number to the right is 17 and the fifth extra number to the left is 1.

For each of the other two puzzles, find the fifth extra number to the right and the fifth extra number to the left.

8. Julie and Trevor were timing the traffic signal of one of the lights at the intersection by their school. They drew a timeline of their data.

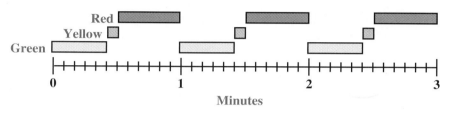

a. How many seconds is one complete cycle?

b. Julie and Trevor noticed that when the light in one direction turned red, there was a delay of about 2 s before the light in the other direction turned green. Why do you think this delay is built into the sequence?

c. Given a delay of 2 s, how would the timeline for the light in the other direction look? That is, when would it be red, green, and yellow, compared to the timeline for the first light? Create a timeline to show this light sequence. Assume the yellow lights stay on for the same amount of time in each direction.

9. Astronomy Halley's Comet is a regular visitor to our skies. It completes its elongated orbit about once every 76 years. Its last visit to Earth was in 1986.

a. When will the next visit be? How old do you expect to be then?

b. Find the years of the three appearances of Halley's comet prior to 1986.

c. Scientists can use patterns to do "detective work" and explain past events. In about 1304, the Italian artist Giotto referred in a painting to a wonderful comet he had seen in 1301. Could this have been Halley's comet? Why or why not?

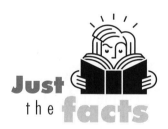

Just the **facts**

Halley's comet is named for Edmond Halley (1656–1742), a British astronomer. He discovered the pattern of its visits and predicted its return in 1758—but didn't live to see the comet that bears his name.

10. Ecology Many animal species depend on other animals for food. For example, a fox's main source of nourishment is often rabbit. The fox is a *predator,* and the rabbit is its *prey.*

The populations over time of the foxes and rabbits in one forest might be modeled by the graphs below. Compare the two graphs.

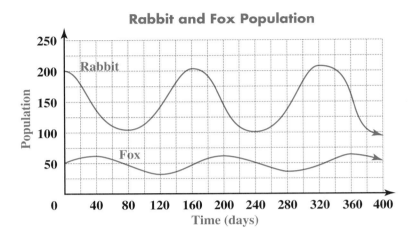

Rabbit and Fox Population

a. How often do the patterns repeat?

b. How are the graphs alike? How are they different?

c. Find the times when the fox population is changing fastest, and look at the rabbit population at those times. What do you notice?

d. Challenge Notice that when the fox population is dropping fastest, the rabbit population is at its lowest point.

   i. What does a drop in the fox population do to the rabbit population? Why?

   ii. What would a low rabbit population mean for the foxes?

   iii. Use Parts i and ii to explain what is happening in the part of the graph where the fox population is dropping.

11. **Earth Science** You may have heard of the El Niño weather effect. El Niño is a warming of the central and eastern Pacific Ocean. It affects weather in many countries, including the United States.

One climate measure is the Southern Oscillation Index (SOI), which is based on the difference in air pressure taken at the islands of Tahiti and in Darwin, Australia. When the index is negative, the West Coast of America usually experiences wet weather, and Australia usually experiences dry weather. When the index is positive, it tends to be drier in the Americas and wetter in Australia. Sustained negative values for the SOI accompany El Niño.

The table lists the approximate SOI measures every month from 1996 to 1998.

SOI Measures

	1996	1997	1998
January	8.4	4.1	⁻23.5
February	1.1	13.3	⁻19.2
March	6.2	⁻8.5	⁻28.5
April	7.8	⁻16.2	⁻24.4
May	1.3	⁻22.4	0.5
June	13.9	⁻24.1	9.9
July	6.8	⁻9.5	14.6
August	4.6	⁻19.8	9.8
September	6.9	⁻14.8	11.1
October	4.2	⁻17.8	10.9
November	⁻0.1	⁻15.2	12.5
December	7.2	⁻9.1	13.3

Source: SOI archives, Australia Bureau of Meteorology, *www.bom.gov.au*

a. Draw a graph of these data to show the pattern. Put the month on the horizontal axis.

b. A significant El Niño effect hit in 1997 and 1998 when the SOI was negative for a long period. During that time, Australia suffered a drought. It was extremely wet in the Americas, causing massive mud slides in California, among other problems. Between what approximate dates (month and year) did this occur?

In y o u r
own
words

What characteristics do the graphs of repeating patterns share? What do these kinds of graphs allow you to do with data points that may not be given?

12. A city has kept temperature records for over 100 years. From the records, someone calculated these statistics for the month of January:

> *Average daily temperature:* 26°C
>
> *Percent of days that vary from the average:*
> - within 5°C of average: 60%
> - within 10°C of average: 85%
> - within 15°C of average: 100%

a. Based on these long-term patterns for January, how many days would you expect the temperature to be from 21°C to 31°C?

b. How many days would you expect the temperature to be from 16°C to 36°C?

c. How many days would you expect the temperature to be above 36°C or below 16°C?

d. Do you think the temperature in January could ever be 40°C? Explain your answer.

Mixed Review

Evaluate each expression for $k = 5$.

13. $\frac{1}{k} + \frac{k}{3}$
14. 1.2×10^k
15. $\sqrt{k \cdot k}$

16. $3k^k$
17. $k^{2k} \times k^{-k}$
18. $\frac{4^k}{4^{k-2}}$

19. Geometry If you drew a quadrilateral using Points *A–D* as the vertices, what would the figure's perimeter be?

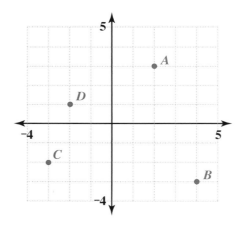

20. Rob has a free-throw average of 51%. In his next basketball game, he makes 4 out of 7 free-throws he attempts. Did he do better or worse than his average? Explain.

21. Social Studies The landlocked South American country of Paraguay is divided into 17 regions called *departments*. The table lists the area and population of each department.

Department	Area (km^2)	Population (2002 census)
Alto Paraguay	82,349	15,008
Alto Paraná	14,895	563,042
Amambay	12,933	113,888
Boquerón	91,669	45,617
Caaguazú	11,474	448,983
Caazapá	9,496	139,241
Canendiyú	14,667	140,551
Central	2,465	1,363,399
Concepción	18,051	180,277
Cordillera	4,948	234,805
Guairá	3,846	176,933
Itapúa	16,525	463,410
Misiones	9,556	103,633
Ñeembucú	12,147	76,738
Paraguarí	8,705	226,514
Presidente Hayes	72,907	81,876
San Pedro	20,002	318,787

Source: www.citypopulation.de

a. Which department has the greatest population density (people per square kilometer)? Which has the least?

b. How many times greater is the department with the most area than the department with the least?

c. Which department has the median area? What is that area?

d. Which department has the median population? What is that population?

22. The Paraguay currency unit is the *guaraní*. In June 2003, 1 U.S. dollar was worth about 6,119 guaraní.

a. How many guaraní would you have received in exchange for 40 U.S. dollars?

b. How many U.S. dollars would you have received in exchange for 1,000,000 guaraní?

Chapter Summary

VOCABULARY
line graph
multiplicative inverse

In this chapter, you worked with several types of graphs. First you looked at graphs that show how one variable changes over time. Then you saw how relationships with a square, with an inverse, or with a variable as an exponent look when graphed.

You used graphs to solve problems for different situations. You also saw situations—including some related to weather—that involve repeating patterns. Some graphs, such as those for inverse relationships and bank-account balances, had points that could not be connected.

MATERIALS
graph paper

Strategies and Applications

The questions in this section will help you review and apply the important ideas and strategies developed in this chapter.

Interpreting graphs over time

1. Luis's birthday party began around 2:00 P.M. Kasinda noticed that the noise level changed as the party progressed. Use the following events to sketch a graph that may show how the noise level rose and fell over time.

   • Most of the guests arrived between 2:00 P.M. and 2:35 P.M.

   • The guests sang "Happy Birthday" to Luis at about 2:55 P.M.

   • Luis blew out the candles on his cake at about 2:58 P.M.

   • Everyone applauded after Luis blew out the candles.

   • At 3:30 P.M., Kasinda turned up the music so people could dance.

2. Kenyon dropped a ball from a height of about 10 meters. Lucetta carefully measured the maximum heights of the ball as it bounced, and Kenyon timed when it reached each maximum height. Using what they know about the way dropped objects fall, they created this graph.

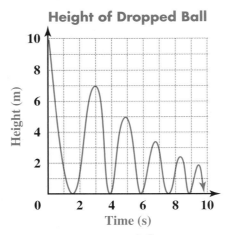

Height of Dropped Ball

impactmath.com/chapter_test

a. Explain why the graph looks as it does.

b. About what height does the ball reach on its first bounce?

c. About how long does the second bounce last? That is, how much time passes between the second time the ball hits the ground and the third time?

Recognizing graphs of square, inverse, and exponential relationships

3. Think about the similarities and the differences among the formulas in the box.

$s = 9.8t$	the speed s in meters per second of a falling body after t seconds
$A = \pi r^2$	the area A of a circle with radius r
$d = 4.9t^2$	the distance d in meters a falling body has fallen after t seconds
$B = 100(1.04^n)$	the balance B in a bank account after n years if \$100 is left in the account and the account pays 4% interest compounded annually
$W = \frac{36}{L}$	the width W of a rectangle with length L and area 36 square units
$S = 6L^2$	the surface area S of a cube with edge length L
$B = \frac{1}{4}F$	the amount B of butter needed in a recipe for F units of flour
$b = 3(4^t)$	the number of bacteria b in a petri dish after t hours, if the dish held 3 bacteria to begin and they quadruple every hour
$Q = 150 - 10t$	the quantity Q of water in liters remaining in a tank after t minutes, if the tank holds 150 liters and is drained at a rate of 10 liters per minute
$V = \frac{500}{P}$	the volume V of a certain amount of gas at a certain temperature as pressure P varies

a. Organize the formulas into four groups:

   i. those you think have graphs that are straight lines

   ii. those you think have graphs most like the graph of $y = x^2$

   iii. those you think have graphs most like the graph of $y = \frac{1}{x}$

   iv. those you think have graphs most like the graph of $y = 2^x$

b. Choose four formulas, one from each category in Part a. Find several pairs of values, and draw a graph for each formula.

Using graphs to solve problems and equations

4. The graph shows how the height of a dropped ball changed over time.

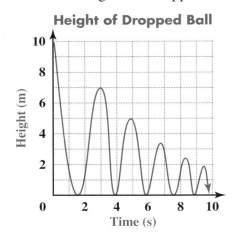

Height of Dropped Ball

a. When did the ball have a height of 6 meters?

b. When was the last time the ball reached a height of 4.9 meters?

Recognizing and using repeating patterns in graphs

5. Although a large portion of Arizona is warm or even hot year-round, some areas have a more varied climate. The table shows the average minimum temperatures in Flagstaff, Arizona, for three years.

Month	Average Minimum Temperature (°F)		
	2000	2001	2002
Jan.	18.7	15.0	17.9
Feb.	22.9	17.0	17.3
Mar.	22.6	24.7	20.0
Apr.	28.7	29.9	31.4
May	37.0	37.4	33.1
Jun.	44.5	43.1	42.8
Jul.	49.1	50.3	55.5
Aug.	51.0	49.5	49.7
Sep.	42.9	41.3	43.2
Oct.	32.6	33.2	31.5
Nov.	17.7	24.7	25.2
Dec.	20.5	12.9	15.4

Source: www.wrcc.dri.edu

a. Plot the Flagstaff, Arizona, temperature data on axes like those shown below. Connect the points using line segments.

Average Minimum Temperature

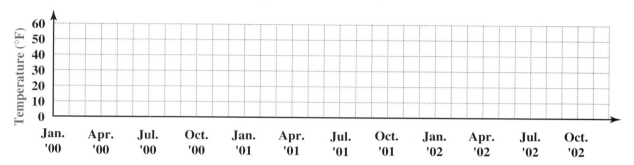

b. Describe the pattern in your graph.

Demonstrating Skills

Draw a rough sketch to show how the graph of each equation will look.

6. $y = \frac{3}{x}$

7. $y = {}^-3x + 5$

8. $y = 3^x$

9. $y = x^2 + 5$

CHAPTER 10

Data and Probability

Real-Life Math

What Do Most Americans Think? "Most Americans are optimistic about their futures." Have you ever wondered where statements like this come from? After all, how would anyone know what "most Americans" think? To uncover the American public's opinions, a group called the Gallup Organization conducts surveys on a variety of topics. This group, which has been conducting surveys for more than 60 years, polls a small sample of the American public and draws conclusions about the entire American population based on the responses. The Gallup Organization says the margin of error in their survey results is plus or minus 3 percentage points.

Think About It Suppose you were to conduct a survey to determine the favorite flavors of ice cream of students in your class. If you were to conduct the same survey of all students in your school, do you think the results would be the same?

Family Letter

Dear Student and Family Members,

Many games we play at home rely on chance or probability. In the next chapter, our class will study the basics of probability. We will begin by considering marbles drawn from a bag. For example, if there are 6 green and 4 yellow marbles in a bag, we can answer questions like these:

- If one marble is drawn at random, which color is it most likely to be?

- What are the chances that the marble drawn will be green?

- Suppose a yellow marble is drawn, put in a pocket, and then another marble is drawn. What are the chances of drawing a green marble?

We will use probability to test whether games of chance are fair. We'll also play a game that uses probability to develop a winning strategy.

Another common use of probability is in statistical sampling. In statistical sampling, a small group is chosen at random and used to draw conclusions about the entire population. Can you think of surveys or statistics that you have seen which may have been based on statistical sampling? Do you think this is a fair and accurate way to draw conclusions about a large population?

However used, finding probabilities depends on having data. We will review and study many ways to display data, including box-and-whisker plots. Then we will discuss when it is most appropriate to use those displays.

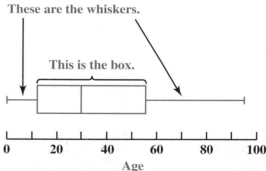

Vocabulary
Along the way, we'll learn about these new vocabulary terms:

population	**representative sample**
quartile	**sample**

What can you do at home?
You can join your student in our study of probability by playing different games of chance together. He or she can teach you the games we play at school; other games may include card games where it helps to know what cards you are likely, or very unlikely, to draw next. You can also talk about probabilities in our daily lives, such as the chance of rain.

10.1 Dependence

Many board and card games rely on *chance,* or *probability,* to make them fun. With some games, whether you win depends on nothing more than the cards you choose or the dice rolls you make. In this lesson, you will explore situations and games involving probability.

Think & Discuss

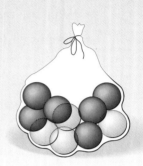

There are 10 marbles in a bag, 6 blue and 4 purple. If you draw out one marble at random, which color is it most likely to be?

The chances that the marble you draw will be purple are 4 out of 10. What are the chances that the marble will be blue?

Suppose you draw a blue marble and put it in your pocket. Then you draw another marble. What are the chances it will also be blue?

Investigation Combinations and Probability

You have probably heard people talk about probabilities in many ways. The Think & Discuss, for example, states that the chances of the marble being purple are 4 out of 10. You might also say that the chances of choosing purple are 4 in 10, 4:10, $\frac{4}{10}$, 0.4, or 40%.

Problem Set A

A bag holds six blocks, numbered 1 to 6. Imagine drawing one block from the bag.

1. What are the chances of drawing the number 2? Explain how you found your answer.

2. What are the chances of drawing an odd number? Explain.

3. What are the chances of drawing 2, 3, or 5?

4. What are the chances of drawing an odd number that is 2, 3, or 5?

5. Suppose a friend tells you she drew 2, 3, or 5. What is the probability that she drew an odd number?

Problem Set B

Suppose you take a block from the bag of six blocks and record its number. Then you return the block to the bag, mix the blocks, and draw another.

1. What is the probability that the number on the second block is 2? Compare your answer to Problem 1 of Problem Set A.

2. What is the probability that the second number is odd? Compare your answer to Problem 2 of Problem Set A.

3. Suppose the first block you drew was 3. You might have drawn 1, 2, 3, 4, 5, or 6 for the second block. That gives six ways you could have drawn two blocks: 3-1, 3-2, 3-3, 3-4, 3-5, or 3-6.

   a. List *all* the possible ways you could have drawn two blocks. How many are there?

   b. In how many of these ways are both numbers odd? What is the probability that you will draw two odd numbers?

   c. What is the probability that you will draw the same number twice?

Now imagine taking a block from the bag, putting it in your pocket, and then drawing a second block.

4. What is the probability that the first number is odd?

5. Now you will find the probability that the second number is odd.

   a. List all the possible ways you could have drawn the two blocks. How many ways are there?

   b. In how many of these ways is the second number odd? What is the probability that you will draw an odd number the second time?

   c. Compare the probability you found in Part b to the one you found in Problem 4. What do you notice? (Hint: Write both probabilities as ratios to compare them.)

6. What is the probability that both numbers will be odd?

7. Now suppose that, before you choose a second block, you look at the number on the first block.

   a. If the first number is odd, what is the probability that the second number will also be odd? You might want to look at your list from Problem 5 to help you answer this.

   b. If the first number is even, what is the probability that the second number will be odd?

Just the facts

Many people—such as meteorologists, engineers, insurance specialists, and geneticists—use the concepts of probability every day.

Share & Summarize

1. Suppose you draw one block from the bag of 6 blocks and then—without putting it back—draw another. What is the probability of drawing the same number twice? Compare your answer to Part c of Problem 3 in Problem Set B, and explain the reasons for any differences.

2. Compare the probabilities you calculated for Part b of Problem 3 and for Problem 6. Explain any differences.

3. Look again at Problems 5 and 7 of Problem Set B.

   a. Compare your answers to Parts a and b of Problem 7. Try to explain any differences.

   b. Now compare your answers for Part b of Problem 5 with those for Problem 7, and try to explain any differences.

Investigation 2 Heads or Tails?

Shaunda just tossed a coin twice and got heads both times.

Think & Discuss

Whose reasoning do you agree with, Shaunda's or Zach's? Explain.

Shaunda and Zach decided to conduct an experiment to test their ideas.

Shaunda suggested, "We can toss a coin until we get two heads in a row. Then we can see what the next toss is."

Zach says, "If we only do it once, it won't prove anything. We need to do the experiment lots of times."

◣ MATERIALS

coin

Just the facts

One way to simulate tossing a coin is to have a computer or calculator give a random number between 0 and 1. If the number is less than 0.5, the result is heads. If the number is 0.5 or greater, the result is tails.

Problem Set C

1. Follow Zach's suggestion and try the experiment at least 10 times. Keep a tally of how many times you get tails after two heads, and how many times you get heads after two heads.

2. Do your results support Shaunda's argument (tails are more likely) or Zach's (heads are just as likely as tails)? Why?

Shaunda and Zach thought they needed more than 10 results, but they were tired of tossing coins. They created a computer program to simulate tossing a coin many times. Here are the results of 200 tosses.

```
T T T T H H T H H H H T H H H H T H T H T H H T H T
H T T T H T H T T T H T H T T T H H T H H T H T H H T T
T H T T T H H H T H T T T T H H T T T T T T H H H T
T T T H T H H H H T T T T H T T T T T H H T H T T H
H T T H T T T H T T H H T H H H T H T H T H H H T H
T H H H H T H T H T H T H T T T T T T H T H H T
T T H H H H H H H H T T H T T T H T H H H H T T T
H T T H H H H T T T H H T H H T H H T H H T T T H H H
```

3. Starting from the first toss, reading left to right, find the first instance of two heads in a row. Tally the result of the next toss, and then continue the process, looking for the next instance of two heads in a row.

4. Do another experiment using the 200 tosses, this time looking for two tails in a row and tallying the results of the next toss.

5. Comment on whose reasoning your results support, Shaunda's or Zach's.

6. Shaunda and Zach ran another 500 tosses on their computer. For two heads in a row, the next toss was heads 38 times and tails 31 times. For two tails in a row, the next toss was heads 30 times and tails 38 times. Do these results support Shaunda or Zach?

Zach remembered that he had once tossed four heads in a row. "That surprised me. How often do you think that would happen?" Shaunda suggested drawing a tree diagram to find out.

Problem Set D

Trying 16 times to get four heads is not enough to test Shaunda's conclusion. More data are needed, but tossing coins is tedious. Shaunda and Zach used their computer program to generate another 500 coin tosses.

```
H H H T H T T H H H T H H H H T H T T H H H H H T
T T T H T H H H T T T T T H H T T T T T T T H H T
T T T H T H T H H H H H H T T H T H T H H T T H H
T T T H H H T T H H H T H H H H T H H T T H T H H
T H T T T T H T T H T H T T T H T T T T T T T H H
H T T T T T T H T H T T T H T T H T H T H T T T T
T T H T T H H T H T T T T H T H T T H H H H H H T
T H T H T T H H T T T T H T T H H T H T H T H H T
T T T T T T H T H H H H H T H T H T T H H H H H H
H T H T H H H H T H H T T T H T H T H T H T H H H
T H T T T T T T H T T T H H T H T H T T T T T H T
T H H T T H T T H T H H T T H H T H H H H H H H H
T H T H T T H H H H T H H T T H T H H H H H T T H
T T T T T H T H H H H T H H H T T T T H H T H H T
T T T T T T H T H H H H T T H H H T T T T H H T H
H T T T T H T H H H H H H H H H H T H T T H H H H
H H H H T H T H H H H H T H T T T T T T H H H T T
H T H H T H T T T H T H T H H H H H H H H H H T T
T H T H H H T H H T T H T H H T T T T T H T H H T
T T T H H H T H H H H T H H T T T H H H T T T H T
```

Choose a starting point anywhere in this set of tosses, and look at the next four tosses. If they are all heads, make a note. Then look at the four tosses after that four and note whether they are all heads. Examine at least 48 sets of four tosses. (If you reach the end of the set, start over at the beginning.)

1. How many times were all four tosses heads? Does your result support Shaunda's conclusion?

2. In Shaunda's tree diagram, how many times does the combination of two heads and two tails—in any order—occur? This is a prediction of how many times out of 16 you could expect to get two heads and two tails.

3. Use the tree diagram to work out the chances of all the combinations of heads and tails from four tosses. Enter the results in a table.

Combination	4H	3H 1T	2H 2T	1H 3T	4T
Probability					

4. Choose one of the predictions from your table. Then obtain some evidence by using the chart of 500 coin tosses. Comment on whether your evidence supports the prediction.

Just the facts

When you are considering only final outcomes (not their order), tossing a coin five times in a row is the same mathematical situation as tossing five coins at once.

Share & Summarize

1. If you are trying to estimate how likely something is, would it be better to run an experiment 1 time or 100 times? Explain your reasoning.

2. Jonah made a tree diagram for tossing a coin 5 times. From his diagram, he estimated he would get four heads and one tail 5 times out of every 32 tries. He then tossed five coins 64 times.

   a. What is the most likely number of times Jonah will toss four heads and one tail?

   b. Jonah actually got four heads and one tail 8 times. Does this contradict your answer to Part a? Explain.

On Your Own Exercises

1. Martin and Jill play a game with eight marbles numbered 1 to 8. They mix the marbles in a cup, and then draw two.

   a. List all the pairs of numbers that are possible to draw from the cup. (Note: Drawing 1 and 2 is the same as drawing 2 and 1.)

   b. What are the chances that the two numbers are 4 and 7?

   c. What are the chances that the two numbers have a sum of 11?

   d. What are the chances that the two numbers include an even number?

   e. What are the chances that the two numbers have an even sum?

2. Mr. Richards, a math teacher, works as a circus clown in the summer. His clown shirts come in three colors: yellow, green, and red. He also has four pairs of clown pants: yellow, purple, orange, and green.

   a. List all 12 possible combinations of Mr. Richards' shirt and pants.

   b. Mr. Richards chooses his clothes at random, so that each outfit has the same chances of being selected. What is the probability that he will select a yellow shirt and orange pants?

   c. What is the probability that at least one of the items will be yellow?

   d. What is the probability that he will choose purple pants?

   e. What is the probability that his pants and shirt will be different colors?

3. Jerry has pulled all the numbered spades from a deck of cards, so he now has nine cards: 2, 3, 4, 5, 6, 7, 8, 9, and 10 of spades. He mixes them up and then holds them out for Alanah to choose one. After Alanah chooses her card, Jerry holds them out for Marquez to choose one.

   a. What is the probability that Alanah chose a number less than 8?

   b. What is the probability that Alanah chose an even number?

   c. List all the possible ways Alanah and Marquez could have chosen their cards. For example, if Alanah chose 5 and Marquez chose 3, the combination is (5, 3).

   d. What is the probability that Alanah chose a higher card than Marquez?

   e. If Alanah drew the 8, what is the probability that she chose a higher card than Marquez?

4. Ms. Cooper has three children. Assume she is just as likely to have a boy as a girl.

    a. Make a tree diagram you could use to find probabilities in this situation, such as how likely it is that all three of her children are boys.

    b. What is the probability that all three children are boys?

    c. What is the probability that Ms. Cooper has two girls and one boy?

    d. Suppose you know that Ms. Cooper has at least one girl. Now what is the probability that she has two girls and one boy?

5. A bookshelf contains three books, *After the Bell Rings, Beware of the Frog,* and *Cornered!* When Malik accidentally knocked the books on the floor, he put them back on the shelf without paying attention to their order.

    a. Run this experiment at least 12 times:

    Label three identical pieces of paper A for *After the Bell Rings,* B for *Beware of the Frog,* and C for *Cornered!* Fold each piece so you can't see the labels. Mix them up, and then drop them on the floor and pick them up one by one. How often do you pick them up in alphabetical order?

    b. List all the ways the books could be put on the shelf.

    c. In how many of these ways would the books be in alphabetical order? Does this agree with your experimental results from Part a?

6. Zoe has a cube with faces numbered as follows:

    • Each face has a whole number.

    • Two faces have the same number.

    • The sum of all six numbers is 41.

    Zoe rolls the cube 10 times, with these results:

$$3 \quad 11 \quad 12 \quad 4 \quad 11 \quad 12 \quad 3 \quad 4 \quad 4 \quad 4$$

    a. Obviously 3, 4, 11, and 12 are on four of the faces. How many possibilities are there for the other two faces? What are they?

    b. Which of these possibilities do you think is most likely?

7. Preview A fairground entices customers with this game:

a. Draw a tree diagram to work out the chances of winning this game.

b. What is the probability that a player will win? What is the probability that a player will lose?

c. The prizes for the winners cost the fairground $2 each. If 500 people play, how much money would the fair expect to make? Explain.

8. Cary ran an experiment. Using a computer program, he examined the number of heads in various numbers of coin tosses. For example, he simulated tossing a coin 10 times, and got 4 heads. For 100 tosses, he got 46 heads. He made a table of his results.

Remember

The absolute value of a number is its distance from 0 on a number line. For example:

$|3| = 3$

$|{-3}| = 3$

Number of Tosses	Actual Number of Heads	Expected Number of Heads	Difference |Expected − Actual|
10	4		
100	46		
1,000	513		
10,000	5,087		

a. Complete the last two columns of Cary's table.

b. Cary noticed that the numbers in the "Difference" column were increasing. He claimed that this shows that as the number of trials grows, the results get further from the actual probabilities of the situation. Explain why his reasoning is incorrect.

Mixed Review

Rewrite each expression as simply as possible.

9. $t \times t \times t^{-2}$

10. $\dfrac{2.4(r^3 - 3r^3)}{0.3}$

11. $(a^p)^{2p}$

Order each group of numbers from least to greatest.

12. $-\frac{1}{3}$, $^-0.6$, $^-0.32$, $-\frac{2}{3}$, $^-0.65$

13. 3.2×10^{-6}, 0.2×10^{-5}, 0.2×10^{-7}, 1.2×10^{-6}, 0.02×10^{-5}

14. Joy took an inventory of the trees on her property and made a table of her findings.

Tree	Cedar	Redwood	Eucalyptus	Apple	Oak
Number	3	7	1	2	4

a. Joy made a circle graph of her data. Explain how she found the percentage for each section. Then explain how she determined the size of each section. Give an example of each calculation.

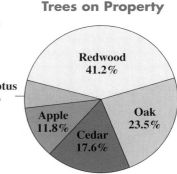

Trees on Property

b. Joy also made a pictograph of her data, but something is wrong with her pictograph. What is the problem, and how could she correct it?

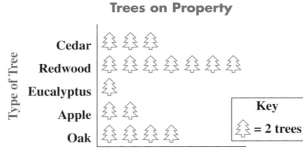

Trees on Property

Key
🌲 = 2 trees

c. Joy made a third graph of her data. Does it make sense to connect the points on her graph? Why or why not?

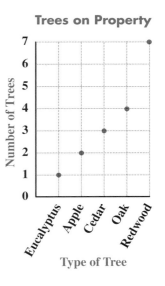

Trees on Property

10.2 Applying Probability

VOCABULARY
outcomes

Have you ever wondered whether the games you play are really fair? Determining the number of possible results, or **outcomes,** of a game can often help you judge its fairness.

MATERIALS
2 coins

Explore

Suppose you and your friend play a game in which you toss two coins. If the coins show the same faces, you score 1 point. If they show different faces, your friend scores 2 points.

Is this a fair game? Play 10 rounds of the game to see what happens.

Investigation 1 ▶ Is It Fair?

In the problems that follow, you will test whether games are fair by determining probabilities for them.

Problem Set A

Taylor and Manuel have three disks. Two are blue on both sides, and one is blue on one side and red on the other.

The friends take turns tossing all three disks. Taylor scores 1 point if they all turn up blue, and Manuel scores 2 points if they are not all blue.

1. Is this a fair game? Why or why not?

2. If the game is not fair, assign points in a different way to make it fair.

Problem Set B

Three people are playing the game *1, 2, 3, Show!* On the count of three, each player shows 1, 2, or 3 fingers. Points are scored according to these rules:

- Player A receives 1 point if *exactly two* players show the same number of fingers.

- Player B receives 1 point if all three players show a *different* number of fingers.

- Player C receives 1 point if all three players show the *same* number of fingers.

The winner is the person with the most points after nine rounds.

1. Predict which player will win.

2. Play *1, 2, 3, Show!* in groups of four. Three people should be Players A, B, and C, and the fourth should record the results to estimate the probabilities of winning. Play an entire game (nine rounds).

   Complete a table like the following for your game.

Player	Tally of Points Scored	Estimated Probability
A		
B		
C		

3. Which player was the winner? Was your prediction correct?

4. Now collect the data for the entire class. What percentage of the time did each player win?

5. Using a tree diagram or a chart, list all the possible outcomes for a round of *1, 2, 3, Show!* For example, 122 is the outcome of Player A showing 1 finger and Players B and C each showing 2 fingers.

6. Given the outcomes you listed in Problem 5, what is the probability of each player winning a round?

7. Is this a fair game? Explain.

8. How would you assign points to make the game fair? Explain why you think your system is fair.

Share & Summarize

A certain game has four outcomes. Outcome A has probability $\frac{1}{10}$, Outcome B has probability $\frac{1}{5}$, Outcome C has probability $\frac{3}{10}$, and Outcome D has probability $\frac{2}{5}$.

Four players choose outcomes and score a certain number of points when their outcomes arise. How should points be assigned to each outcome to make the game fair?

Investigation 2 ► What's the Difference?

As you read the rules for the game *What's the Difference?* think about whether the game is fair.

Rules for *What's the Difference?*

- Two players each roll a single die.
- The players find the difference by subtracting the lower number rolled from the higher.
- Player A scores 1 point if the difference is 2 or less.
- Player B scores 1 point if the difference is 3 or more.
- The first player to reach 5 points wins.

You will now look at the game more closely and decide whether it is fair.

MATERIALS

2 dice

Problem Set C

1. Play six games of *What's the Difference?* with a partner. One person should be Player A for all six games. Record your scores in a table.

Game	1	2	3	4	5	6	Total
Player A							
Player B							

2. Describe any patterns you notice in the results.

3. Write your totals on the board. Then compare your totals with the others found in your class.

4. Does this game seem fair? Explain.

Problem Set D

In their analysis of *What's the Difference?* Luis and Kate noticed that some differences occurred more often than others.

Luis said, "A difference of 5 doesn't happen very often. If you roll 6, I would have to roll 1. If I roll 6, you would have to roll 1. That's only two ways."

1. Luis and Kate decided to find how many outcomes give a difference of 1. They started to work out the possible ways by using a table. Complete their table, adding more columns if necessary.

Outcomes with a Difference of 1

Die 1	6	5	5	4				
Die 2	5	6	4	5				

2. How many outcomes have a difference of 1?

3. Kate decided they should count how many outcomes give each difference, not just differences of 5 and 1. Complete the table below. Begin by putting your answer to Problem 2 in the "Number of Outcomes" row for a difference of 1.

Difference	0	1	2	3	4	5
Number of Outcomes						2

4. Look again at the rules for *What's the Difference?* Use your table to explain whether the game is fair.

MATERIALS

2 dice

Problem Set E

Luis and Kate are convinced that *What's the Difference?* is quite unfair.

Kate said, "What if we were to change the scoring system? Could we make the game fair?"

1. Luis suggested these new rules:

   • Player A scores 1 point for differences of 0, 2, 3, 4, 5, and 6.

   • Player B scores 1 point for a difference of 1.

   Are these rules fair? Why or why not?

2. Using your table from Problem 3 of Problem Set D, write new rules that will make the game fair. Play a few games with your new rules, and tell whether the results support your suggestion.

3. Challenge If a player can receive only 1 point for a roll, there are exactly three ways to assign the differences to give the players an equal chance of winning. Try to find all three.

4. Zander joined Kate and Luis. He suggested these rules for three players, still rolling two dice:

- Player A scores 1 point if the difference is 0 or 1.

- Player B scores 1 point if the difference is 2 or 3.

- Player C scores 1 point if the difference is 4 or 5.

a. Which player do you think has the advantage in this game? Explain.

b. Find a way to assign the differences that is fair for all three players.

c. Play a few games with two classmates to see whether the results show that the rules are fair.

Share & Summarize

1. Make up an *unfair* game in which players score points based on the results of two tosses of a coin.

2. Now rewrite the rules to make your game *fair*. Explain how you know the game is fair.

Investigation 3 The Hidden Prize

In the *Hidden Prize* game show, a prize is hidden under one of three boxes. Only the game show host knows where the prize is located.

Carmen has won the right to try for a prize. The host asks Carmen to name the box she thinks the prize is under: A, B, or C.

Once Carmen has named a box, the host will show her that one of the unchosen boxes is empty. The host will then ask her, "Would you like to change your mind?"

Think & Discuss

The challenge in this situation is to advise Carmen whether to change her mind once the host has revealed one of the empty boxes.

What is your initial reaction? Would you advise Carmen to change her mind or to stay with her first guess?

To solve the hidden-prize problem, you will now run an experiment and collect data.

Problem Set F

Work in pairs to set up and run the following experiment.

Write the word *Prize* on one of three identical sheets of paper. Then either fold all three sheets—being careful that you can't tell them apart—or just place them on a desk with the "prize" face down.

Step 1   One person pretends to be the quiz show host and the other the contestant. While the contestant looks away, the host rearranges the papers, being careful to note where the prize is.

Step 2   The contestant indicates the paper he or she thinks is the prize.

Step 3   The host turns over one of the unchosen sheets of paper. If the contestant did not choose the prize, the host must be careful to turn over the unchosen paper that is *not* the prize.

Step 4   The contestant decides whether to change his or her mind.

1. First see what happens when the contestant decides *not* to change his or her mind. Play the game 10 times, always staying with the first choice. Record the number of wins in the middle column of a table like the one shown below.

Games	Win or Lose?	
	Do Not Change Your Mind	**Change Your Mind**
1		
2		
3		
4		
5		
6		
7		
8		
9		
10		

2. Now see what happens when the contestant *always* changes his or her mind. Play another 10 times, always changing to the unchosen paper in Step 4. Record the results in the last column of your table.

3. Based on your results, do you think the contestant should or should not change his or her mind?

4. Does your initial guess agree with the result you found through your experiment?

5. Try to explain why the advice you would now give Carmen is correct.

Share & Summarize

In trying to decide which strategy to use when playing a game, why is it useful to conduct an experiment?

Will an experiment always give you the best strategy? Explain.

Investigation 4 ▶ The Greatest-Number Game

The game you will explore next involves chance, but a high score is also based on having a good strategy.

Maya and Simon are playing a place-value game. The object is to form the greatest three-digit numbers possible. They each draw a playing board like the one below. There are three columns, one each for hundreds, tens, and ones.

Greatest Number Playing Board

	H	T	O
Round 1			
Round 2			
Round 3			
Round 4			
Grand Total			

Rules for *Greatest Number*

1. Put six blocks, numbered 1 to 6, in a bag.

2. For each round, players take turns creating a three-digit number in this way:

   • Draw a block from the bag.

   • Write the number in one of the three columns on the playing board. Once it has been placed, a number cannot be moved.

   • Without returning the first block, draw a second block. Write the number in one of the remaining two columns.

   • Without returning either block to the bag, draw a third block. Write the number in the remaining column.

3. Each game consists of four rounds. At the end of the game, each player has four three-digit numbers.

4. Add the four three-digit numbers to get the grand total.

5. The winner is the player with the highest grand total.

For example, Maya drew 3 and placed it in the tens column. She then drew 1 and placed it in the ones column. Finally she drew 6 and put it in the hundreds column. Her three-digit number is 631.

Problem Set **G**

Play *Greatest Number* with a partner once to see who gets the highest grand total.

1. Calculate the highest and lowest possible grand totals.

2. The greatest three-digit number possible is 654. One way to get 654 is to draw 6, 5, and then 4. Calculate how often you can expect to draw these numbers in that order.

3. In Parts a–f, the player has placed the first number and has just drawn a second number. What advice would you give the player about where to place the second number? State your reasons.

   a. ☐ 4 ☐   Second draw: 5

   b. ☐ ☐ 2   Second draw: 4

   c. ☐ 3 ☐   Second draw: 4

   d. ☐ 5 ☐   Second draw: 2

   e. ☐ ☐ 1   Second draw: 5

   f. 6 ☐ ☐   Second draw: 3

4. Now play the game twice more. Are your scores better or worse than the first time you played? Do you think your strategy has improved? If so, explain why you think you are making better decisions.

Maya and Simon were trying to find the best strategy for playing *Greatest Number*. Simon drew 3 first and put it in the tens place, and then drew 4.

☐ 3 ☐

They discussed where to place the 4. One suggestion was to finish the round many times. Maya said, "I'll put the 4 in the ones place. Then I can draw the third block 10 times to see what scores I get." Simon agreed to put the 4 in the hundreds place and also draw 10 times.

Maya   Simon

☐ 3 4   4 3 ☐

Problem Set **H**

Remove the 3 and 4 blocks from your bag, and mix the remaining blocks.

1. Draw one block from the bag. Record what Maya's number would have been, and then return the block to the bag. Do this 10 times.

2. Now draw a block 10 times and record what Simon's number would have been each time.

Remember

In a stem-and-leaf plot, the "stem" gives the place values just to the left of the "leaves." For example, the second line of this plot

23 | 39, 54, 81

means three of the scores were 2,339, 2,354, and 2,381. This is indicated by the key shown at the bottom right of the plot.

3. Find the totals for the two strategies. Which strategy do you think is better?

4. Maya and Simon suggested this fund-raising game for the school fair:

Fairgoers donate $1 to play Greatest Number. If they score over a certain grand total, they win a prize costing about $5.

Maya and Simon want to figure out what score a player should get to win a prize. To gather information about the game, they asked 50 people at the mall to play it. They made a stem-and-leaf plot of their scores.

Stem	Leaf
24	12
23	39, 54, 81
22	17, 22, 23, 40, 60, 72, 99
21	08, 17, 23, 24, 35, 37, 51, 59, 67, 78, 90
20	03, 06, 31, 46, 54, 55, 72, 82
19	07, 19, 27, 32, 50, 72, 74, 83
18	07, 11, 22, 47, 53, 59, 60
17	39, 45, 70
16	11, 82 Key: 23	39 = 2,339

a. What was the highest score for these 50 people? What was the lowest score?

b. Create a set of four scores that could have given the high score. Do you think your scores are likely or unlikely for someone to get?

c. Suppose Maya and Simon decide that a fairgoer must score 2,350 or above to win the prize. How many of these 50 people would have won?

d. Remember that fairgoers pay $1 to play, and that each prize costs the school $5. How much money would be raised or lost by the fund-raising game if these 50 people had played and winning scores are 2,350 or higher? Explain.

e. How much money would be raised or lost from these 50 games if winning scores are 2,100 or higher? Explain.

f. From these 50 scores, what do you think is a good cutoff for the winning score? Give your reasons.

Share & Summarize

James, a sixth grader, plays *Greatest Number* and usually loses. What advice would you give him for a winning strategy for the game?

On Your Own Exercises

Practice & Apply

1. Alejandro and Lamond have two chips. One is red on both sides; the other is red on one side and blue on the other. Each boy tosses a chip. Lamond scores 1 point if the chips match, and Alejandro scores 1 point if they don't match.

   a. Is the game fair? Explain how you know.

   b. The boys get a third chip, red on one side and blue on the other. Make up a fair game that involves tossing all three chips.

2. Suppose you spin the spinner below and then toss a cube that is numbered 1, 2, 3, 4, 5, and 6. You and another player add the two results. You score 2 points if the sum is 5, and your friend scores 3 points if the sum is 3. Is this a fair game? Why or why not?

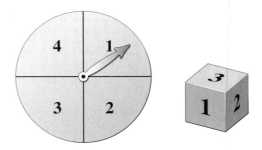

3. Rashard and Kellie are getting ready to play checkers, and they devise a competition to decide who will make the first move. They will put three red and two black checkers in a bag, and then each draw one. If the two checkers are the same color, Rashard wins; if they are different colors, Kellie wins.

   a. Design an experiment that could help you decide whether the competition is fair. Run the experiment 20 times. Describe your experiment and your results.

   b. List all the possible pairs of checkers that could be chosen. Hint: There are 10 possible pairs, all equally likely. Use R1, R2, and R3 to represent the red checkers and B1 and B2 for the black checkers. For example, one possible pair is R1, B2.

   c. In how many of the 10 possible pairs are the two colors the same? Does this agree with your experimental results?

   d. Is the game fair? Explain. If it is not fair, who has the advantage?

impactmath.com/self_check_quiz

4. Timoteo and Jin Lee have cubes with different numbers on their faces.

Jin Lee's cube: 1, 3, 5, 7, 8, 9   Timoteo's cube: 4, 4, 4, 5, 6, 8

They are playing a game in which they each roll a cube, and the highest roll scores a point. Ties are ignored. Is this game fair? Explain.

Hint: Suppose Jin Lee rolls a 1 and then Timoteo rolls. List all the possibilities. Then suppose Jin Lee rolls a 3 and list all the possible results for Timoteo. Do this for all the numbers Jin Lee can roll.

5. For a bonus round in the *Hidden Prize* game show, Carmen is shown four cards, numbered 1 to 4. The cards are shuffled, and Carmen must choose two of them without looking.

The game host looks at the cards Carmen has chosen and then shows one of them to her—being sure it is *not* the card numbered 1. Carmen can then keep the other card or trade it for one of the two unchosen cards. Carmen will win the bonus prize if the final card she chooses is numbered 1.

a. Design an experiment to decide whether it's better for Carmen to choose again or to keep her original card. Describe your experiment.

b. Which is better, keeping the original card or trading it? You may want to run your experiment to help you answer this question.

c. Explain why your answer to Part b is correct.

6. Here are the scores of 50 people who played *Greatest Number*.

Stem	Leaf
24	03
23	15, 26, 38
22	02, 13, 29, 32, 44, 61, 83
21	06, 11, 17, 22, 36, 40, 43, 51, 62, 81, 99
20	09, 25, 28, 36, 41, 55, 70
19	14, 20, 32, 47, 48, 53, 69, 88, 91
18	00, 17, 24, 40, 61, 76, 78
17	03, 44, 62
16	21, 27 Key: 23|15 = 2,315

a. From these data, how many players out of 50 would you expect to score over 2,300?

b. How often would you expect a player to score from 1,800 to 2,200?

c. How often would you expect a player to score under 2,100?

7. Consuela and Sarah decided to play *Greatest Number* with their younger brothers and sisters. They used blocks numbered from 1 to 5, and playing boards with only the tens and ones columns. Suppose a player draws 3 first. Which place is better for the number, the ones column or the tens column? Explain.

Connect Extend

8. Mai, Keenan, and Andrés are playing a dart game with this target.

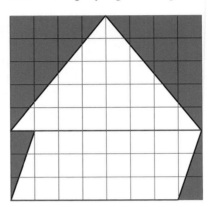

Mai scores 1 point if she hits the white triangle. Keenan scores 1 point if he hits the parallelogram. Andrés must hit a purple region to score a point. Is this game fair? Why or why not?

9. Erika and Eneas tossed a coin 20 times. Eneas scored 1 point if it landed heads, and Erika scored 1 point if it landed tails. At the end of 20 tosses, there were 14 heads and 6 tails. Erika said that was impossible, since if you toss a coin 20 times, it should come out 10 heads and 10 tails. Is Erika right? Why or why not?

10. At the school carnival, students play games and collect points they can trade in for prizes.

- In Game A, they toss 2 coins and score 2 points if the coins show the same thing.

- In Game B, they roll a die and score 2 points if the number is 3 or 4.

- In Game C, they roll two dice and score 4 points if the numbers are the same.

a. Ching-Li decides to play only one game. He wants to choose the game that gives him the best chance of collecting many points. Which game should he choose? Why?

b. How could you reassign points to the games so it doesn't matter which game Ching-Li chooses?

11. **Challenge** Invent a game using three dice that could be played at a fund-raising carnival. The player pays a set amount to play, and receives a prize worth a certain amount, in cents, that is calculated using the numbers on the dice.

Explore different ways the three dice might be used to calculate the prizes. For example, a player might multiply any two of the numbers and subtract the third.

What might be a reasonable cost to play your game so that the carnival doesn't lose money?

12. Latrell and Nykesha are playing darts with this target.

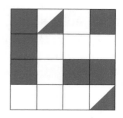

Latrell scores 1 point for every dart that lands in a purple region. Nykesha scores 1 point for every dart that lands in a white region. Is this game fair? Explain.

13. Two friends, Deanna and Rodrigo, were on a game show. For their bonus round, Rodrigo was taken offstage. Deanna was given two boxes, numbered 1 and 2, and two paper bills, marked $500 and $1,000.

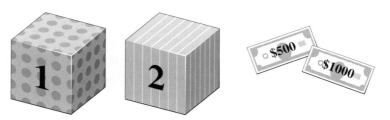

Deanna was told to place the bills in the boxes however she wanted. Rodrigo would be brought back onstage to choose a box and—if at least one bill was in that box—to pull a bill from it without looking. The friends would win the amount on the bill multiplied by the number of the box.

a. How many ways can Deanna place the bills in the boxes? Describe them.

b. How should Deanna place the bills in the boxes? Explain your reasoning. (Hint: For each way you described in Part a, imagine how much money the friends would win if they could play several times.)

14. Carin and DJ are playing a game. Each player starts with five cards, and the first player to correctly turn over all five cards wins.

The first player turns over the first card, and then either guesses whether the next card will be higher or lower or passes the turn to the other player. If the player guesses and is correct, the player gets another turn. If the player guesses and is incorrect, all the player's cards are removed and he or she must start over with five new cards on his or her next turn.

If the player passes, the other player gets a turn. If he or she passes back to the first player, that player replaces his or her last card with a new one from the deck. For example, Carin's first card was a 3, so she guessed higher and turned over the next card, a 7. She wasn't sure what to guess next, so she passed to DJ.

DJ's first card was a queen, so he guessed lower. The next card was a jack, so he guessed lower again. The third card was an 8, and he passed back to Carin.

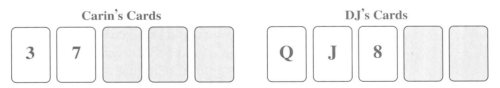

Carin took a new card, an ace, from the deck and put it over her 7. Nothing is lower than an ace, so she guessed higher.

Play the game a few times, using a real deck of cards or a deck made from paper. Use the results to develop a strategy for playing. Explain why your strategy is a good one.

Mixed Review

Find the positive square root of each number.

15. 36   **16.** 100   **17.** 121   **18.** 0.04   **19.** 1

Solve each equation.

20. $2(3a - 2.4) = 14a - 24.8$

21. $3b - 10 = 3^4 + 17$

22. $\frac{k - 0.01k}{0.03} = 0.7k + 3.23$

23. $97b \div 3.4 = 5b$

Evaluate without using your calculator. Write your answers in scientific notation.

24. $3.2 \times 10^{-6} + 0.2 \times 10^{-7}$

25. $5 \times 10^9 - 3 \times 10^9$

Algebra In Exercises 26−28, write an equation to represent the situation.

26. the number of houses, T, Geoff sells in a year if he sells h houses per month

27. the distance d traveled in h hours by a plane averaging 200 miles per hour

28. the width of a room if the ratio of width w to length l of the room is 3:2

29. Geometry Identify all the pairs of similar triangles.

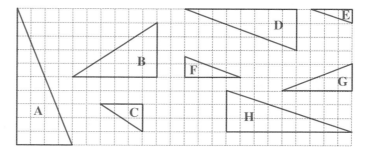

30. A 24-oz box of corn flakes costs $2.99 and a 32-oz box costs $4.19. Which costs the least per ounce? Why?

31. Janine wants to wallpaper a 6 meter by 9 meter room with 3-meter-high ceilings. The room has two 1.5-meter-square windows and one entranceway, 80 cm wide by 2.4 meters high. How many square meters of wallpaper will she need (including a little extra to practice with)?

Suppose you ask 10 people in your class to name the all-time greatest professional basketball player, and 7 say Michael Jordan. If your school had 1,000 students, you might predict that close to 700 of them would also name Michael Jordan. What if you were trying to predict for a group of 1,000 people in another area of the country? The seventh graders in Chicago—where Michael Jordan played for the Bulls—might not represent the opinions of people in Boston, where the Celtics play.

Understanding how to evaluate groups—and how to use that information to make predictions about larger groups—is the focus of this lesson.

MATERIALS
- colored tiles
- paper bag

Explore

Work in groups of two or three to play *What's in the Bag?* Each group will need about 20 colored tiles. From these 20 tiles, your group should select a total of 10 tiles. Put the 10 tiles into a paper bag. Then, without revealing your colors, exchange bags with another group.

Someone in your group should draw 4 tiles from the bag you are given. Discuss how to predict, based on those 4 tiles, the number of tiles of each color in the bag. For example, if the tiles chosen are RBBY—1 red, 2 blue, and 1 yellow—someone might say, "I think there are likely to be more blues than the others, so my guess is 5B, 2R, 2Y, 1G."

Return the tiles to the bag, shake the bag, and draw 4 tiles again. Record the tile colors in a table like the one below, adding a second prediction about the contents of the bag. As you make your new prediction, keep in mind the first group of tiles.

Attempt	Tiles Drawn	Prediction
1	RBBY	4B, 3R, 3Y, 0G
2	BBBG	6B, 2R, 1Y, 1G

Draw a third group of 4 tiles, and make another prediction. Finally, look at all the tiles in the bag. How close were your predictions?

Investigation What's in the Bag?

Medical researchers often need to predict how well a new drug will work by testing it on part of the population. Advertisers test the effectiveness of a TV commercial on part of the viewing population before they spend large amounts of money airing it. And, in Lesson 8.4, you learned about the capture-tag-recapture method biologists use to estimate how many of a particular type of animal live in a certain area.

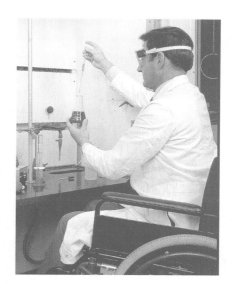

VOCABULARY
population
sample

These people are all using a technique called *sampling*. The smaller group is called a **sample.** The larger group from which the sample is taken is called the **population.** In the Explore activity, the population was the 10 tiles in the bag, and you took samples of 4 tiles.

Problem Set A

Each of the following are actual *What's in the Bag?* experiments conducted by students. Look at their samples, and make your own prediction about the contents of the bags. Then check your prediction against the actual contents of the bag, which your teacher will reveal.

1. In this game, there were 20 blocks in a bag. The blocks come in yellow, blue, red, and green. Dana and Sabina drew three samples of 5 blocks. After drawing each sample, they replaced the blocks.

- Sample 1: YBRGG

- Sample 2: GYYBR

- Sample 3: RGRBY

a. Predict the colors of the 20 blocks.

b. Dana and Sabina have seen three samples of 5, or 15 data items in all. Since there are 20 tiles in the bag, they may have seen some tiles twice, and certainly some tiles were not drawn at all. Should Dana and Sabina be fairly confident about their prediction? Explain.

c. Check your predictions against the actual contents of the bag as given by your teacher.

2. Andrea and Tyson played a game with red, green, blue, and yellow disks. They had a bag with 10 colored disks. They took five samples of 4 disks at a time.

- Sample 1: RRGG

- Sample 2: GYRR

- Sample 3: RBRG

- Sample 4: RYRR

- Sample 5: BRRR

a. What colors do you predict were in the group of 10?

b. Check your prediction against the actual contents of the bag.

3. In which problem, Problem 1 or 2, did you make a more accurate prediction? In which would you *expect* to make a more accurate prediction? Why?

Share & Summarize

Sook Leng and Peter are playing *What's in the Bag?* with 100 colored blocks. They took Sample 1, returned the blocks to the bag, and then took Sample 2.

- Sample 1: RRGYGBRRGR

- Sample 2: RGGRBRBGBY

Explain how you would use their sample data to make a prediction about the colors of the 100 blocks in the bag.

Investigation 2 Exploring Sample Sizes

People often use a sample to get a sense of the characteristics of a population—particularly when it would be time-consuming and costly to test the whole population. If you have a good sample, the proportion of the sample with a certain characteristic or holding a certain belief can be used to make a prediction about the population.

In a poem, for example, one characteristic you might look at is word length. The following poem has long words, such as *interrupted,* which has 11 letters. It also has words with only 2 letters. The population in this situation is the 109 words of the poem.

Acquainted with the Night
by Robert Frost

I have been one acquainted with the night.
I have walked out in rain—and back in rain.
I have outwalked the furthest city light.

I have looked down the saddest city lane.
I have passed by the watchman on his beat
And dropped my eyes, unwilling to explain.

I have stood still and stopped the sound of feet
When far away an interrupted cry
Came over houses from another street,

But not to call me back or say good-by;
And further still at an unearthly height
One luminary clock against the sky

Proclaimed the time was neither wrong nor right.
I have been one acquainted with the night.

Just the facts

The famous American poet Robert Frost (1874–1963) is best known for his verse about life in New England and his love of nature.

Problem Set B

1. Before reading on, estimate or guess the mean word length in the poem.

The *certain* way to find the mean word length is to count all the letters in the poem and divide by the number of words. However, your task here is to see how well you can *predict* the answer by using a sample.

2. The poem has 14 lines.

   a. For each line, count the numbers of letters and words. Record your totals in a table like the one shown here.

Line	1	2	3	4	5	6	7	8	9	10	11	12	13	14
Letters														
Words														

   b. Find the mean word length for Line 6.

   c. Find the mean word length for Line 10.

   d. Do you think you can estimate the mean word length for the poem from a one-line sample?

3. Make a table for recording the results from different sample sizes.

Lines	Letters	Words	Mean Word Length
2			
4			
6			
8			
10			

Randomly choose 2 lines from the poem. Add the numbers of letters for the 2 lines, and then add the numbers of words. Calculate the mean word length, and write your results in the first row of your table.

Next choose a sample of any 4 lines, and record the results. Then choose a sample of any 6 lines, then any 8 lines, and then any 10 lines. Enter the data you collect into your table.

4. Do all samples give the same mean word length? Discuss with your partner which sample you think gives the most accurate prediction.

5. Find the mean word length in the poem by dividing the total number of letters in the poem by the total number of words. How close was your prediction?

6. If you wanted to choose a one-line sample to bias the prediction toward longer words, which line would you choose? Explain.

Your results may be different from those of other students in your class, depending on which lines you chose.

7. Create a class data set of the number of lines in the samples that gave the closest prediction. Each student should report how many lines he or she used for the sample that gave the best prediction.

   a. Count the number of times each sample size was reported.

   b. Which sample size seems to be most reliable for predicting the mean word length in the poem?

Just the facts

One way teachers check the "readability" of a piece of writing for their students is by using a similar sampling technique. They count the number of sentences and syllables in three randomly chosen 100-word passages, and plot their results on a chart that gives them the approximate grade level of the reading material.

Share & Summarize

1. Count the numbers of letters in the first names of the six students sitting nearest to you. What is the mean word length for these data?

2. Do you think your group of six students is a reasonable sample for predicting the mean number of letters in the first names of everyone in your class?

3. Describe a situation for which six students is a good sample size and a situation for which six students is not a good sample size.

Investigation ▶3▶ Making Predictions from Samples

For a statistics project, Alison's group wanted to predict which of the following after-school activities were the favorites of seventh graders in the four schools in their district.

- playing sports
- playing music
- listening to music
- watching TV

- going to the mall
- playing video games
- reading
- drawing or painting

Think & Discuss

How would you recommend that Alison's group collect a sample that would help them make an accurate prediction? Make a list of suggestions.

As you have learned, a good sample must be large enough to give accurate results. In addition, it must be *representative* of the population.

VOCABULARY
representative sample

A **representative sample** has approximately the same proportions as the population with respect to the characteristic being studied. For example, Alison's group will want their sample to have about the same proportion as the district population of males and females, because boys might be more likely (or less likely) than girls to engage in a particular activity.

Finally, anyone conducting a survey using a sample will want a method that is practical to carry out.

That gives three important questions to ask when examining whether a particular survey method is a good one.

- Is the sample *large enough* to give accurate results?
- Is the sample *representative* of the population?
- Is the survey method *practical*?

Problem Set C

Alison's group proposed five strategies for conducting their survey.

Strategy 1   Ask all 600 seventh graders in the four schools.

Strategy 2   Hand out questionnaires to the 130 seventh graders at the district band competition next week.

Strategy 3   Get a list of all the seventh graders in the district and call 5% at random from each school.

Strategy 4   Survey every third seventh grader out of the 125 in our school as they enter their homerooms.

Strategy 5   Survey the 12 seventh graders on one student's bus.

1. Which strategy or strategies do you think would give the least representative data? The most representative data?

2. Which strategy would give the smallest sample size? The largest sample size?

3. Which strategy seems the least practical? The most practical?

4. Which strategy do you think is the best? Explain your reasoning.

Problem Set D

M A T E R I A L S

protractor

Alison's group decided to survey every third student to enter each seventh grade homeroom in their school. The five homerooms average 25 students each. They displayed their survey results in a circle graph.

Favorite After-School Activities

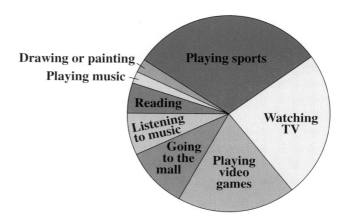

1. How large is the sample of students surveyed?

2. How many students in the sample prefer playing sports after school?

3. What is the measure of the angle (in degrees) of the section that represents the 10% of students who prefer going to the mall?

4. Based on the sample, predict how many in a district of 600 seventh graders would prefer reading.

5. Suppose you could choose 1 seventh grader from the 600 in Alison's school district at random. What is the probability that he or she prefers playing video games? Explain how you found your answer.

6. If Alison was predicting the favorite leisure activities of all students in the district—from kindergarten to twelfth grade—would sampling just seventh graders give an accurate picture of the population? Explain.

Share & Summarize

1. Name some situations in which you are more likely to predict from a sample rather than measure an entire population.

2. Look back over your class's suggestions for the Think & Discuss on page 697. For each suggestion, remark on these three questions:

- Is the sample large enough?
- Is the sample representative?
- Is the survey method practical?

On Your Own Exercises

Practice & Apply

1. Tobias and Kareem were playing *What's in the Bag?* From 12 hidden tiles, they took five samples of 3.

   • Sample 1: BYG

   • Sample 2: YBG

   • Sample 3: BGB

   • Sample 4: GBB

   • Sample 5: RBG

   What colors do you predict were in the group of 12 tiles?

2. Jemma and Desiree were playing *What's in the Bag?* There were 20 hidden colors, and they drew four samples of 5.

   • Sample 1: RYRGR

   • Sample 2: YBGYB

   • Sample 3: BYGYB

   • Sample 4: RYGRY

   What colors do you predict were in the group of 20?

3. Juan and Kelly were playing a computer simulation of *What's in the Bag?* The software creates a group of 200 tiles in a bag with a combination of four colors. Users enter the sample size and the number of samples.

   Juan and Kelly selected four samples of 10.

   • Sample 1: RBBGGGGYYY

   • Sample 2: RRBGGGGGYY

   • Sample 3: RRRGGGGYYY

   • Sample 4: RRBBGGGGGY

   a. From this sample, what do you think the software chose for the percentage of each color?

   b. Is the sample large enough for you to feel confident about your prediction?

impactmath.com/self_check_quiz

4. Akiko and Tankia are playing *What's in the Bag?* The bag has 15 marbles, a mixture of red, blue, yellow, and green. The girls decide to take five samples of 4.

- Sample 1: GBGR
- Sample 2: BRGG
- Sample 3: RBGG
- Sample 4: BGRR
- Sample 5: RYGG

a. Predict what colors make up the group of 15 marbles.

b. Describe a method for making a prediction.

5. The local ice cream shop, 5 Flavors, had the following sales in one week:

- Vanilla: 300 scoops
- Chocolate: 350 scoops
- Strawberry: 180 scoops
- Lemon: 70 scoops
- Raspberry: 100 scoops

a. 5 Flavors is one of a chain of five similar shops throughout the county. Assuming the other shops have similar sales, predict how many scoops of strawberry ice cream would be sold per week in the county.

b. Do you think these data are enough to predict for all five shops for the year? Explain.

6. Literature This poem was written by Emily Dickinson.

> Bee! I'm expecting you!
> Was saying Yesterday
> To Somebody you know
> That you were due—
>
> The Frogs got Home last Week—
> Are settled, and at work—
> Birds, mostly back—
> The Clover warm and thick—
>
> You'll get my Letter by
> The seventeenth; Reply
> Or better, be with me—
> Yours, Fly.

a. This poem has 12 lines. How many lines do you think you should use as a sample to predict the mean word length in the poem?

b. Choose a sample of lines at random, and predict the mean word length. Indicate which lines you used.

c. Now find the actual mean word length in the poem. How close was your prediction?

7. Choose a page from a novel, a story you are reading for English class, or a book for a social studies class.

a. Look at the first full paragraph on a typical page. (If the first line on the page is a continuation from the previous page, skip to the next paragraph.) Count the number of words in the paragraph, and divide that by the number of sentences in the paragraph. What is the mean sentence length?

b. Do you think you can now accurately predict the mean sentence length for the entire book? Explain.

c. Find the mean sentence length for at least 10 paragraphs chosen at random from 10 different pages.

d. Would you now have more confidence making a prediction for the mean sentence length in the book? Do you think your prediction would be accurate? Explain.

8. Jenny and Carmelita were trying to predict the outcome of an election for student president in which 350 students were eligible to vote. Jenny sampled 50 students at a party given for the football team. Carmelita sampled the first 30 people entering the school after 8.00 A.M.

Whose sample is more likely to be representative? Why?

9. Ten students surveyed from a class of 100 seventh graders were asked this question: "What is the first fruit you think of?"

a. Of the 10 students surveyed, 6 said oranges, 2 said apples, 1 said grapefruit, and 1 said lemons. Use the results of this sample to predict the results for the whole class.

b. Comment on the accuracy of your prediction.

c. The school these students attend is in Florida. Suppose the school had been in Hawaii. Do you think the students would be likely to name the same fruits? Explain.

10. The drama club at Marilyn Middle School is planning to host a movie night to raise money for costumes and props. They want to know what type of movie would draw the most interest: comedy, romance, or action/adventure. The club members decide to conduct a survey, and they have three suggestions for who to ask:

• all seventh graders

• all students in one homeroom from each grade

• the drama club

a. For each suggestion, comment on the three issues of representativeness, sample size, and practicality.

b. Which suggestion do you think is best?

Connect & Extend

Remember

The *mean* is the sum of all the numbers divided by how many numbers there are.

11. Grace records her points in each basketball game she plays. For the first four games, her results are 3, 3, 5, and 7.

a. What is the mean of this set of points?

b. Assuming these scores are typical for Grace, predict the total number of points she will score in the next 10 games.

c. How accurate do you think your prediction is? Explain.

12. Peter keeps a record of the runs he makes in each baseball game he plays. His results for the first 16 games of the season are as follows:

2 1 1 1 0 2 3 2 3 4 3 2 2 4 0 5

a. What is the mean of Peter's runs?

b. Use Peter's first 16 games to predict the total number of runs he is likely to score in the season, which lasts for 40 games.

c. How accurate do you think your prediction is? Explain.

13. In Investigation 2 of Lesson 8.4, you used proportions to estimate animal populations in a capture-tag-recapture experiment. For example, you considered these situations:

• To estimate the number of bald eagles in the United States, ornithologists tagged and released 240 eagles from across the United States. A couple of months later, they caught 100 birds and found that 3 of them had tags.

• To estimate the number of red wolves in certain forests, biologists caught 20 wolves, tagged them, and then freed them. Of the 15 wolves they later caught, 5 had tags.

a. Estimate the number of bald eagles in the United States and the number of red wolves in the forests in the study.

b. Calculate the percentages of captured eagles to total estimated population; do the same for wolves. Then compare the percentages. Do they make sense? Explain.

14. In the following situations, a sample must be taken to estimate information about the whole population. Analyze each situation by doing the following:

• Identify the population.

• Discuss what factors might be considered when deciding how large a sample to choose.

• Discuss how a sample might be chosen.

• Discuss how the sample data would be used to make a prediction for the population.

a. Clear Light Company wants to develop a new light bulb. It hopes to sell 1,000,000 a year and to guarantee customers the bulbs will last an average of 200 hours before failing. To test a light bulb, it must be left on until it fails.

b. Authorities in a country of 25 million people need to test a new vaccination for effectiveness and side effects. The sample size of volunteers must be large enough for the authorities to be confident of the results before they let the vaccine be used.

c. ACME toy makers have a new electronic game that runs on batteries. They want to be confident the game will last 30 hours before it needs new batteries. They hope to sell 10,000 a year.

15. On a sheet of paper, write the letters of the alphabet, one letter per line. Then choose some reading material, such as a book, magazine, or newspaper.

Randomly select a line of text from the reading material. Carefully go through the line one letter at a time. For each letter, make a tally mark on your alphabet list next to that letter. For example, for the word *would,* you would make a tally mark next to *W, O, U, L,* and *D.*

a. From that one line of text, what are the three most common consonants? What is the most common vowel?

b. Choose another line at random, and add the letter counts for that line to the count for the first line. Now what are the three most common consonants and the most common vowel?

c. Continue choosing single lines at random until the most common letters are the same for three lines in a row. What are the letters?

d. How many lines did you have to test?

e. Do you think your sample is large enough to predict the most common letters in your reading material?

f. Do you think your letters will be the most common letters in the language that your reading material is written in?

16. Suppose you want to find out the favorite food of all the students in your school. Devise a strategy for finding a sample. What do you think is the minimum size you need to get a reasonably accurate prediction?

17. Zara conducted a survey of eye color for the 25 students in her homeroom and put the results in a circle graph.

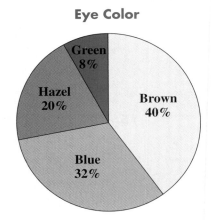

Eye Color

Green 8%
Hazel 20%
Brown 40%
Blue 32%

   a. 40% of the students had brown eyes. How many students is this in the class of 25?

   b. How many students had the other eye colors?

   c. Draw a pictograph to present the data in another form. Remember, there are 25 students in the class.

   d. Draw a line graph of the data.

   e. Which type of data display do you feel is best for conveying the data?

   f. Which type of data display do you feel conveys the data least well? Explain.

18. A school wishing to run a road-safety campaign surveyed a group of 200 seventh graders in an urban area. The survey found that the students traveled to school in five ways. Bus and car travel were the two most common methods. The survey also found that 35 students take the subway, 30 walk, and 25 ride a bicycle.

   a. How many students altogether travel to school in a bus or a car?

   b. What is one possible result the survey would find for the number of students who travel by car?

   c. Challenge Find all the possible solutions for the number of students who travel by car and the number who travel by bus.

Just the facts

Gallup polls are named after George Horace Gallup (1901–1984). Born in Iowa, he is best known for his method of surveying public opinion on politics.

19. Social Studies Gallup polls use a type of sampling to predict the outcome of elections. There is a way to determine how accurate these predictions are, based on the population and the sample size.

Suppose 1,000 people are chosen at random from a voting population of 1,000,000 and asked which of two candidates they voted for. The *margin of error* for the poll is 4%. That means it's very likely that the percentage of votes for each candidate predicted by the poll will be within 4 percentage points of the real figure.

a. Why would a Gallup poll choose people at random?

b. If a sample of 1,000 people shows that 47% intend to vote for Candidate A, what would you expect the range of percentage votes for Candidate A to be on the actual election day?

c. It is unlikely that 47% is exactly correct. What are the least and greatest numbers of votes you would expect Candidate A to receive? Is it possible the candidate would get more than 50% of the votes, winning the election? Explain.

Mixed Review

Evaluate each expression.

20. $23 \cdot 2 - 5(3.1)$

21. $23 \cdot [2 - 5(3.1)]$

22. $23 \cdot (2 - 5)(3.1)$

23. $3^3 \cdot \frac{4}{3} - 2$

24. $3^3 \cdot (\frac{4}{3} - 2)$

25. $\frac{3^3 \cdot 4}{3} - 2$

26. Geometry Match each angle to one of the measurements below.

27°   78°   240°   90°   167°   115°

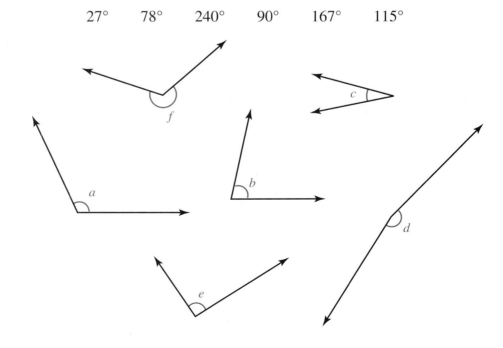

27. Which of these figures are nets for a cube? That is, which of them will fold into a cube?

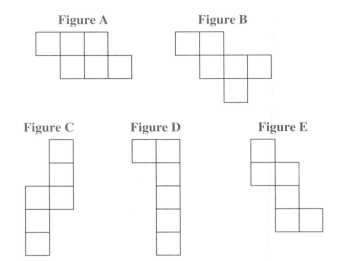

Figure A   Figure B

Figure C   Figure D   Figure E

Supply the missing information for each stretching or shrinking machine.

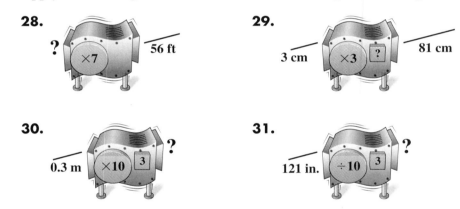

28. ?   ×7   56 ft

29. 3 cm   ×3   ?   81 cm

30. 0.3 m   ×10   3   ?

31. 121 in.   ÷10   3   ?

If a 1-cm stick of gum is sent through each super machine, how long will it be when it exits?

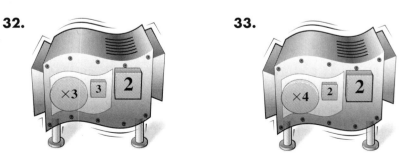

32. ×3   3   2

33. ×4   2   2

10.4 Statistical Tools and Graphs

Here is a list of the ages of all 64 residents of the town of Smallville.

1	1	2	3	3	4	5	6	6	6
7	9	11	12	12	12	12	13	15	15
15	16	18	18	20	22	24	28	28	28
29	30	30	31	33	34	35	35	36	39
40	41	42	42	46	48	51	53	58	60
62	66	73	75	77	78	81	81	81	86
86	88	92	95						

Imagine that you work for the local newspaper and want to publish these data in a graph. You have three choices for the type of graph to use.

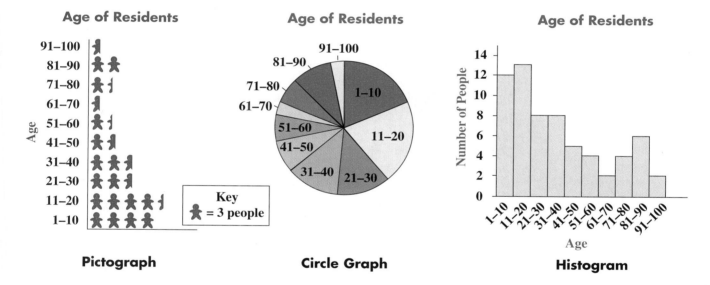

Pictograph **Circle Graph** **Histogram**

Think & Discuss

Which type of graph do you think best conveys the data? Why?

Investigation Box-and-Whisker Plots

By now you are familiar with several types of graphs, such as pictographs, line graphs, circle graphs, bar graphs, stem-and-leaf plots, and histograms. Below is another type of graph, a *box-and-whisker plot,* for the Smallville population.

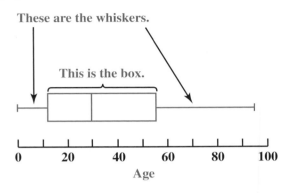

VOCABULARY
quartile

A box-and-whisker plot separates the data into four sections: two whiskers and two parts of the box. Each section represents about 25% of the data. The points that divide the sections are called **quartiles.**

The following problems will help you understand these statistics and how they are displayed in a box-and-whisker plot.

Problem Set A

Every spring Mark receives compliments on the tulips in his front yard. To monitor the flowers' growth one year, he measured the length of the longest leaf on each flower. On one day, a bed of nine tulips had the following leaf lengths, in centimeters:

12.6   13.8   16.0   16.1   18.6   23.3   24.4   27.4   32.5

Mark made a box-and-whisker plot of these data.

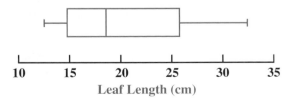

1. Two important points in box-and-whisker plots are the *minimum* (lowest data point) and *maximum* (highest data point).

   a. What are the minimum and maximum for this data set?

   b. How are these values shown on the graph?

2. One of the quartiles is the *median* of the data set.

   a. What is the median of this data set?

   b. How is this value shown on the graph?

3. To find the remaining two quartiles, you can consider the data below the median as one data set and the data above the median as another data set.

   12.6   13.8   16.0   16.1   18.6   23.3   24.4   27.4   32.5

   Lower Data Set   Median   Upper Data Set

   The median of the lower data set is the *first quartile,* and the median of the upper data set is the *third quartile.* The median of the entire data set is also the *second quartile.*

   a. What is the first quartile? What is the third quartile?

   b. How are these values shown on the graph?

4. The *range* of a set of values is the difference between the maximum and minimum values. What is the range of heights in this data set?

Problem Set B

Here are the data and the box-and-whisker plot for the ages of the 64 Smallville residents.

1	1	2	3	3	4	5	6	6	6
7	9	11	12	12	12	12	13	15	15
15	16	18	18	20	22	24	28	28	28
29	30	30	31	33	34	35	35	36	39
40	41	42	42	46	48	51	53	58	60
62	66	73	75	77	78	81	81	81	86
86	88	92	95						

The important statistics have been labeled in the graph. Notice that there are five important points: the minimum, the maximum, and the first, second, and third quartiles.

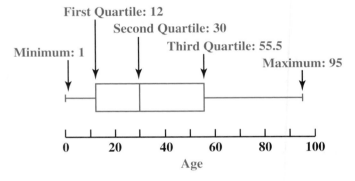

1. Copy the data set onto your paper. Draw a slash (/) to divide the two halves on either side of the median. Then draw two more slashes, dividing each of those halves.

2. Your slashes represent the quartiles in the data set.

   a. How many values are represented by each whisker? What percentages of the town's population are these?

   b. How many values are represented by each section of the box? What percentages of the town's population are these?

3. Why is the left-hand whisker shorter than the right-hand whisker?

4. Why is the median not in the exact center of the box?

5. Name another type of graph that could be useful for displaying this set of data.

Problem Set **C**

Here are two box-and-whisker plots for the ages of the citizens of two towns.

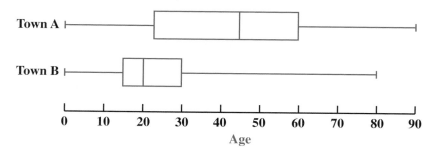

1. What is the second quartile, or median, for Town A?

2. Which town has a greater range of ages?

3. About what percentage of the population of Town B is between 15 and 80 years old?

4. One of the towns is a university town. Which do you think it is? Why?

5. If these data were displayed in a stem-and-leaf plot, what additional information would you know?

Now you will draw your own box-and-whisker plots.

Problem Set **D**

Rhiannon kept a record of the points she scored throughout the basketball season. In 11 games, she scored these numbers of points:

    20   15   23   14   18   12   25   10   24   12   17

1. Write the scores in order.

2. What is the median of this data set?

3. What are the first and third quartiles?

4. What is the range of points scored?

5. Draw a box-and-whisker plot of Rhiannon's scores.

Simona and Derek scored the following runs in several games of baseball.

Game	1	2	3	4	5	6	7	8	9	10
Simona	7	2	5	6	4	5	9	3	2	1
Derek	6	1	2	1	8	7	9	8	7	8

6. On the same axis, draw a box-and-whisker plot for each player.

7. Compare the ranges of each player's scores.

8. Who do you think is a better scorer? Use the graphs you made to explain your answer.

Share & Summarize

Suppose you have a data set with 18 values, ordered from least to greatest. Explain as completely as you can how to create a box-and-whisker plot of the data set.

Investigation ▶ 2 Choosing a Graph for Displaying Data

When you have a choice of graph for organizing and displaying data, what do you do? The "best" graph for a situation is the one that most clearly shows the data for the purpose you need.

These are some of the types of graphs that you have learned about and can choose from.

- line graphs
- histograms
- bar graphs

- circle graphs
- stem-and-leaf plots
- box-and-whisker plots

For example, the data below—which show the average monthly temperatures for Knoxville, Tennessee, in degrees Fahrenheit—were used to create the six graphs on the next page.

Jan	Feb	Mar	Apr	May	Jun	Jul	Aug	Sep	Oct	Nov	Dec
36	40	49	58	65	73	77	76	70	58	49	40

Reprinted with permission from *The World Almanac and Book of Facts.*

Monthly Temperature

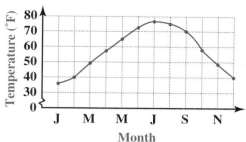

Line Graph

Monthly Temperature

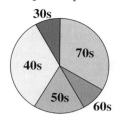

Circle Graph

Monthly Temperature

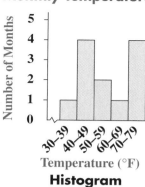

Histogram

Monthly Temperature

Stem	Leaf
7	0 3 6 7
6	5
5	8 8
4	0 0 9 9
3	6

Key: 5 | 8 = 58

Stem-and-Leaf Plot

Monthly Temperature

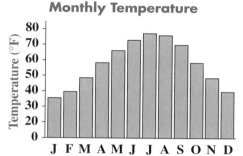

Bar Graph

Monthly Temperature

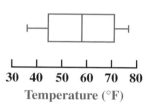

Box-and-Whisker Plot

Remember

In a histogram, the height of a bar tells how many data values are in a particular range of values. In a bar graph, the height of a bar gives a particular data value.

Problem Set E

1. Which graphs give a sense of how the temperature changes over the year?

2. From the graphs you named in Problem 1, describe the average monthly temperatures in Knoxville.

3. Consider the graphs you didn't mention in Problem 1.

   a. Why don't these graphs give a sense of how the temperature changes over the year?

   b. Suppose a newspaper printed the circle graph but gave no other graphs or data. What would a reader be able to say about Knoxville's average monthly temperatures?

   c. Suppose a newspaper printed just the box-and-whisker plot. What would a reader be able to say about Knoxville's average monthly temperatures?

4. Which graphs seem most useful for displaying the temperature data? Explain.

5. Are the other graphs useless for displaying the data? Explain.

Problem Set F

Kyung and Isandro gathered some data on pets for a school project.

Most Popular Pets

Kyung surveyed the 27 students in his class about their favorite pets. Here are his findings:

Dogs	Cats	Birds	Fish	Mice
8	9	5	4	1

Tropical Fish

Isandro found 10 students who kept tropical fish and asked how many fish they each had in their tanks. The responses were as follows:

4   9   11   13   15   16   16   16   18   20

Cost per Week

Kyung and Isandro gathered information on how much it costs students to care for their pets each week, on average. Here are their results:

- under $5: 23%

- between $5 and $10: 48%

- over $10: 29%

Life Expectancy of Dogs

Isandro researched the following data for how long pet dogs live on average:

Years	0 to 4	5 to 8	9 to 12	13 to 16	over 16
Percentage	17%	8%	26%	37%	12%

1. The boys wanted to use graphs to effectively display their data. For each of the four types of data they collected, choose two of the following graph types you think would be useful and one that would be inappropriate. Explain your choices.

   - line graphs
   - histograms
   - bar graphs
   - circle graphs
   - stem-and-leaf plots
   - box-and-whisker plots

2. For each data set, choose an appropriate graph type and make a graph to display the data.

On Your Own Exercises

1. Two ice cream shops, Scoops and Sundae Sunday, report the number of ice cream cones sold per day for a week.

Day	Mon	Tue	Wed	Thurs	Fri	Sat	Sun
Scoops	125	140	180	130	150	160	260
Sundae Sunday	130	155	120	140	160	170	250

   a. On one axis, draw a box-and-whisker plot for each shop.

   b. Which shop has the greater range of ice cream sales?

   c. Which shop has the higher median sales?

2. For 4 weeks (20 nights), four students recorded the number of minutes they spent on homework each night.

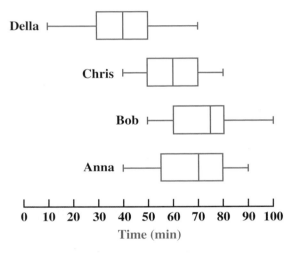

   a. Which student has the highest median time? Suggest some possible reasons for this.

   b. If the school suggests that a median study time of 70 minutes is adequate, which students are putting in an adequate amount of study time?

   c. Which student studied the longest in one night?

   d. Which student studied the least in one night?

   e. Which student has the least range of study times?

   f. Which student has the least amount of study time above 50 minutes?

   g. How many of the students worked for more than 2 hours in any one evening?

 impactmath.com/self_check_quiz

3. During football season, Albert, Bettina, and Paul sold soft drinks at the home games from drink carts. They kept records of how many drinks they sold each game and drew the following box plots. They want to use the data to find the best locations for their carts.

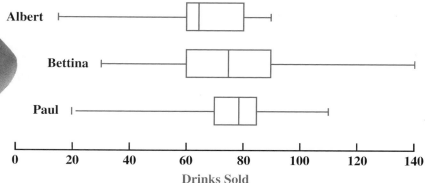

Drinks Sold

a. Who had the highest median?

b. Who sold the least number of soft drinks in a game?

c. Who had the highest proportion of sales above 30 drinks per game?

d. Who had the greatest range of sales?

e. Who had 75% of his or her sales above 70 drinks per game?

f. Who sold more than 140 drinks during a game?

g. What information about the best location can you get from these graphs?

4. A farmer recorded the temperature every 2 hours during daylight.

Time	6 A.M.	8 A.M.	10 A.M.	12 noon	2 P.M.	4 P.M.	6 P.M.	8 P.M.
Temp (°F)	45	50	60	65	80	75	60	50

a. Draw a graph showing the change in the temperature during the day.

b. Use your graph to estimate the temperature at 11 A.M.

c. During which hours of the day was the temperature above 60°F?

d. It is recommended that crops be irrigated when the temperature is between 50°F and 60°F. From these data, at what times should the farmer irrigate the crops?

5. Alexis lives on a farm. For a math project, she measured the growth of several chicks in their first week of life. The table shows the averages for her chicks. What type of graph would you use to display these data? Explain.

Day	0	1	2	3	4	5	6	7
Mass (grams)	25	28	31	34	38	43	48	54

6. Craig counted the colors in a package of candy. What type of graph do you suggest he use to display his data? Explain.

Color	Red	Green	Yellow	Brown	Blue	Orange
Number	8	3	10	18	5	13

7. Ms. Isaacs polled her class about their birthday months. What type of graph might she use to display these data? Explain.

Month	Jan	Feb	Mar	Apr	May	Jun	Jul	Aug	Sep	Oct	Nov	Dec
Students	5	3	2	3	3	2	0	6	3	2	2	1

8. Mr. Ritter polled his class on the kinds of exercise they prefer. Of 24 students, 10 said running, 5 said walking, 4 said swimming, and 3 said biking. The other 2 said they never exercise. What kind of graph do you suggest he use to display these data? Explain.

9. Ben works as a florist's assistant. He sorted a collection of long-stem red and yellow roses and then measured their lengths. To compare the two varieties, he drew box plots.

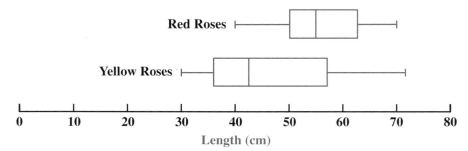

a. What is the median length of the red roses?

b. What is the median length of the yellow roses?

c. What lengths are the middle 50% of red roses between? What lengths are the middle 50% of yellow roses between?

d. Which color rose would you buy for a 40-cm-tall vase? Why?

e. What percentage of the red roses were longer than 63 cm?

f. What percentage of the yellow roses were less than 57 cm long?

g. Challenge Red roses cost $3, and yellow roses cost $2. A customer spent exactly $25 on roses. What combinations might she have bought?

10. Box-and-whisker plots use the median as a measure of the center of a data set. The mean is another measure of center.

Lana surveyed two groups of 13 students each. She asked the students how many posters they had hanging in their bedrooms. She made two box-and-whisker plots of her findings.

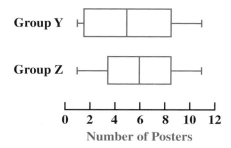

The data set for Group Z is as follows:

1 2 3 4 5 6 6 6 7 8 9 10 11

a. How will the data set for Group Y be different from the data set for Group Z? Explain.

b. Find the mean for Group Z's data.

c. Is the mean for Group Y higher or lower than the mean for Group Z? Explain.

d. The mean and median for Group Z are equal, but that is not true for all data sets. For example, find the mean and median of this data set:

1 1 1 1 11

e. Consider your answers for Parts c and d. The median for Group Y is 5. Estimate the mean for Group Y's data. Explain how you made your estimate.

11. Julian measured the heights in inches of 15 carrots he had grown. Draw a box-and-whisker plot of his data.

10.5	10.9	11.1	11.2	11.3
11.4	11.5	11.8	11.9	11.9
12.1	12.8	12.3	12.8	14.0

12. Megan rolled two dice 100 times and found the difference between the two numbers each time. Here are her results.

Difference	0	1	2	3	4	5
Number of Times	20	29	18	15	13	5

a. Which difference occurred most often? Why do you think this happened?

b. Draw a box-and-whisker plot of Megan's results.

c. Draw a bar graph of Megan's results.

d. Which graph do you feel best displays the data? Explain.

13. Earth Science The maximum distances between the nine planets in our solar system and the sun are given below.

Planet	Distance from Sun (millions of miles)
Mercury	43.4
Venus	67.7
Earth	94.5
Mars	154.8
Jupiter	507.0
Saturn	936.0
Uranus	1,867.0
Neptune	2,818.0
Pluto	4,586.0

a. Display these data using any kind of graph you like. Explain your choice of graph.

b. Write a few sentences describing any observations you can make about the planets from your graph.

Tell whether each set of number pairs could describe a linear relationship.

14. (4, 15); (7, 24); (3, 12); (5, 18); (6, 21)

15. (0, ⁻5); (2, 1); (1, ⁻2); (3, 4); (⁻1, ⁻8)

16. (⁻1, ⁻0.5); (1, 4.5); (3, 9.5); (2, 7); (0, 2)

Determine the slope for each relationship.

17. $y = 2x + 1.7$ **18.** $y = {}^-x - 1.2$ **19.** $2y = 6x + 9$

Give the coordinates of the y-intercept for each relationship.

20. $y = 2x + 1.7$ **21.** $y = {}^-x - 1.2$ **22.** $2y = 6x + 9$

23. Draw a flowchart for this rule: *output* $= 1.1(n + 3.2)$. Use it to find the output for the input 0.8.

24. From the following list, find all the pairs of fractions with a sum of 1.

$$\frac{12}{21} \qquad \frac{12}{14} \qquad \frac{10}{14} \qquad \frac{6}{12} \qquad \frac{2}{7} \qquad \frac{3}{7} \qquad \frac{1}{6} \qquad \frac{15}{18} \qquad \frac{1}{7} \qquad \frac{1}{2}$$

Geometry Use the Pythagorean Theorem to find each missing side length.

25.

26.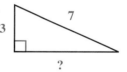

Find a single machine to do the same work as the given hookup.

27.

28.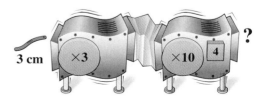

29. How long will the gummy worm be when it exits this hookup? Give your answer in meters.

Chapter Summary

In this chapter, you learned more about probability. You studied situations in which outcomes were influenced by previous outcomes and situations in which each outcome was independent of the others. You also determined the fairness of several games of chance and devised strategies, based on probabilities, to increase your chances of winning.

The basics of sampling a population and how to make predictions from samples were introduced in this chapter. You learned that the size of a sample and whether it is *representative* are important factors in sampling.

Finally, you encountered another kind of statistical graph, a *box-and-whisker plot,* which uses the maximum, minimum, and *quartiles* of a data set. Then you reviewed other kinds of graphs and considered which might be best to display different types of data.

Strategies and Applications

The questions in this section will help you review and apply the important ideas and strategies developed in this chapter.

Recognizing when previous outcomes influence later ones

1. Luis has two quarters, one dime, and four pennies in a pouch of his backpack. He needs a quarter for a vending machine. Suppose he reaches into his pack without looking and pulls out the first coin he touches.

   a. What is the probability Luis will select a quarter?

   b. The first coin Luis pulled out was a penny, so he will reach in again and pull out the first coin he touches. Which would give him a better chance of choosing a quarter: putting the penny back into the pack first, or leaving it out? Explain your reasoning, and support your answer with actual probabilities.

   c. Luis's hands were full, so he put the penny back before pulling out a second coin. He pulled out another penny. He put it back into the pack and then pulled out yet another penny. He put it back and tried once more, muttering, "I have to get something other than a penny this time!"

   Is this true? That is, is Luis more likely to get a quarter or dime than a penny? Explain.

Identifying whether a game is fair

2. Ben and Gabriela are playing a game with two dimes and two nickels. They toss the four coins. If one of the pairs of coins lands heads up, Ben scores 1 point. If both pairs land heads up, Ben scores 2 points. On every toss for which Ben does not score, Gabriela is awarded 1 point. They play to 5 points.

Determine whether this game is fair. If it isn't fair, identify who has the advantage. Explain or show how you found your answer.

Using probability to make decisions and create strategies

3. Three boards and two spinners are used to play the game *Cover Up*.

Board A		
1	4	2
2	9	2
9	2	1

Board B		
3	9	4
1	3	9
3	4	3

Board C		
2	2	3
3	6	2
2	3	6

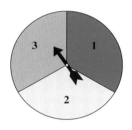 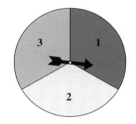

To play the game, each of two players chooses a board. They take turns. On each turn, a player spins both spinners, computes the product of the two numbers, and then covers a space on the board with that product, if there is one. The winner is the first player to cover all the spaces.

a. Which board gives the best chance of winning *Cover Up*? Explain your choice.

b. Of the remaining two boards, which is better? Explain.

Analyzing the appropriateness of a sample or a sampling process

4. A home-decorating magazine wanted to determine whether Americans think a living room should be painted or wallpapered. The people in charge of the survey considered different ways to gather this information.

For each suggestion, comment on whether the sample would be a good one and whether the process seems practical. Explain your answers.

a. One suggestion was to run a poll in the magazine with two toll-free numbers. If you prefer wallpaper, you call one number. If you prefer paint, you call the other number.

b. Another suggestion was to get a listing of home purchases over the past month in 100 cities across the country, chosen at random. The surveyors would call 20% of the new homeowners and ask them what they prefer.

c. The third suggestion was to call a random sample of 100 telephone numbers from across the country and ask the people who answer.

d. A fourth suggestion was to call a random sample of 10,000 people across the country for their opinions.

Creating and interpreting box-and-whisker plots

5. These box-and-whisker plots give information about the heights of adult men and women in a particular city. There are about 56,000 women and 52,000 men in the city.

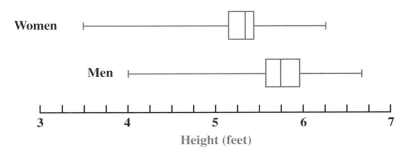

a. About how many adult women in this city are between 5'2" and 5'5"?

b. About how many adult men in this city are taller than 5'11"?

c. Estimate the third quartile for women's heights in this city. What does that number tell you?

d. Estimate the first quartile for men's heights in this city. What does that number tell you?

e. Estimate the median for men's heights and for women's heights.

6. Jamie opened 20 snack bags of peanuts and counted the nuts in each bag. His findings are shown below. Make a box-and-whisker plot of Jamie's data.

13	13	13	13	13	14	14	14	14	14
14	14	15	15	15	16	16	16	17	18

Choosing appropriate displays for data sets

7. For each data set described, indicate which types of graphs (line graph, stem-and-leaf graph, circle graph, bar graph, histogram, or box-and-whisker plot) are most appropriate for displaying the information.

a. average score on a standardized test each year for the past 10 years

b. scores on a standardized test for a group of 80 students

c. numbers or percentages of students in a large district receiving *exemplary, satisfactory, unsatisfactory,* and *poor* results on a standardized test

Demonstrating Skills

Jin Lee has a bag with three marbles in it, one orange and two yellow. She draws marbles from the bag one at a time.

8. Suppose Jin Lee keeps each marble she draws before drawing the next one.

   a. Find the probability that she will draw the orange marble first.

   b. Suppose Jin Lee drew a yellow marble first. Find the probability that she will draw the orange marble second.

9. Now suppose Jin Lee returns each marble to the bag before drawing the next one.

   a. Find the probability that she will draw the orange marble first.

   b. Suppose Jin Lee drew a yellow marble first. Find the probability that she will draw the orange marble second.

   c. Malik ran a simulation of Jin Lee's marble drawing by rolling a die. Rolling a 1 or 2 corresponds to drawing the orange marble; 3, 4, 5, and 6 correspond to drawing the yellow marble.

   2 1 2 1 2 5 6 2 1 2 2 4 1 3 3 4 6 5 4 4

   4 6 6 5 2 5 4 2 5 1 5 1 2 4 3 1 4 4 4 2

   1 3 1 4 3 1 5 4 1 2 3 6 2 6 5 1 1 3 4 4

   4 4 5 3 1 1 5 5 1 2 3 3 4 2 4 3 1 2 6 3

   5 1 3 2 3 1 3 3 4 6 4 2 2 1 1 1 3 4 1

   Use Malik's simulation to estimate the probability of Jin Lee drawing the orange marble on the third of three draws. Start at the top left of the table and read from left to right. When you reach the end of a row, return to the beginning of the next row. Use the entire table; you should have 33 sets of three rolls.

10. Suppose 300 students at a large school were randomly surveyed. The survey found that 180 were planning to attend the last home football game of the year. The school has 2,400 students. How many students would you expect to go to the game?

English	Español

algebraic expression A rule written with numbers and symbols. Examples: $n + n + n + 2$, $3n + 2$. [page 5]

backtracking A method of solving algebraic equations by working backwards from the known answer to figure out the value of the variable. [page 20]

base

1: The parallel faces of a prism. [page 112]

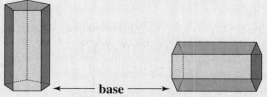

2: The number in an exponential expression that is multiplied by itself. For example, in t^3, t is the *base*; in 10^4, 10 is the *base*. [page 149]

coefficient A number that is multiplied by a variable. For example, in $y = 3x + 2$, 3 is the *coefficient* of the variable x. [page 367]

congruent Having the same size and the same shape. [page 450]

conjecture A statement that is believed to be true but has not yet been proven. [page 422]

constant term A number that stands by itself in an expression or equation. For example, in $y = 3x + 2$, 2 is the *constant term*. [page 367]

expresión algebraica Regla escrita con números y símbolos. Ejemplos: $n + n + n + 2$, $3n + 2$.

vuelta atrás Método en que se empieza con la respuesta y se trabaja de atrás hacia adelante para despejar la variable y resolver ecuaciones algebraicas.

base

1: Las caras paralelas de un prisma.

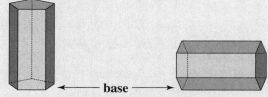

2: El número en una expresión exponencial que se multiplica por sí misma. Por ejemplo: en t^3, t es la *base*; en 10^4, 10 es la *base*.

coeficiente Un número que se multiplica por una variable. Por ejemplo: en $y = 3x + 2$, 3 es el *coeficiente* de la variable x.

congruente Que tiene el mismo tamaño y la misma forma.

conjetura Enunciado que se cree que es verdadero, pero el cual no se ha probado todavía.

término constante Número que no cambia en una expresión o ecuación. Por ejemplo: en $y = 3x + 2$, 2 es el *término constante*.

English

corresponding angles Angles of two similar figures that are located in the same place in each figure. For example, in the figure angle *B* and angle *E* are *corresponding angles*. [page 457]

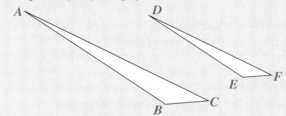

corresponding sides Sides of two similar figures that are located in the same place in each figure. For example, in the figure above, sides *AB* and *DE* are *corresponding sides*. [page 457]

counterexample In testing a conjecture, an example for which the conjecture is not true. [page 453]

cube A three-dimensional figure with six square sides, or faces. [page 81]

cylinder A figure that is like a prism, but its two bases are circles. [page 113]

dilation A process that creates a figure similar, but not necessarily congruent, to an original figure. [page 482]

distance formula The symbolic rule for calculating the distance between any two points, (x_1, y_1) and (x_2, y_2), in the coordinate plane:
distance $= \sqrt{(x_2 - x_1)^2 + (y_2 - y_1)^2}$. [page 274]

distributive property The *distributive property of multiplication over addition* states that for any numbers *n*, *a*, and *b*, $n(a + b) = na + nb$. The *distributive property of multiplication over subtraction* states that for any numbers *n*, *a*, and *b*, $n(a - b) = na - nb$. [page 62]

equivalent expressions Expressions that always give the same result when the same values are substituted for the variables. For example, $2K + 6$ is equivalent to $2(K + 3)$. [page 57]

Español

ángulos correspondientes Ángulos de dos figuras semejantes ubicadas en el mismo lugar en cada figura. Por ejemplo: En la figura, el ángulo *B* y el ángulo *E* son *ángulos correspondientes*.

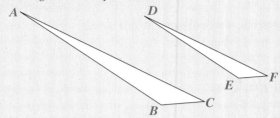

lados correspondientes Los lados de dos figuras semejantes que están ubicados en el mismo lugar en cada figura. Por ejemplo: En la figura anterior, los lados *AB* y *DE* son *lados correspondientes*.

contraejemplo Al probar una conjetura, un ejemplo para el cual la conjetura no es verdadera.

cubo Figura tridimensional con seis lados cuadrados, o caras.

cilindro Figura que parece un prisma, pero que tiene un par de bases circulares.

homotecia Proceso mediante el cual se crea una figura semejante, pero no necesariamente congruente, a la figura original.

fórmula de la distancia Regla simbólica para calcular la distancia entre cualquier par de puntos, (x_1, y_1) y (x_2, y_2), en el plano the coordenadas:
distancia $= \sqrt{(x_2 - x_1)^2 + (y_2 - y_1)^2}$.

propiedad distributiva La *propiedad distributiva de la multiplicación sobre la adición* establece que para cualquier número *n*, *a* y *b*, $n(a + b) = na + nb$. La *propiedad distributiva de la multiplicación sobre la sustracción* establece que para cualquier número *n*, *a*, y *b*, $n(a - b) = na - nb$.

expresiones equivalentes Expresiones que siempre dan el mismo resultado cuando los mismos valores se reemplazan con las variables. Por ejemplo: $2K + 6$ es equivalente a $2(K + 3)$.

English	Español
equivalent ratios Two different ratios that represent the same relationship. For example, 1:3 and 4:12 are *equivalent ratios*. [page 457]	**razones equivalentes** Dos razones diferentes que representan la misma relación. Por ejemplo: 1:3 y 4:12 son *razones equivalentes.*
expand To use the distributive property to remove parentheses. [page 64]	**expandir** Use la propiedad distributiva para eliminar los paréntesis.
exponent A symbol written above and to the right of a quantity that tells how many times the quantity is multiplied by itself. For example: $t \times t \times t$ is written t^3. [pages 13 and 149]	**exponente** Símbolo que se escribe arriba y a la derecha de una cantidad y el cual indica cuántas veces la cantidad se multiplica por sí misma. Por ejemplo: $t \times t \times t$ se escribe t^3.
exponential decay A decreasing pattern of change in which a quantity is repeatedly multiplied by a number less than 1 and greater than 0. [page 182]	**descomposición exponencial** Patrón de cambio decreciente en que una cantidad se multiplica repetidas veces por un número menor que 1 y mayor que 0.
exponential decrease See *exponential decay.*	**disminución exponencial** Ver *descomposición exponencial.*
exponential growth An increasing pattern of change in which a quantity is repeatedly multiplied by a number greater than 1. [page 179]	**crecimiento exponencial** Patrón de cambio creciente en que una cantidad se multiplica repetidas veces por un número mayor que 1.
factor To use the distributive property to insert parentheses. [page 64]	**factor** El uso de la propiedad distributiva para agregar paréntesis.
flowchart A visual diagram that shows each step in evaluating an algebraic expression. [page 19]	**flujograma** Diagrama visual que muestra cada paso en la evaluación de una expresión algebraica.

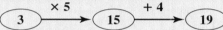

English	Español
formula An algebraic "recipe" that shows how to calculate a particular quantity. For example, $F = \frac{9}{5}C + 32$ is the *formula* for converting Celsius temperatures to Fahrenheit temperatures. [page 37]	**fórmula** Una "receta" algebraica que muestra cómo calcular una cantidad dada. Por ejemplo: $F = \frac{9}{5}C + 32$ es la *fórmula* para convertir temperaturas Celsius en temperaturas Fahrenheit.
linear relationship A relationship whose graph is a straight line. *Linear relationships* have a constant rate of change. As one variable changes by 1 unit, the other variable changes by a set amount. For example, $m = 5t$ shows that m changes 5 units per 1-unit change in t. [page 303]	**relación lineal** Relación cuya gráfica es una recta. Las *relaciones lineales* muestran una tasa constante de cambio. A medida que una variable cambia en 1 unidad, la otra variable cambia en la misma cantidad. Por ejemplo: $m = 5t$ muestra que m cambia 5 unidades por 1 unidad de cambio en t.
line graph A graph on which points are connected by line segments. [page 605]	**gráfica lineal** Gráfica cuyos puntos se conectan con segmentos de recta.

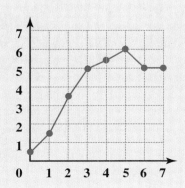

model Something that has the key characteristics of something else. For example, in mathematics you could use a balance to *model* an equation. [page 395]

modelo Algo que tiene una característica clave de algo más. Por ejemplo: en matemáticas podrías usar una balanza para hacer un *modelo* de una ecuación.

multiplicative inverse The number by which another number is multiplied to get 1. For example, the *multiplicative inverse* of 4 is $\frac{1}{4}$. [page 627]

inverso multiplicativo El número por el cual se multiplica otro número para obtener 1. Por ejemplo: el *inverso multiplicativo* de 4 es $\frac{1}{4}$.

net A flat figure that can be folded to form a closed, three-dimensional object called a solid. [page 129]

red Figura plana que al doblarse forma un cuerpo tridimensional cerrado llamado sólido.

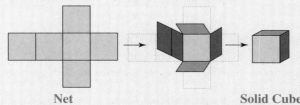

Net   Solid Cube

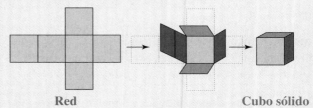

Red   Cubo sólido

outcomes The possible results of an experiment. For example, 4 is an outcome when a number cube is rolled. [page 676]

resultado Uno de los resultados posibles de un experimento. Por ejemplo, 4 es un resultado posible cuando se lanza un dado.

population A larger group from which a sample is taken. [page 693]

población Grupo grande del cual se toma una muestra.

power A number that is written using an exponent. [page 149]

potencia Número que se escribe usando un exponente.

prism A figure that has two identical, parallel faces that are polygons, and other faces that are parallelograms. [page 109]

prisma Figura con dos caras paralelas idénticas, las cuales son polígonos, y otras dos caras que son paralelogramos.

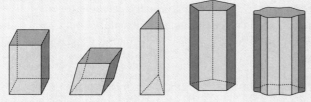

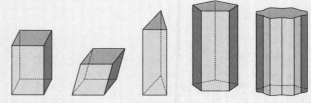

proportion An equation that states that two ratios are equal. For example, 2:3 = 6:9. [page 545]

proporción Ecuación que establece que dos razones son iguales. Por ejemplo: 2:3 = 6:9.

proportional Used to describe the relationship between two variables in which, when the value of one variable is multiplied by a number, the value of the other variable is multiplied by the same number. For example, when someone is paid an hourly rate and works double the hours, that person gets double the pay. When someone works triple the hours, that person gets triple the pay. The hours worked and rate of pay per hour are *proportional* to each other. [page 309]

proporcional Se usa para describir la relación entre dos variables en las cuales, cuando el valor de una variable se multiplica por un número, el valor de la otra variable se multiplica por el mismo número. Por ejemplo: cuando a alguien se le paga un sueldo a cierta tasa por hora y esa persona trabaja el doble del número de horas, dicha persona obtiene el doble del pago. Cuando alguien trabaja el triple de las horas, esa persona obtiene el triple del pago. Las horas trabajadas y la tasa de pago por hora son *proporcionales* entre sí.

English	**Español**

quadrant One of the four sections created by the axes on the coordinate plane. [page 259]

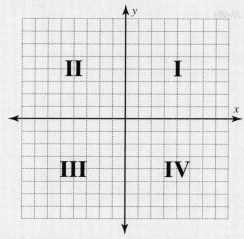

cuadrante Una de las cuatro secciones creadas por los ejes en el plano de coordenadas.

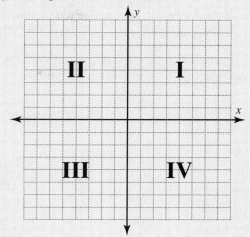

quartile The values that divide a set of data into four parts; each of which includes about 25% of the data. In a box-and-whisker plot, these values are represented by the ends of the box and a segment inside the box. [page 710]

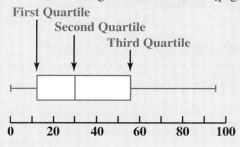

cuartil Los valores que dividen un conjunto de datos en cuatro partes; cada una de las cuales incluye aproximadamente un 25% de los datos. En un diagrama de caja y patillas, estos valores se representan con los extremos de la caja y un segmento dentro de la caja.

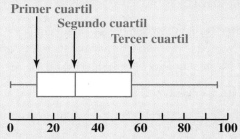

rate Describes how two unlike quantities are related or how they can be compared. [page 300]

tasa Describe la relación entre dos cantidades diferentes o la manera de comparar dichas cantidades.

ratio A way to compare two numbers. For example, when one segment is twice as long as another, the *ratio* of the length of the longer segment to the length of the shorter segment is 2 to 1, or 2:1. [page 456]

razón Una manera de comparar dos números. Por ejemplo: cuando un segmento es el doble de largo que otro segmento, la *razón* de la longitud del segmento más largo al segmento más corto es 2 a 1 ó 2:1.

representative sample A part of a population that has approximately the same proportions as the whole population with respect to the characteristic being studied. [page 698]

muestra representativa Una parte de la población que tiene aproximadamente las mismas proporciones que la población entera con respecto a las características bajo estudio.

sample A smaller group taken from a population that is used to represent the larger group. [page 693]

muestra Un grupo más pequeño que se toma de la población y el cual se usa para representar el grupo más grande.

scale factor The number by which you multiply the side lengths of one figure to get the side lengths of a similar figure. [page 482]

factor de escala Número por el que multiplicas las longitudes de los lados de una figura para obtener las longitudes de los lados de una figura semejante.

English	Español

scientific notation A number that is expressed as the product of a number greater than or equal to 1 but less than 10, and a power of 10. For example, 5,000,000 written in *scientific notation* is 5×10^6. [page 196]

notación científica Número que se expresa como el producto de un número mayor que o igual a 1, pero menor que 10 y una potencia de 10. Por ejemplo: 5,000,000 escrito en *notación científica* es 5×10^6.

similar Having the same shape but possibly different sizes. [page 450]

semejante Que tiene la misma forma, pero posiblemente tamaños diferentes.

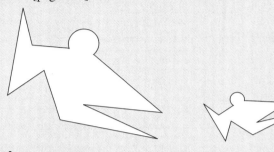
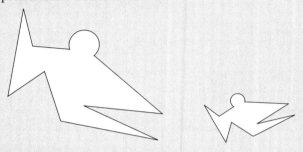

slope The steepness of a line. [page 324]

pendiente El grado de inclinación de una recta.

speed How fast an object is going (always positive). [page 329]

rapidez El grado de velocidad con que viaja un cuerpo (es siempre un número positivo).

surface area The area of the exterior surface of an object, measured in square units. [page 98]

área de superficie El área de las superficies exteriores de un cuerpo, medida en unidades cuadradas.

unit rate Term used when one of two quantities being compared is given in terms of one unit. Example: 65 miles per hour or $1.99 per lb. [page 530]

tasa unitaria Término que se usa cuando una de dos cantidades bajo comparación se da en términos de una unidad. Ejemplo: 65 millas por hora o $1.99 por lb.

variable A quantity that can change or vary, or an unknown quantity. [pages 5 and 367]

variable Cantidad que puede cambiar o variar, o una cantidad desconocida.

velocity The rate at which an object is moving from or toward a designated point (can be positive or negative). [page 329]

velocidad Tasa a la cual se mueve un cuerpo desde un punto o hacia un punto designado (puede ser positiva o negativa).

volume The space inside a three-dimensional object, measured in cubic units. [page 98]

volumen El espacio dentro de un cuerpo tridimensional, medido en unidades cúbicas.

y-intercept The point at which a graph intersects the *y*-axis. [page 333]

intersección y El punto en que una gráfica interseca el eje *y*.

INDEX

PHOTO CREDITS

Cover Paddy Grass, Toyohiro Yamada/Getty Images;

Front Matter **v** Getty Images; **vi** NASA; **vii** Aaron Haupt; **viii** PhotoDisc; **1** Timothy Fuller;

Chapter 1 **2** (t)File Photo, (b)Getty Images; **2-3** Getty Images; **12** (t)Doug Martin, (b)Alan Schein/CORBIS; **22** MAK-I; **28** Aaron Haupt; **29** (t)Getty Images, (b)CORBIS; **41** Getty Images; **43** Aaron Haupt; **46** Richard Hutchings; **50** Reuters NewMedia/CORBIS; **69** CORBIS;

Chapter 2 **76** (t, inset)Amanita Pictures, (b)Rick Weber; **76-77** MAK-I; **90** Morton & White; **92-93** Tim Courlas; **96** DigitalVision/PictureQuest; **114** Aaron Haupt; **124** Digital Vision/PictureQuest; **126** (l)American Airlines, (r)Doug Martin; **128** CORBIS;

Chapter 3 **144-145** NASA; **145** NASA; **146** CORBIS; **171** Doug Martin; **176** Scott Cunningham; **182** Matt Meadows; **194** Doug Martin; **202** Geoff Butler;

Chapter 4 **216** (t)CORBIS, (b)Galen Rowell/CORBIS; **216-217** Digital Stock; **223** Doug Martin; **224** Mark Romesser; **228** Doug Martin; **230** Carl & Ann Purcell/CORBIS; **241** Doug Martin; **242** Johnny Johnson; **249** Geri Murphy; **253** CORBIS; **260** Tim Courlas; **265** Getty Images; **267** CORBIS; **268** James L. Amos/CORBIS; **273** Eliot Cohen; **278** Getty Images; **294** Tom Stewart/CORBIS; **296** CORBIS;

Chapter 5 **298** (t)Christine Osborne/CORBIS, (b)Bill Ross/CORBIS; **298-299** Digital Stock; **300** CORBIS; **302** Todd Anderson/Photo Op; **303** Getty Images; **306** Matt Meadows; **308** Allen Zak; **310** Getty Images; **315** Geoff Butler; **316** CORBIS; **319** Kenji Kerins; **323-324** CORBIS; **325** Aaron Haupt; **330-331** Rudi Von Briel; **339** CORBIS; **352** Getty Images; **357** Janet Adams; **363** Brent Turner; **366** CORBIS; **372** Getty Images; **378** Tim Fuller;

Chapter 6 **382** (t)Getty Images, (b)Courtesy Cedar Point Amusement Park/Photo by Dan Feicht; **382-383** Aaron Haupt; **386** Geoff Butler; **390** Courtesy Apple Computers; **393-396** CORBIS; **397** PhotoDisc; **400** Aaron Haupt; **402** Doug Martin; **406** Getty Images; **407** K.S. Studios; **413** PhotoDisc; **416** Getty Images; **420** (t, cr)Mark Burnett, (cl)Milo Stewart, Jr./National Baseball Hall of Fame Library, Cooperstown, NY, (b)file photo; **428** Corbis; **429** File Photo; **430** Matt Meadows; **446** CORBIS;

Chapter 7 **448-449** Getty Images; **449** LEGOLAND, California; **452** Getty Images; **472** CORBIS; **486** Elaine Comer Shay; **495** James W. Richardson/Visuals Unlimited; **500** Larry Hamill;

Chapter 8 **518** Tim Fuller; **518-519** Tim Courlas; **528** StudiOhio; **532-533** Getty Images; **546** van Gogh, Vincent. The Starry Night. (1889) Oil on canvas, 29 X 36 1/4 " (73.7 x 92.1 cm). The Museum of Modern Art, New York; **548-552** Getty Images; **566** Doug Martin; **571** Getty Images; **575** Steven Ferry; **587** Rudi Von Briel; **588** Getty Images;

Chapter 9 **600** Timothy Fuller; **600-601** Getty Images; **603** John Evans; **610** Getty Images; **616** Larry Hamill; **624** Matt Meadows; **625** CORBIS; **629** file photo; **630-633** Getty Images; **636** file photo; **647** Mark Burnett;

Chapter 10 **664** Ted Rice; **664-665** CORBIS; **673** Timothy Fuller; **688** Getty Images; **693** Matt Meadows; **697** Doug Martin; **701** CORBIS; **706** MAK I; **711** Getty Images; **717** Amanita Pictures; **718** Doug Martin; **721** Getty Images; **726** CORBIS.

Unlisted photographs are property of Glencoe/McGraw-Hill.